# PUTTING KNOWLEDGE TO WORK

# PUTTING KNOWLEDGE TO WORK:

## From Knowledge Transfer to Knowledge Exchange

Edited by

*Amare Desta, Mentesnot Mengesha & Mammo Muchie*

AFRICA WORLD PRESS

TRENTON | LONDON | CAPE TOWN | NAIROBI | ADDIS ABABA | ASMARA | IBADAN | NEW DELHI

AFRICA WORLD PRESS
541 West Ingham Avenue | Suite B
Trenton, New Jersey 08638



Book design: Dawid Kahts

Cover design: Ashraful Haque

Library of Congress Cataloging-in-Publication Data

Names: Desta, Amare, 1964- editor. | Mengesha, Mentesnot, 1959- editor. | Muchie, Mammo, editor. | Global Knowledge Exchange Network. Conference (1st : 2011 : London, England) | Global Knowledge Exchange Network. Conference (2nd : 2012 : London, England)
Title: Putting knowledge to work : from knowledge transfer to knowledge exchange / edited by Amare Desta, Mentesnot Mengesha & Mammo Muchie.
Other titles: Title should read: Putting knowledge to work
Description: Trenton, New Jersey : Africa World Press, 2016. | Selected papers from the 1st and 2nd yearly conferences of GKEN (Global Knowledge Exchange Network), held 2011 and 2012 in London, England. | Includes bibliographical references and index.
Identifiers: LCCN 2016009850| ISBN 9781569024805 (pb) | ISBN 9781569024799 (hb : alk. paper)
Subjects: LCSH: Africa--Intellectual life--Congresses. | African diaspora--Intellectual life--Congresses. | Knowledge management--Africa--Congresses. | Technology transfer--Africa--Congresses.
Classification: LCC DT14 .P88 2016 | DDC 306.42096--dc23
LC record available at http://lccn.loc.gov/2016009850

# CONTENTS

# ACKNOWLEDGMENTS

We recognize and appreciate the support given to GKEN by the Chair of the DST/NRF Research SARChI on Innovation Studies at Tshwane University of Technology, Pretoria, South Africa to publish the first book.

Some of the papers used in this and the other GKEN books to follow have first appeared in the Ethiopian Electronic Journal for Research and Innovation Forsight (Ee-JRIF). founded by the Network of Ethiopian Scholars(NES).

http://www.nesglobal.org/eejrif4/index.php?journal=admin

We would like also to acknowledge the encouragement and support we received from the African Universities' Research Approaches (AURA) in the publication of this first GKEN book

# ACRONYMS

- AAU - Addis Ababa University
- ABIDE - Alliance for Brain Gain and Innovative Development
- ADLI - Agricultural Development Led Industrialization
- AEHHPS - Association of Ethiopian Health and Health Associated Professionals in Sweden
- ACRD - Agriculture-Centred Rural Development
- AGDP – Agricultural Gross Domestic Product
- AHEAD - Association for Higher Education and Development
- ASALs - Arid and Semi-Arid Lands
- AU – African Union
- AWDM - Agricultural Water Development and Management
- BoA - Bureau of Agriculture
- BDU - Bahir Dar University
- Bo-EPLUA - Bureau of Environmental Protection Land Use and Administration
- BoWR - Bureau of Water Resource
- CAADP - Comprehensive African Agricultural Development Programme
- CPD – Continuing Professional Development
- CSA – Central Statistics Agency
- CSO – Civil Society Organisation
- DCC - Decoding Communication Competence
- DfID – The British Department for International Development
- DWD -Directorate of Water Development
- ECC - Encoding Communication Competence
- EKTTS - Ethiopian Knowledge and Technology Transfer Society

- ENAHPA - Ethiopian North American Health Professionals Association
- EPA - Environmental Protection Authority
- EPA-NA - Ethiopian Physicist Association in North America
- EPH - Ethiopian Professionals in Huston
- EPRDF – Ethiopian People Revolutionary Democratic Front
- ESAi - Books for Africa, Ethiopian Students' Association, international
- ETFF - Ethiopian Tree Fund Foundation
- EWDCA - Ethiopian Wildlife Development and Conservation Authority
- FAO – Food and Agricultural Organisation
- FDI - Foreign Direct Investment
- FDRE – Federal Democratic Republic of Ethiopia
- FIELD - Forum International for Ethiopians Living in the Diaspora
- GAP-A - Group Against Poverty and AIDS
- GDP – Gross Domestic Product
- GIS – Geographical Information System
- GMB - Grain Marketing Boards
- GR – Green Revolution
- GTP – Growth and Transformation Plan
- HE – Higher Education
- HEI – Higher Education Institution
- HERI - Higher Education, Research and Innovation
- HPRs - The House of Peoples' Representatives
- IBC - Institute of Biodiversity Conservation
- ICT- Information Communication Technology
- IEAR - Institute of Ethiopian Agricultural Research
- IFAP – UN Information for All Progemmes
- IGAD – Inter-Governmental Authority for Development
- IK - Indigenous Knowledge
- IP – Intellectual Property
- IS – Information System
- ITC - IT Competence
- KBE – Knowledge Based Economy

- KE – Knowledge Exchange
- KM - Knowledge Management
- KSB - Knowledge Sharing Behaviour
- LMS - Learning Management Systems
- MDGs – Millennium Development Goals
- MEA - Millennium Ecosystem Assessment
- MENA - Middle East and North Africa
- MoA – Ministry of Agriculture
- MoARD- Ministry of Agriculture and Rural Development
- MoFED – Ministry of Finance and Economic Development
- NBE – National Ban of Ethiopia
- NEPID - Network of Ethiopian Professionals In the Diaspora
- NEPAD - New Partnership for Africa's Development
- NGO – Non-Governmental Organisation
- NNI - National Nanotechnology Initiative
- NRM - Natural resource management
- NPEAP - National Poverty Eradication Action Plan
- NWP - National Water Policy
- ODA - British Overseas Development Administration
- OECD - Organization for Economic Cooperation and Development
- P2P - People to People
- PASDEP -Plan for Accelerated and Sustained Development to End Poverty
- PBC - Perceived Behavioural Control
- PESTEL - Political, Economical, Social, Technological, Environmental and Legal
- PRSP - Poverty Reduction Strategy Paper
- PSP - Private Sector Participation
- R&D- Research and Development
- RDPS – Rural Development Policy and Strategy
- RDPSI - Rural Development Policies, Strategies and Instruments
- RoSS - Republic of South Sudan
- SDPRP – Sustainable Development and Poverty Reduction Programme

- SN - Subjective Norm
- S&T- Science and Technology
- SWAP - Sector Wide Approach to Planning
- TNCs - Trans-national companies
- TOKTEN - Transfer of Knowledge Through Expatriate Nationals
- TPB - Theory of Planned Behaviour
- TPLF - Tigrean People's Liberation Front
- USAID - United States Agency for International Development
- UNCTAD - UN Conference on Trade and Development
- UNDEP – United Nation's Environment Programme
- UNESCO – United Nations Educational, Scientific and Cultural Organization
- WAP - Water Action Plan
- WB – World Bank
- WWF - World Wildlife Fund
- ZANU-PF - Zimbabwe African National- Patriotic Front
- ZINATHA - Zimbabwe National Traditional Healers' Association
- ZISCO - Zimbabwe Iron and Steel Company

# FORWARD

*Amare Desta, Mammo Muchie, and Mentesnot Mengesha*

The idea behind this book came out of discussions between the GKEN (Global Knowledge Exchange Network) founders Amare Desta, Mentesnot Mengesha, Mammo Muchie, Abiy Negussie and Behailu Nekatebeb in London UK in 2014. GKEN was established in December 2010 owing to the realisation that there is a large number of Africans all over the world attained hallmarks of excellence in their respective fields of specialization and many have made significant contributions to scientific research and innovation in the country of their origin and also across many other countries. However this massive and valuable resource has not been organised and communicated in a systematic way in order to make a meaningful contribution to the wider African and global communities.

It is in light of the above considerations that GKEN has taken the first step to connect the human and social capital of those learned Africans and friends of Africa knowledge workers (i.e. academics, researchers and practitioners) bearing in mind, the availability of massive potential of African scientific knowledge scattered in the world on one hand and the challenges to unlock it on the other

GKEN as a knowledge exchange platform has now created a visible space and continued to organize a series of yearly conferences since 2011 in order to promote the culture of knowledge, good practice and valuable information sharing between the aforementioned knowledge workers and civil society in array of disciplines by organizing conferences and publishing their evidence based research papers via its journals.

Two years has passed since the first and second conferences took place and the connection we have created helped us to find such great and diverse authors who submitted papers to the two unique conferences GKEN organised. However the idea to publish this book became a realistic option only when three of the five founders Amare, Mentesnot & Mammo joined up in the winter 2014. Had it not been for Mentesnot editorial effort it would not have been possible to finalise the book publishing project. In the end, we believe that three of us worked as one person and this arrangements provided us the perfect blend of knowledge and skills that went into editing this first book titled "Putting Knowledge to Work" which we be believe is a great asset to the African and the world community at large.

The two conferences were supported by the London South Bank University (LSBU) and partly funded by Globelics and by further encouragement from NES-Global. We are very grateful indeed to those authors, reviewers and keynote speakers who freely shared and continued to share their knowledge and expertise.

We thank all present and past conference participants for their strong support. We are also grateful to our friends, colleagues & family members who are always very supportive and encouraging. We therefore dedicate this book to them for their unreserved support and helping us to realise the noble vision of Global Knowledge Exchange Network (GKEN).

While we are thanking the authors and the reviewers and facilitators for their invaluable contributions and support, we also look forward to writing the second and third books to strengthen knowledge communications across Africa and beyond.

# INTRODUCTION

*Mentesnot Mengesha, Amare Desta and Mammo Muchie*

Knowledge exchange as a process between two or more entities (i.e individuals, groups or institutions) is one of the ways to replicate and spread the wealth of knowledge beyond one's own fortress to a new territory, very often, for mutual benefits. Knowledge, be it tacit or explicit, is a dynamic route to reinforce our comprehensive understanding, description and benefit from things and situations around us and beyond. It is therefore, that the value of knowledge exchange is dependent on how it is disseminated, reciprocally used between both the knowledge producer and the recipients. Knowledge is the powerful tool for socio-economic development and a means of remedy to the challenges faced by humanity in a day in day out manner. When a net of indigenous and non-indigenous knowledge is connected and complimented each other, indeed, it produces more powerful instrument in tackling the ills of society.

The advancement of information technology has also contributed to the gulf of a knowledge divide. The discussion of knowledge exchange is partly to narrow the gap and partly to nurture the indigenous knowledge to the best and effective use of it.

As various authors also demonstrated in this book knowledge exchange is a dynamic process by which a new set of knowledge is produced or built on through the transaction of approaches, skills and theories in a continuing way.

The linkage and harmonisation of indigenous and non-indigenous knowledge are matters that instigated discussions among the circle of academics and practitioners in social and natural sciences. The knowledge that humanity possesses throughout history has helped her through adversities and challenges and how to overcome it. As various literature also suggest before penicillin was invented to cure many broad spectrum bacteria local medicinal knowledge had helped to relieve many of the ailments people faced. The basis of many of the advancement of technologies and its innovations has linkage with indigenous and local knowledge.

This book has been compiled to address a range of ideas and views in four main parts and eighteen chapters. Part one addresses topics on harmonising indigenous with scientific knowledge, while part two is about linking knowledge exchange and indigenous knowing to everyday challenges. On part three, the book discusses the themes of both competition and cooperation and the future of knowledge exchange in the world of science and innovation. Part four deliberates and revisits the emergence of debates about the role of indigenous knowledge in higher learning institutions and the contribution and visible impact they make.

In chapter one Amare Desta addresses the heightened global concern regarding climate change, poverty and hunger and stresses that the natural resource management is becoming increasingly important. However, Amare argues that in many developing countries, the local custodians of natural resources are indigenous farmers who have been working the land for generations. He further argues over time, they have built up a large amount of informal indigenous knowledge concerning sustainability in the local environment that could be disseminated more widely. Amare reviews previous research on indigenous knowledge and discusses the possible contributions that the disciplines of information systems and innovation studies could provide.

However Tewabech Bishaw in chapter two discusses that most developing countries like Ethiopia aspire to fast-track their development processes to alleviate poverty and achieve the Millennium Development Goals. Tewabech argues that these aspirations are dependent on a nation's human capital and the progress it makes in education, science technology and innovations.

She also further asserted that, for developing nations like Ethiopia, engaging the brain power of its professionals fully and effectively is imperative to achieve the quest for rapid development and sustainable social change. In this regard, the author argued that Ethiopia needs to create an enabling environment and establish strategic collaborations and partnerships with all its development partners and sectors.

The chapter reiterates that the Ethiopian Diaspora, including the second and third generation, is among the country's critical intellectual resource whose maximum mobilization and engagement is indispensable. Finally, the author concludes by saying the Diaspora has not been effectively mobilized, networked and engaged. She further highlights the enormity of the problem of brain-drain by underlining the Diaspora's role in knowledge exchange and recommends the need to increase the viability of Diaspora professionals.

Medicine Magocha investigates how prior to the post-modern industrial eras the African states had different methods, ways, approaches and systems of doing things for survival and sustainability. The author highlights knowledge systems could be maintained, improved, implemented and passed on from generation to generation through oral tradition data banks, written records and archaeological packages. Skills and expertise though they appeared individualistic they were used to benefit many people for example the "cow horn formation" was developed from the existing knowledge systems of that time. Medicine discusses in pre-historic eras people could practice business principles like warehousing, marketing, accounting, human resources management, operations management, economics and purchasing and supply to mention just a few; that was through knowledge systems of the moment. The chapter suggests how knowledge systems can be capitalized towards the development of sustainable future. There are a lot of developments which were generated from the indigenous knowledge systems for example the issue of education in Africa has a direct relationship to the folklores of the traditional societies. Medicine argues that African societies are losing popularity because they lack contributions in terms of technological developments, product developments and innovations. So capitalizing on their indigenous knowledge system (IKS) would come as a springboard towards development of Africa.

In chapter four Demese Chanyalew writes about the outcome of a desk study how Ethiopia is one of the few African countries which recorded an agriculture sector growth and budgetary appropriation more than the targets set for the Comprehensive African Agricultural Development Programme (CAADP). Since 2004/05 the average annual growth rate of the agriculture sector in Ethiopia was about 8.4%, exceeding the CAADP target of 6%. Despite the positive changes in the economy, poverty and hunger remained the daunting challenges in Ethiopia. Some politicians and academicians relate the problems of poverty and hunger to the agricultural policies and strategies of the EPRDF government.

Demese uses trend analysis based on the supposition that in recent years indigenous knowledge based policy making has made a difference in the agriculture sector performance of Ethiopia. Demese reviews the strategy known as ADLI and the principles of "a foot on the ground," in order to comprehend the relevance of indigenous knowledge in agricultural policy formulation. The paper concludes that in Ethiopia centuries' old high dependence on agriculture is changing. Agriculture, and the economy at large, is experiencing remarkable growth. On the other hand poverty and hunger remained discouraging to the government and people of Ethiopia.

In chapter five Bedru B. Balana examines the impact of information/knowledge exchange on human wellbeing and highlights the salient features of the current situation of information flow in Ethiopia. The author argues that knowledge, both technical know-how and information about attributes, makes a difference from the impact of 'green revolution' on food security in India to the Grameen Bank's 'credit revolution' in Bangladesh to the spectacular health changes in Costa Rica. Berdu further argues that many developing countries still suffer from knowledge gaps and information problems. In this regard Ethiopia is a case point. Institutional rigidity, heavy-handed government, extensive poverty, high illiteracy, and cultural barriers can explain the problem in Ethiopia. Bedru finally concludes that Ethiopians and friends of Ethiopians (especially those who are capable to contribute in knowledge exchange) need to work hand in hand to make a difference via knowledge exchange in Ethiopia.

Samuel Galiwango in chapter six discusses the water sector as one of the priority sectors in Uganda, as it directly impacts on the quality of life of the people and overall productivity of the population. Water supply and sanitation are among the key issues emphasized under the Millennium Development Goals (MDGs, Goal 7 section c) and National Poverty Eradication Action Plan (PEAP), which is the key government framework for ensuring poverty eradication through creation of an enabling environment for rapid economic development and social transformation.

Samuel argues that water is a key strategic resource, vital for sustaining life, promoting development and maintaining the environment. Access to clean and safe water and improved sanitation facilities and practices are pre-requisites to a health population and therefore have a direct impact on the quality of life and productivity of the population. Besides domestic water supply, water is also vital for: Livestock Water Supply, Industrial Water Supply, Hydropower generation, Agriculture, Marine Transport, Fisheries, Waste Discharge, Tourism, and Environmental Conservation. Water, therefore, as Samuel argues significantly contributes to the national socio-economic development and thus poverty eradication.

Samuel reiterates despite Uganda's being well endowed with significant freshwater resources, the challenges of rapid population growth, increased urbanization and industrialization, uncontrolled environmental degradation and pollution are leading to accelerated depletion and degradation of the available water resources. Uganda is also faced with the challenge of low safe water coverage (59% rural and 65% urban, as of December 2003).

In order to meet the above challenges, government initiated reforms in the water sector in 1997 to ensure that water services are provided and managed with increased efficiency and cost effectiveness. Samuel concludes by answering questions like what percentage of Ugandans use protected water sources; how much time is spent while collecting water from the water source; and what reasons do households give for not using protected water sources?

In chapter seven Dessalegn Molla Ketema, Nick Chisholm and Pat Enright investigate the trade-off between environmental protection and development in most dynamic and complex ecosystems. They took the catchment areas of Lake Tana of Ethiopia. They argue despite the positive impact on the country's drive towards food security and poverty alleviation, development activities that have been taking place around Lake Tana catchment are casting doubt on the sustainability of the region. They further discuss unless timely, protective policy intervention measures put in place, there might be a serious environmental degradation which ultimately ends up with a complete collapse of the whole ecosystem. The authors try to analyse the importance of integrating different knowledge systems and transferring research knowledge into policy framework and immediate intervention in the Lake Tana ecosystems by which researchers, policy makers at governmental level and user groups at community level can achieve desired outcomes through the sustainable resource use, development and application of innovative approaches.

Zeremariam Fre and Bereket T. Tsegay discuss a broad context to the current status of pastoralism and agro-pastoralism in the Horn of Africa from a food security and smallholder livelihoods perspectives. They argue that the significant contribution pastoral and agro pastoral livestock producers make to domestic, national and regional food security is often misrepresented and such contribution to food security is regularly unappreciated by policy makers at State level. They further argue that such an attitude is reflected in national and regional government policy perspectives which often do not apportion adequate financial and human resources to help the sector develop to its fullest potential.

Zeremariam Fre and Bereket T. Tsegay discuss that pastoralism and agro-pastoralism (i.e. as an environmentally/socially sustainable livelihood system) is going through a serious self-generated and externally driven evolving process which may be irreversible. Thus a change of the pastoralist production system from subsistence to more commercial orientation may be feasible. According to the authors this reality is so fundamental that government planners, researchers, international organizations and funders must recognise this if they wish to contribute to the wellbeing of pastoral peoples from a food security dimension in the Horn region. Drawing from authors' experiences from Eastern Sudan, Western Ethiopia and Western Eritrea, they demonstrate the significant economic contribution small producers make to food security within the Region as well as the Middle East, and recommend ways in which such contribution could be enhanced at national and regional levels.

In chapter nine Amare Desta and Mammo Muchie discuss the existence of increasing national and international concerns and uncertainties caused by the forces of globalization on local communities in the developing world. They focus on related and pressing issues on how to identify, accommodate, modify or fully accept practices associated with globalized and localized knowledge. The author hope that the issues they are proposing within the paper can contribute to the ongoing debate on globalization vs. localization in both economically developed and the developing world. Given the increasing international concerns about the impacts of globalization on local communities, how to manage the realities and practices of globalization and localization, the author proposes five different scenarios in order to appropriately foster meaningful and sustainable knowledge in contemporary world.

Hailemichael Teshome writes that humanity is at the cusp of a technological revolution transitioning from the 'Age of Discovery' to the 'Age of Mastery'. According to Hailemechael nanotechnology is central to this revolution, hence there is an air of realistic optimism about the nanotechnology revolution. He further writes those who missed out on the information technology revolution, resulting in the once yawning 'digital divide', will not miss out this time due to the cost or nature of nanotechnology (Khan, 2006, p.104). He says some even argue that such modern technologies should not just be affordable but also extremely affordable to create 'value for money and for many' (Mashelkar and Borde, 2010). Hailemichael discusses to capitalise on this optimism and argues that Africa must renegotiate the governance of nanotechnology to ensure that the social solidarity, inclusionary and redistributive ethos are duly and effectively engaged at the global level. The author highlights what Africa needs to do on its part besides demanding a renegotiation of the terms of the governance of the technology.

In chapter eleven Semahagn Gashu highlights the significance of good governance in facilitating the transfer of knowledge and capital to Ethiopia, where there is pressing need to make reforms aimed at improving governance. Specifically, he writes that there is a need to change the repressive culture that has passed from generation to generation, having representative government with well function democratic institutions and doing away with failed Marxist oriented policies. According to the author, in order to achieve this goal, there has to be proper checks and balances among the three branches of the state and efficient executive branch that implement government policies effectively. He further argues that there has to be also clear demarcation between party and government activities and the civil service need to be out of political influence and employment need to be made only based on merit rather than political loyalty. The author reiterates the existence of an independent judicial organ that uphold rule of law is also critical in terms of building confidence among investors and the Diaspora to freely enter into the country and engage in various activities that facilitates transfer of capital and knowledge.

Tadesse Biru Kersmo in his part argues contrary to conventional wisdom, lightweight corporations could beat heavyweight conglomerates. He writes that market competitions are not necessarily a two-person zero-sum games in which the winner takes all and heavyweight conglomerates have always the highest probability of success. The author furthers this by saying that too often, market competitions are games played by more than two players, in which the number of players makes a difference. He reiterates as the number of players increases, the probability that heavyweight conglomerates will triumph over lightweight corporations diminishes while that of lightweight corporations increases. Market competitions are positive-sum games in which both competition and cooperation go together; and even the loser gets a stake. Finally the author based on a three-person three-way duel, demonstrates mathematically the proposition that heavy weight conglomerates do not always win.

In chapter thirteen Yared Haile-Meskel discusses how the Middle East and the North African (MENA) countries have been going through uprising and changes. These changes were triggered on December 17, 2010, when a street vendor set himself on fire to protest against confiscation of his merchandise and humiliation at the hands of a provincial official in Tunisia. According to Yared since then it had engulfed the whole region in to turmoil driving radical changes and reformation of the political landscape. These uprisings have caught everyone unprepared, including strategic analysts, government's officials, migrant workers and foreign and local investors, leading to loss of investment due to force majeure or loss of market and customers. The author analyses macroeconomic data of MENA countries to understand what could have predicted the uprising and how relevant it is for other sub-Saharan countries. The analysis result shows that it is not poverty, Facebook, income inequality or corruption that caused the uprising. Yared concludes it is more likely to be explained by the hierarchy of needs when a rapid improvement of living standard was not matched by political liberalisation and democratic rights. He suggests that these same events may be repeated in countries like China.

In chapter fourteen Dessalegn Molla Ketema, Nickolas Chisholm and Patrick Enright focus on exploring an overview of the status and governance problems of Lake Tana fishery sector. Using household survey, focused group discussions, key informant interviews case studies, stakeholder identification and analysis workshop; the authors show that there were no legally binding rules that govern fisher's and other stakeholder's behaviour at user level. Though fishery proclamation and regulation were enacted at Federal and Regional level, directives to enforce proclamations at the user level are not yet prepared. The authors through their stakeholder identification and analysis reveal that, Bureau of Agriculture, Bureau of Environmental Protection Land Use and Administration and Bureau of Water Resource, ranked high on stake value-power level and totalled the highest score. They had a particularly strong and legitimate claim to govern the fishery sector. Above all, the power to control and monitor the fishery resource management and governance was vested to Bureau of Agriculture. However, the authors argue that these stakeholders were not working synergistically and discharging their responsibilities to resolve the fishery governance problems. As a result too much fishing pressure and illegal fishing activities were believed to be causing overexploitation of endemic fish stocks, threatening sustainable fish production, marketing and the livelihood of local fishermen. The authors suggest enacting legislations at the constitutional level could not be a panacea to protect the fish resource from depletion; rather the process of rulemaking should be participatory to understand the local context, rules and regulations should be integrated with the local institutions and it has to be communicated with relevant stakeholders in the action arena.

Elizabeth Ayalew writes that advanced economies are now moving from a resource-based to a knowledge-based mode of production. National governments of many developing economies also have come to recognize knowledge and innovation as significant driving forces of economic growth and social development. However, according to Elizabeth they have many challenges to handle before they pave the way to exploit the knowledge resource they have. Elizabeth argues in light of this view, Higher Education Institutions are required to play a pivotal role in driving this vision by complementing their traditional role – teaching and research – with the development of an effective knowledge exchange system with the private and public sectors. To be able to do this, according to the author they need to develop an innovative approach that allows them exploit the knowledge that exists outside their reach for further knowledge reconstruction. This vision however requires clear deliberation on important strategic as well as technical issues.

The author further highlights the challenges of the 21st century in the context of the university to set the background of discussion. Then she discusses the rationale for knowledge exchange, identifies the challenges and opportunities the local context offers and concludes by identifying priority issues that need the attention of such institutions while considering the implementation of knowledge exchange in Ethiopian universities.

In chapter sixteen Rahel Bekele and Ermias Abebe discuss the striding increase of Universities and Higher Learning Institutions in Ethiopia. These institutions the authors write have knowledge repositories both electronic and paper based. The institutions are also expected to facilitate the sharing of available knowledge among the academic community – a situation which needs particular attention. Based on a survey, the authors review of available documents and discussions with relevant individuals, the paper presents the existing level of staff awareness and practice of knowledge sharing among higher learning institutions in Ethiopia, as well as the extent of organizational and infrastructural support. The chapter also makes recommendations on strategies and mechanisms that need to be devised in order to overcome knowledge sharing barriers and promote effective utilization of ICT for knowledge exchange.

Mariye Yigzaw Workineh on his part recognises that knowledge creation has been practiced since the beginning of mankind. Communities have been creating and using indigenous knowledge for centuries. However, indigenous knowledge and indigenous techniques of knowledge creation have been neglected for decades and there has been a divide between the modern (scientific) and traditional worlds. The author identifies that despite advances in knowledge creation, there is a fear that knowledge management can become another management fad that promised much and delivered little. Besides, application of modern knowledge is limited especially in third world communities like Ethiopia. Such communities are heavily dependent on traditional knowledge for solving problems. Thus, Mariye argues studying traditional practices may lend an additional lens to better understand and facilitate knowledge creation. By leveraging the past and incorporating the traditional practices in knowledge creation, one can further enable today's knowledge creation efforts. The chapter explores the potentials of indigenous knowledge creation techniques of instantaneously creating knowledge, which has been in use by the Orthodox Church communities of Ethiopia (and Eritrea) since the 5th Century. Using an interpretive field study, the author collected qualitative data from the indigenous Poetry schools through participant observation, unstructured and semi-structured interviews, and document analysis. The analysis indicated that the indigenous techniques have important implications related to promoting creative problem solving, creative education, tolerance, ethics, inspiration, and entertainment, expressing the unsayable, and predicting the future. Besides helping to preserve the heritage of the accumulated wisdom, the author plays a role in narrowing the divide between tradition and modernity regarding knowledge creation and is useful to researchers in creative education, communication, natural language, and information retrieval.

In chapter eighteen Fekadu Fullas and Mammo Muchie discuss Intellectual Property Rights (IPR) has been a subject of voluminous literature. Much has been written about its origin, implications and dispositions. Yet, the subject of IPRs is fraught with murky issues, especially as they apply to developing countries. This article examines IPRs within the context of developing countries, in particular those in Africa. It cited examples in the field of traditional medicine/medicinal plants to show how IPRs have been applied in the past. The paper has tried to chart a few pointers into the future to turn the tide in favour of developing countries in Africa by focusing more on potential collaboration among developing countries.

By and large, the chapters in this volume have brought a range of insights to enrich the debates on knowledge exchange in the form of indigenous and non-indigenous and challenge its linkage and harmonisation to our day to day lives. In the current fast moving technological advancement the place for knowledge exchange is increasingly in demand to narrow the knowledge gap. Globalisation has stridden to bring together the latest innovations and technologies that are meant to resolve many of the challenges humanity currently faced. I hope the chapters in this volume have contributed to the debates in knowledge exchange.

# PART I:
# LINKING INDIGENOUS KNOWLEDGE WITH SCIENTIFIC KNOWLEDGE

# CHAPTER 1

## Indigenous Knowledge in the Context of Natural Resources Management: An Information Systems Perspective

*Amare Desta*

### 1.0 Introduction

Over recent decades knowledge management has become the mantra of most contemporary organizations while knowledge associated with local communities has been largely ignored. However, there has been a growing recognition of the role that local communities' knowledge, especially indigenous knowledge (IK), plays in the management of natural resources. At the same time, worries about food production and global hunger have been enhanced by increased public concern over the rapid deterioration of the Earth's ecosystems.

A large proportion of the world's population depends upon IK to meet their food and medicinal needs, especially in developing countries. This paper reviews the existing indigenous knowledge (IK) literature and relates it to mainstream information systems and innovation studies issues, within the context of natural resource management (NRM).

### 2.0 Natural Resource Management (NRM)

According to The World Bank (1997), natural resources refer to a broad spectrum of 'environmental assets', including air, water, land, plants, animals and microorganisms. These assets are linked together to form natural systems, such as rivers and forests, and NRM reflects these linkages within and between natural systems. It integrates the management of social, economic and environmental values by involving the community in planning and other activities. NRM is fundamentally about people as its success is ultimately determined by the level of community involvement and the adoption of ecologically sustainable practices across the community (Ashley, 2000).

The extensive literature on NRM (e.g. Norfolk et al, 2003; Pritchard & Sanderson, 2002), highlights the importance of participatory development and knowledge management. There is widespread recognition (Bessette, 2004) that participatory development is critical for achieving sound resource management but this means empowering local communities. Information describing the natural resources forms the base upon which sustainable development is built (Tabor & Hutchinson, 1994) and hence it is important to manage knowledge resources effectively. However, the local knowledge resources in many communities in developing countries are not codified in Western scientific terms but instead comprise what is known as 'indigenous knowledge'.

Natural resource exploitation provides the livelihoods for a high proportion of the world's population (Pimental, et al., 2002). Since the 1992 Earth Summit in Rio, and culminating in the Copenhagen Conference in 2009, there have been increasing concerns about climate change and the sustainability of the world's natural resources. As human activity is the major destructive force in nature, improving NRM primarily requires changing human behaviour at 'grassroots' level (Röling, 2000). Today it is widely agreed that local people's perspectives need to be at the centre of research efforts for development and that innovations need to be 'owned' by the local land users, if changes in decision-making and behaviour leading to a positive impact are to be achieved. Such ownership can be created effectively through development and implementation of innovations by local people themselves in cooperation with outsiders (Hagmann & Chuma, 2002).

## 3.0 Indigenous knowledge (IK)

Various scholars (e.g. Murdoch & Clark, 2005; Norgaard, 2003) argue that IK plays an important role in the sustainable management of natural resources. At the same time, there has been a realization that scientific knowledge may have contributed little to the development of particular communities; rather, it may have hastened the depletion of their social and natural resources (FAO 1999).

However, what is meant by `indigenous knowledge' is by no means clear and Antweiler (2004) identifies no fewer than twenty-two similar terms, such as 'local knowledge' and 'folk knowledge', which are often used interchangeably. Sillitoe & Marzano (2009, p14) find that IK "varies within and between societies, comes from a range of sources and is a dynamic mix of past tradition and present innovation. It is heterogeneous and complicated which is an inconvenience for development." It is also diffused 'skills as knowledge', held by various people within a society and communicated through various symbols, myths and rites in an apparently piecemeal everyday fashion. They argue that "it is neither static nor uniform but ever-changing and subject to continual negotiation between people ... it is a process featuring the acquisition and integration of current information and experience" (p15).

Berkes & Berkes (2009) emphasise the relationship between IK and the local natural world and note that IK comprises institutions, in terms of rules and norms, about how to treat the environment, as well as comprising a particular worldview that influences how they make sense of this natural world. They also emphasise the holistic nature of IK, compared to Western reductionism. Thrupp (1989) points out that, while at first IK was seen as a potentially useful source of mere 'technical ideas', it also extends to "non-technical insights, wisdom, ideas, perceptions, and innovative capabilities which pertain to ecological, biological, geographical and physical phenomena" (p.15).

An IK system therefore provides the basis for decision-making, which is operationalised through indigenous organizations, which provide the foundation for local innovations and experimentation. IK systems are therefore adaptive skills of local people, usually derived from many years of experience, which have been communicated through oral traditions and learned through family members and generations. Local people, including farmers, landless labourers and rural artisans are all stakeholders of IK systems.

Dewalt (1994) identifies the following features of IK, which are relevant to NRM:

*locally appropriate*: IK represents a way of life that has evolved with the local environment.

*restraint in resource exploitation*: production is for subsistence needs only.

*diversified production systems*: there is no overexploitation of a single resource.

*respect for nature:* a "conservation ethic" often exists. The land is considered sacred, humans are dependent on nature for survival; all species are interconnected.

*flexible*: IK is able to adapt to new conditions and incorporate outside knowledge

*social responsibility*: there are strong family and community ties, and with them feelings of obligation and responsibility to preserve the land for future generations.

Furthermore, Grenier (1998) points out that IK:

Is usually parochial, confined to a small area, and limited to what rural people can sense, observe, and comprehend using their own terms and concepts.

Is not uniformly spread. Individuals vary in their aptitude for learning, storing, and generating knowledge.

Specialized knowledge often belongs to certain groups or individuals; for example, male elders, midwives, traditional healers (Eythorsson, 2000).

Includes both explicit and implicit knowledge, some of it intuitively practised through cultural rituals or revealed through stories and legends.

Is embedded in culture, can be complex. Attempts to "scientize" IK by removing it from its owners will tend to compromise the subtle nuances of this knowledge (Thrupp 1989).

Turnbull (2009), quoting a position paper from the World Summit on the Information Society in Geneva in 2003, argues that IK is the basis of people's cultures, identities, institutions and value systems and cannot be separated from their spiritual and material relationships with their lands. Furthermore, these cultures provide the rules for sharing and applying this knowledge. Berkes & Berkes (2009) see IK as "a body of knowledge built up by a group of people through generations of living in close contact with nature" (p7). IK is therefore understood to be the starting point for NRM in rural communities.

## 3.1 Indigenous knowledge and scientific knowledge

It is very difficult to discuss different systems of knowledge and different cultures without considering the realities of political and economic power. Bryan (2009, p24) notes that "the very concepts used to identify certain kinds of knowledge as indigenous remain steeped in colonial power relations". He discusses the production of maps and, in particular, the difficulties of 'indigenous mapping' where the traditional relationships between a people and the land are often considered to be 'unmappable'. Nevertheless, he argues that indigenous people are in the position of having to "map or be mapped" (p24). Maffie (2009) critiques the notion that Western hegemony, reflected in the triumph of the Gatling machine gun, somehow demonstrates the superiority of Western epistemology. As he argues: "indigenous knowledges have been defeated, not disproven, by Western technology" (p56).

IK is seen to be different from scientific knowledge and conventional wisdom has been that scientific knowledge is somehow more advanced and global than IK. However, the onset of 'global warming' and adverse climate change raises questions as to how advanced Western science actually is. Turnbull (2009) makes the point that scientific knowledge itself is 'local', based on the sociological notion that science is 'what scientists do' and is based on highly situated practices. Both knowledges are based on observation, some form of experimentation and the desire to create order out of apparent disorder (Berkes & Berkes, 2009) and "in some sense we are all indigenous and all knowledge including science is local" (Turnbull, 2009, p.3).

Similarly all knowledges are "the product of human movement, actions, practices and protocols. … [they] are dynamic, heterogeneous, social and distributed" (p.3). The knowledges have different epistemologies, with science based on evidence, repeatability and quantification while IK is often more related to spiritual and religious practices. However, there is no meaningful meta-theory to compare the different varieties. IK may be lacking in terms of scientific (positivist) epistemology but it rests on a very different epistemology. Furthermore, Sillitoe & Marzano (2009) argue that the distinction between IK and science is 'misleading' as, in practice, they borrow from each other.

In trying to 'square the circle' between IK and scientific knowledge, authors offer various solutions in terms of providing a space for different knowledges. Green (2009) talks about a 'duality' of IK and science, suggesting that different epistemologies, based on different 'moral economies' should be accepted, such that different knowledges are not seen as mutually exclusive. She argues for a 'reflective equilibrium' to compare the different epistemologies. Berkes & Berkes (2009) note a similarity between IK and 'fuzzy logic', a form of science proposed by Zadeh (1965) which is seen as being highly legitimate within, for example, the artificial intelligence community. Sillitoe & Marzano (2009) argue for a model comprising 'linked spheres of knowledge', in the absence of a single theory of knowledge that would link IK with science, while Maffie (2009) proposes a 'polycentric global epistemology' that would accept such practices as dance, song and ritual performance as legitimate knowledge mechanisms.

Many authors (Labatut & Akhtar, 2005) have stressed the value of IK for development. But IK has its limitations (Bebbington, 1999) and is not in itself capable of addressing all the issues related to sustainable development (Murdoch & Clark, 2005). Sustainable development may well be better served by a system that incorporates both indigenous and scientific knowledge systems (Icamina, 1999) and creating a technological base that includes both traditional and modern approaches to problem-solving (Johnson, 2005).

## 3.2 IK for sustainable NR management

In most developing countries, the majority of the population are small-scale farmers with less than one hectare of land. Their knowledge systems have never been recorded systematically in written form and hence are not easily accessible to researchers. While remaining invisible to the development community, many indigenous organizations (e.g. farmers' associations) operate in rural communities identifying solutions to community problems. NRM planners and policy makers are beginning to recognize the need to understand existing knowledge systems and decision-making processes. There is a general agreement that agricultural innovations based on IK have been tested through time (Warren & Rajasekaran, 1993).

Despite the evidence, there is scepticism about the relevance of IK for NRM, partly because indigenous communities never record their accomplishments, never attach their names and patents to their discoveries and inventions. As a result, the history of natural resource development is written without reference to the main stakeholders (Kajembe & Wiersum, 2004). According to the FAO (1999), NRM has been much more concerned with conserving resources 'without' local communities. Protection of natural resources has at times been seen as necessitating disruption of traditional ways of life.

As documented by IFAP (1990), neglecting IK undermines farmers' confidence in their traditional knowledge, which in turn forces them to become increasingly dependent on outside expertise. IFAP (1990) also asserts that small-scale farmers are often portrayed as backward, obstinately conservative, resistant to change, lacking innovative ability, and even lazy due to:

a lack of understanding of traditional agriculture which further leads to a communication gap between promoters and practitioners giving rise to myths;

the accomplishments of farmers often are not recognized, because they are not recorded;

poor involvement of farmers and their organizations in integrating, consolidating and

disseminating what is already known.

Atte (2004) argues that the under-utilization of IK systems leads to the loss of indigenous acquired knowledge, which results in the inefficient allocation of resources and manpower. With little contact with rural people, planning experts have attempted to implement programs which do not meet the goals of rural people, or affect the structures and processes that perpetuate rural poverty. Human and natural resources in rural areas have remained inefficiently used or not used at all. Planners think they know what is good for these 'poor', 'backward', 'ignorant', and 'primitive' people (Atte, 2004).

Timely attention is now beginning to be paid to incorporating IK into NRM processes as, in North America for example, indigenous perceptions of land use and landscapes have been transcribed into maps (Brody, 1982). Nakashima & Reed (2005) note that IK has been applied to historical climatic research, geophysical research, rural land use and resource management planning. Indeed, the IK of local flora and fauna often exceeds that of western scientists. However, while indigenous peoples have profound knowledge of local ecology, the apparent informality of such information does not sit comfortably with the western scientific tradition. Local knowledge about the land identifies issues of immediate significance and encodes information in a language that local people understand (Ramisch, 2002) unlike much scientific land use information.

Attempts to involve Canadian native indigenous populations in planning through the Northern Land Use Planning Programme and environmental assessment processes were unsuccessful, partly because of a failure to develop adequate frameworks for dealing with indigenous knowledge (Fenge & Rees, 1987). In addition to NRM, IK is widely used in medicine and, according to Kaya (2009), 65% of poor people in sub-Saharan Africa depend on traditional medicine for basic health care. Furthermore, the commercialisation of traditional medicines is an important part of pharmaceutical research and development with world sales of herbal medicines reaching $30 billion in 2000. This raises difficult issues concerning the division of profits and intellectual property rights.

## 3.3 Preservation of IK

IK, which has generally been passed through generations by word of mouth, is in danger of being lost unless it is formally documented and preserved (Warren, 2004). Such a loss would impoverish society because, just as the world needs genetic diversity of species, it needs diversity of knowledge systems (Labelle, 1997). The rapid change in the way of life of local communities has largely accounted for the loss of IK. Younger generations underestimate the utility of IK systems because of the influence of modern technology and education (Ulluwishewa, 1999).

If IK is not recorded and preserved, it may be lost and remain inaccessible to other indigenous systems as well as to development workers. Development projects cannot offer sustainable solutions to local problems without integrating local knowledge (Warren, 1991). IK is the key to local-level development (Schoenhoff, 1999) and ignoring people's knowledge is likely to ensure failure (Brokensha *et al.*, 1997). One should not expect all the expertise for Third World development to come from the West; in the face of dwindling resources, IK could provide vital tools for rural development (Atte, 1989).

Since IK is essential to development, it should be gathered, organized and disseminated, just like Western knowledge (Agrawal, 1995) but this raises issues related to methodology, access, intellectual property rights and the media and formats in which to preserve it (Msuya, 2007). Underlying these challenges is the dilemma of whether to use the Western paradigm for collecting and preserving IK.

Some scholars (Ulluwishewa, 1999) recommend *ex situ* conservation strategies, i.e. isolation, documentation and storage in external archives. In the 1990s this strategy was used to document the healing practices of the

Fulani pastoralists in the north-west province of Cameroon (Nuwanyakpa, 2006). On the other hand, those who advocate maintaining distinctions between scientific knowledge and IK have supported *in situ* preservation of IK (Agrawal, 1995).

Lawas & Luning (1996) point out that the collection of indigenous information is time-consuming and costly. They argue that library and information professionals should design collection development policies but it could be argued that the collection of IK in the field should be left to ethnographers, anthropologists, and related professionals. Instead, information professionals should collaborate with national IK resource centres to enhance access to IK. A contentious issue in the management and preservation of IK is the protection of intellectual property rights. In this regard, the United Nations Draft Declaration on the

Rights of Indigenous Peoples underscore the fact that indigenous peoples have the right to own and control their cultural and intellectual property (Valsala & Kutty, 2002). Although most IK is held in the minds and practices of people, and is commonly held by communities rather than individuals, intellectual property rights that are intended to protect the ownership of the intellectual content of the works of an individual can be applied. In the Western tradition the intellectual property must be tangible, taking the form of a written document, a recording of music, a painting or drawing, and the like. Sometimes IK is tangible. For instance, there are songs, stories, music, statues, paintings, designs, processes and drawings that embody traditional knowledge. These are capable of being protected either individually or communally. Upholding intellectual property rights should benefit indigenous communities by the commercial use of their traditional knowledge. This could be an area where information professionals could contribute.

## 3.4 Challenges and limitations of IK

Although the knowledge of indigenous communities has been found to be very useful, the spread of industrialization threatens its preservation and continued development (Sherpa, 2005). Industrialization, along with its attendant processes of urbanization, exploitation of NR, and increased competition for employment, has set off a problematic chain of events. IK can also be eroded by wider economic and social forces. Pressure on indigenous peoples to integrate with larger societies is often great and, as they become more integrated, the social structures which generate IK and practices can break down. Added to this is the commercial pressure by multinational agro-chemical companies eager to break into new markets (Thrupp, 1989).

As Grenier (1998) puts it:

> *"the growth of national and international markets, the imposition of educational and religious systems and the impact of various development processes are leading more and more to the "homogenisation" of the world's cultures. Consequently, indigenous beliefs, values, customs, know-how and practices may be altered and the resulting knowledge base incomplete."*

As with scientific knowledge, however, IK has its own limitations and these must be recognized. IK is sometimes accepted uncritically because of naive notions that whatever indigenous people do is naturally in harmony with the environment. Thrupp (1989) argues that we should reject "romanticized and idealistic views of local knowledge and traditional societies" (p15). There is historical and contemporary evidence that indigenous peoples have also committed environmental 'sins' through over-grazing, over-hunting, or over-cultivation of the land. It is misleading to think of IK as always being 'good', 'right or 'sustainable'.

Like scientific knowledge, sometimes the knowledge which local people rely on is wrong or even harmful. Practices based on, for example, mistaken beliefs, faulty experimentation, or inaccurate information can be dangerous. Some IK that was once well-adapted and effective becomes inappropriate under conditions of environmental degradation (Thrupp, 1989).

Although IK systems have certain flexibility in adapting to ecological change, when change is particularly rapid or drastic, the knowledge may be rendered unsuitable and possibly damaging. As shown in the above brief review of the IK literature, much of the debate is normative, political and pragmatic. This can be useful in making quick improvements to NRM but adds little to our deeper understanding of IK. Very little of the existing literature has much of a theoretical base and it is conjectured that progress could be made by exploring IK from the perspective of relatively modern disciplines, such as information systems, knowledge management and innovation studies.

## 4.0 Indigenous knowledge, information systems & knowledge management

Apart from knowledge management, IK has been little explored within IS research and development projects, with the possible exception of geographical information systems (GIS), many of which have been concerned with NRM (Mbile et al., 2003). There are important spatial aspects to IK and GIS offer the opportunity to facilitate the management and utilization of IK (Lawas & Luning, 1996). Tabor & Hutchinson (2004) and Gonzalez (1995) describe the advantages of using GIS to document IK. However, as argued by Walsham & Sahay (1999), the use of GIS in developing nations provides a classic example of the utilization and transfer of technology problem, which typically involves the introduction of Western technical systems into developing countries. Furthermore, Sahay & Walsham (1997), in discussing the use of GIS in India, highlight various problems; for example, the development of systems that are not considered relevant by users, the lack of continuity in project management practices; and inappropriate co-ordination between the various agencies. Many approaches to integrating IK into GIS have been participatory in nature. These include Gonzalez (1995) in the Philippines and Rundstrom, (2006) in Nepal. McConchie & McKinnon (2003) pioneered a technique called Mobile Interactive Geographic Information System, developed for integrating IK to produce community based maps for collaborative NRM. While there is an increasing interest in using GIS in a participatory context (Abbot et al., 1998), there are fears that it could be misused, wrongly interpreted, or not used at all and, if poorly designed, it could dis-empower underprivileged groups (Jordan & Shrestha, 2005).

However, rather than within information systems, it is within the field of knowledge management that IK can be more readily discussed. The notion of knowledge management grew from the early predictions that we were entering a post-industrial society (Bell, 1973) which would feature a knowledge economy (Toffler, 1990). Nonaka (1994), one of the most influential theorists, built a theory of organizational knowledge creation, based on Polanyi's (1967) distinction between tacit and explicit knowledge. According to Polanyi, tacit knowledge was based on experience, behaviour and skills, which is held in the brain of the person, whereas explicit knowledge is articulated and can be documented and stored on paper or electronically.

Nonaka (1994) argued that knowledge is created within the firm through modes of interaction between tacit and explicit knowledge and these different modes act together dynamically to form a spiral of knowledge creation. Nonaka's model implicitly views knowledge as an object (Thompson & Walsham, 2004) that is constructed and can then be shared by others. This led to definitions, such as the one by Brooking (1997): "knowledge management is the activity which is concerned with strategy and tactics to manage human centred assets". However, Thompson & Walsham (2004, p.726) argue that "the meaning of any objective

'knowledge' will always remain the subjective product of the person in whose mind this is constituted, always relationally defined, and therefore does not transfer easily to others in a form which may be operationalised to the benefit of the organization".

They also point out that Polanyi himself regarded explicit knowledge as self-contradictory. The alternative approach to knowledge as object is to take a practice-based view (Blackler, 1995) where objective *knowledge* is considered more as an inter-subjective process, resulting in the recipient *knowing*. This also fits better with Weick's (1995) notion of sense-making and Lave & Wenger's (1991) situated learning. Thompson & Walsham (2004) emphasise the importance of the organizational context for these processes and they view *knowing* as "mediated, situated, provisional, pragmatic and contested" (p.743). Habermas (2003) dismisses the notion of knowledge as the representation of reality but regards it rather as a competence to do something successfully in practice. From an actor-oriented perspective, both scientific and IK are fragmentary, partial and temporal. They are constantly being generated and constructed as products of dynamic processes of interaction between various actors with different cultural backgrounds and understandings (Katani, 2005).

Knowledge is not just a commodity but is the outcome of a process which is a result of negotiation on the 'social interface' between multiple actors (Long & Villareal, 1994). From this perspective, local stakeholders (individuals or groups) should be seen as situated agents (Kajembe, 2003). Within the limits of existing information, uncertainty and other constraints (e.g. physical, social and politico-economic), local actors are knowledgeable and capable (Chambers et. al., 1989). They attempt to solve problems, learn how to intervene in the flow of social events around them, and monitor continuously their own actions, observing how others relate to their behaviour and taking note of various contingent circumstances. Human agency, or the capacity to devise ways of coping with life, plays an important role in the way actors create new possibilities.

McAdam & McCreedy (2000) compare what they call the 'social paradigm' of knowledge construction, using Lave & Wenger's (1991) and Demerest's (1997) models, which emphasise practice, interaction and communication, with the 'scientific paradigm' (equivalent to knowledge as object), which produces a "canonical body of facts and rational laws" (p.158). For them, the social paradigm seems more useful within the business context. Sutton (2001) concludes that "knowledge may be codified into texts and artefacts but only functions in people" (p.87).

Alavi & Leidner (2001) discuss the various conceptualizations of knowledge and go on to develop a framework comprising four sets of 'socially enacted knowledge processes':

knowledge creation
knowledge storage/retrieval
knowledge transfer
knowledge application

Lave & Wenger's (1991) work led to a growing research interest into 'communities of practice'. According to them, a community of practice is "a system of relationships between people, activities, and the world: developing with time, and in relation to other tangential and overlapping communities of practice" (1991, p.98). Such communities are seen as being hugely important for knowledge creation as they provide the interaction, the shared basis of understanding and the propagation channels for the creation and sharing of knowledge (Wenger, 1998).

These different conceptualizations of knowledge suggest different epistemologies and Spender & Scherer (2007), among others, argue for a tolerance of these differences and a 'pluralistic conversation' between them. Similarly, Schultze & Leidner (2002) argue that the ambiguity regarding the nature of knowledge, and the different types of knowledge, imply the need for different 'discourses' and they propose the adoption of the four discourse types of Deetz (1996) – normative, interpretive, dialogic and critical - in order to examine knowledge management.

The dominant information management model has been based on acquiring, organizing and preserving recorded and codified knowledge, which is largely generated by researchers, laboratories and research institutions. Such a model has little room for IK, which is not formally codified and resides largely in the minds of local people. Nevertheless, the growing importance of knowledge and knowledge management implies that IK should be accorded a suitable place in the pluralistic conversation of Spender & Scherer (2007).

## 5.0 IK and Systems of Innovation

An innovation study is a fast-emerging multidisciplinary field within the social sciences (Fagerberg & Verspagen 2009) and innovation is high on the agenda of most governments in their attempts to reinvigorate flagging economies. The literature is usually traced back to the work of Schumpeter (1934) who saw innovation as the driving force behind economic and social change. His work, which formed much of the foundation of evolutionary economics, was continued by Freeman et al (1982) and Nelson & Winter (1982).

Mytelka & Smith (2002) emphasise the shift over the years from viewing innovation as a process of discovery to seeing it more as 'a non-linear process of learning', based largely on the evolutionary ideas of Rosenberg (1976). Rogers's (1962) theory of the diffusion of innovation, based on an S-curve, is regularly referred to within the information systems literature. Amabile et al (1996) define innovation as "the successful implementation of creative ideas within an organization" and this reflects the management literature that seeks to enhance the innovative capacity of individual firms. According to Schumpeter (1934), innovation includes the introduction of new products, methods of production, markets, sources of supply and forms of organization.

Another strand of innovation research refers to national systems of innovation (e.g. Edquist, 2004), which is more concerned with the political economy of innovation at the regional, national and supra-national levels. Lundvall (1992) defines them as "elements and relationships which interact in the production, diffusion and use of new and economically useful knowledge ... and are either located within or rooted inside the borders of a nation state".

However, most writers (e.g. Niosi 2002) regard them more specifically as networks of firms, universities and government agencies. These networks include Triple Helix and Globelics, which have grown over the last seven years or so with the purpose of sharing and refining knowledge, learning and development, as well as linking the 'helices' of university-industry-government. It is this stream of research that has largely driven innovation policy and it has little to say about IK or community innovation. Various authors (e.g. Ernst 2002) have discussed innovation systems in developing countries mostly focusing on the networks of formal institutions. There is clearly a considerable overlap between knowledge management and innovation studies. Swan et al (1999) offer a framework that maps process and product innovation against 'cognitive' and 'community' knowledge management. McAdam (2000) views knowledge management as a 'catalyst' for innovation within organizations and he goes on to fit innovation 'drivers' within a framework based on Demerest's (1997) model of knowledge management.

Sorensen & Lundh-Snis (2001) suggest that knowledge classification and codification are means for organizational learning and innovation. Chang & Chen (2004) distinguish between three approaches to national systems of innovation (culture and politics-bounded; technological/sectoral; and regional/local) in terms of knowledge links while Popadiuk & Choo (2006) discuss the relationship between knowledge creation and innovation.

A recent link between innovation and information systems research is the growing interest in 'open innovation', based on the ideas of Chesbrough (2003). Open innovation refers to the notion that, rather than relying on internal sources, organizations should seek innovative ideas and projects externally, particularly through joint ventures and other partnerships with universities, small businesses and individual entrepreneurs. Such partnerships are facilitated by knowledge sharing and improved communication using the Internet. Despite the use of the term 'indigenous innovation' by Lazonick (2004) in discussing the economic development of China, there is very little mention of IK within the innovation literature. As noted above, the emphasis is mostly on formal networks of institutions or the more radical open innovation.

A rare exception is Kaya (2009) who discusses IK and innovation systems in public health in Africa, as well as noting the complementarity between traditional food and traditional medicine. He describes various research and development projects and initiatives in IK and innovation and repeatedly uses the phrase 'IK and innovation systems', suggesting that innovation is an inherent part of IK. He goes on to refer to the process where large multinational pharmaceutical, agricultural and biotechnological corporations patent IK techniques and products as turning "the owners of traditional knowledge into beggars" (p.103).

A little recognised aspect of IK is its experimental and innovative nature. The term IK may create an impression of knowledge that is static, having been handed down through countless generations. However, in reality this knowledge is constantly evolving and being updated with new information. Various authors (e.g. Muchie & Baskaran 2006) emphasise the importance for developing countries to build their own innovation capabilities, rather than relying on the West for innovations that may not be appropriate for the local context. Rhodes & Bebbington (2001) identified three kinds of indigenous farmers experiments: *curiosity experiments* (where farmers experiment simply out of curiosity to test new numbers and sizes of crops); or *problem solving experiments* (where farmers carry out experiments to solve problems); or *adaptation experiments* (where farmers can either test unknown technology in a known environment or test known technology in a new environment).

Studying experiments undertaken by rural people gives an understanding of their 'sense making' activities (Brouwers 2002). Scientists tend to regard an experiment as an enquiry during which all the variables are highly controlled except those under study. Local people differ in the sense that the experiment has to be included in daily circumstances (Kajembe, 2003). Richards (2002) concludes that in recent IK literature the experimenting, innovative, adaptive indigenous farmer is now accepted as the norm, not the exception.

## 6.0 Conclusion

Natural resource management has grown in stature with the increasing concerns stemming from the global problems of climate change, poverty and hunger. Within this context, participatory development and knowledge management have been identified as important factors. In developing countries this raises the issue of indigenous knowledge, which is the knowledge that local people have built up over generations concerning agriculture and the management of local natural resources.

Indigenous knowledge (IK) differs from scientific knowledge in many respects, not the least of which is that it is not systematically recorded. IK is not necessarily inferior, or superior, to scientific knowledge but combining the two knowledge systems is not straightforward. However, IK does offer potential benefits for sustainable natural resource management and there is a strong argument for its preservation, as it is threatened by the forces of globalisation and modernisation. The literature of information systems and knowledge management has largely neglected indigenous knowledge but there is scope for using frameworks and techniques from these disciplines to foster the preservation and continued development of indigenous knowledge. Similarly, although there is considerable evidence that local innovation plays a strong role in indigenous knowledge, the literature on innovation also generally ignores this community innovation.

The author of this paper believes that there is an opportunity to rectify this neglect and to contribute to sustainable natural resource management through the application of the ideas and experience of researchers in information systems and innovation.

## References

Abbot, J., Chambers, R., Dunn, C., Harris T., de Merode, E., Porter G., Townsend, J.,

Weiner, D., (1998) *Participatory GIS: opportunity or oxymoron?* PLA Notes33, October 1998.

Agrawal, A. (1995) Dismantling the divide between indigenous and scientific knowledge. *Development and Change* 26(3): 413–39.

Alavi, M. & Leidner, D.E. (2001) Knowledge management and knowledge management systems: Conceptual foundations and research issues. *MIS Quarterly*, Vol.25, No.1, 107-136

Amabile et al 1996 Assessing the work environment for creativity, *Academy of Management Journal* **39** (5) (1996), pp. 1154–1184

Antweiler, C. (2004) Other knowledge and other ways of knowing. *Journal of Anthropological Research* 50:65-68

Ashley, C. (2000) Applying Livelihood Approaches to Natural Resource Management Initiatives: Experiences in Namibia and Kenya, Working Paper 134.

Atte, O.D. (1989) Indigenous local knowledge – a key to local-level development: possibilities, constraints and planning issues in the context of

Africa. Paper presented at the seminar on 'Reviving local self-reliance: challenges for rural/regional development in eastern and southern Africa.' Arusha, Tanzania, 21–24 February.

Atte, O.D. (2004) Indigenous local knowledge – Paper presented at the seminar on 'Reviving local self-reliance: challenges for rural/regional development in eastern and southern Africa.' Arusha, Tanzania

Bebbington, A. (1999) Modernization from below: An alternative indigenous development', *Economic Geography* 69(3):274-292.

Bell, D. (1973) The Coming of Post-Industrial Society: A Venture in Social Forecasting. New York: Basic Books.

Berkes, F. & Berkes, M.K. (2009) Ecological complexity, fuzzy logic, and holism in indigenous knowledge. *Futures*, 41, 6-12.

Bessette, G. (2004) Facilitating Community Participation, Peinang, Southbound and Ottawa, International Development Research Centre (IDRC).

Blackler, F. (1995) Knowledge, knowledge work and organizations: An overview and interpretation. *Organization Studies*, 16 (6), 1021-1046.

Brody, R. (1982) Problem Solving Concepts and Methods for Community Organizations. Human Sciences Press: New York.

Brokensha, D., Warren, D. and Werner. O. (1997) Indigenous knowledge systems and development. Lanham: *University Press of America.*

Brooking, A. (1997) The management of intellectual capital. *Journal of Long Range Planning*, 30 (3), 364-370.

Brouwers, D. (2002) Privacy, Publicity and Propriety in Congressional Eulogies for Representative Stewart B. McKinney (R-Conn.) *Rhetoric & Public Affairs* -Volume 2 & 7, pp. 191-214

Bryan, J. (2009) Where would be without them? Knowledge, space and power in indigenous politics. *Futures*, 41, 24-32.

Chambers, R., Pacey, A. and Thrupp, L.A. (1989) Farmer First: Farmer innovation and agricultural research. London: *Intermediate Technology Publications.*

Chang, Y-C. & Chen, M-H. (2004) Comparing approaches to systems of innovation: The knowledge perspective. *Technology in Society*, 26, 17-37.

Chesbrough, H (2003) *Open Innovation: The New Imperative for Creating and Profiting from Technology*. Harvard Business School Press, Boston.

Deetz, S. (1996) Describing differences in approaches to organization science: Rethinking Burrell and Morgan and their legacy. *Organization Science*, 7 (2), 191-207.

Desta, A. (2010) PhD Thesis, London School of Economics (LSE), London

Demerest, M. (1997) Understanding knowledge management. *Journal of Long Range Planning*, 30 (3), 374-384.

Dewalt, B. (1994) Using Indigenous Knowledge to Improve Agriculture and Natural Resource Management. *Human Organisation* 53: 123–131, Lanham, Altamira press

Edquist, C. (2004) Systems of innovation: Perspectives and challenges. In: Fagerberg, J. Mowery, D.C. & Nelson, R.R. (eds.) *Oxford Handbook of Innovation*. Oxford University Press, Oxford.

Ernst, D. (2002) Global production networks and the changing geography of innovation systems: Implications for developing countries. *Economics of Innovation and New Technology*, 11 (6), 497-523.

Fagerberg, J. & Verspagen, B. (2009) Innovation studies – The emerging structure of a new scientific field. *Research Policy*, 38, 218-233.

FAO (1999) Environment Report Series 2 - Report No. 99/023 IFAD-RAF

Freeman, C., Clark, J. & Soete, L. (1982) *Unemployment and Technical Innovation: A Study of Long Wave and Economic Development*. Pinter, London.

Fenge, T. and Rees, W.E., (1987) Hinterland or Homeland?: Land-Use Planning in Northern Canada. *Canadian Arctic Resources Committee:* Ottawa

Gonzalez, R. M. (1995) KBS, GIS and documenting indigenous knowledge. *Indigenous Knowledge and Development Monitor 3*(1): 5–7.

Green, L.J.F. (2009) Challenging epistemologies: Exploring knowledge practices in Palikur astronomy. *Futures*, 41, 41-52.

Grenier, L. (1998) Working with Indigenous Knowledge: A Guide for Researchers. IDRC, Ottawa, Canada

Habermas, J. (2003) *Truth and Justification*. MIT Press, Cambridge, MA.

Hagmann, J. and Chuma, E. (2002) Enhancing the adaptive capacity of the resource users in natural resource management'. *Agricultural Systems* Vol. 73, No. 1, pp. 23–39.

Icamina, P. (1999) 'Threads of common knowledge', IDRC Reports 21(1): 14-16. IFAP (1990) Sustainable Farming and the Role of Farmers' Organizations. Wageningen, The Netherlands: International Federation of Agricultural Producers

Johnson, M, (2005) Lore, Capturing Traditional Ecological Knowledge. Pp 190. Dene Cultural Institute (Hay River) and International Development Research Centre (Ottawa).

Jordan, G. and Shrestha, B. (2005) A participatory GIS for community forestry user groups in Nepal: Putting people before the technology. *Participatory learning and action notes 39*. International Institute for Environment and Development.

Kajembe, G.C. and Wiersum, K.F. (2004) Bridging the gap between Indigenous initiative and externally sponsored forest interventions: In Chamshama, S.A.O (ed.). Special Issue, Proceedings of the 1st Annual Forestry Research Workshop. Record No. 67. Faculty of Forestry, Sokoine University of Agriculture Morogoro, Tanzania pp 95 - 105.

Kajembe, G.C. (2003) Perspectives on 'How Local Knowledge can be utilized for Sustainable Natural Resource Management'. Proceedings of the Regional Refresher course for the Alumni of the International Course on the Design of Community Forestry (ICDCF) from Africa held in Nairobi, Kenya.

Katani, J.Z. (2005) Coping Strategies Against Deforestation: Impact of socioeconomic Factors with special attention to Gender-Based Indigenous Knowledge: A case study of Mwanza District, Tanzania.

Kaya, H.O. (2009) Indigenous knowledge and innovation systems for public health in Africa. In: Kalua, F.A., Awotedu, A., Kamwanja, L.A. & Saka, J.D.K. (eds.).*Science, Technology and Innovation for Public Health in Africa*. NEPAD Office of Science and Technology, Pretoria, Republic of South Africa. 95

Labatut, G.M and Akhtar, S. (2005) Traditional environmental knowledge: A resource to manage and share. *Development Journal of the Society for International Development 4*: 24–29.

Labelle, H. (1997) Presidential address. Canadian International Development Agency at the plenary session on global knowledge and local culture of the International Global Knowledge 1997 Conference in Toronto. http://www.kivu.com/

Lave, J. & Wenger, E. (1991) *Situated Learning: Legitimate Peripheral Participation*. Cambridge University Press, Cambridge.

Lawas, C.M. and Luning, H.A. (1996) Farmers' knowledge and GIS. *Indigenous Knowledge and Development Monitor,*4 (1).http://www.nuffic.nl/ciran/ikdm/4_1/articles/lawas.html.

Lazonick, W. (2004) Indigenous innovation and economic development: Lessons from China's leap into the information age. *Industry and Innovation*, 11 (4),273-297.

Long, N and M. Villareal, (1994), The Interweaving of Knowledge and Power in Development Interfaces. In: Scoones, I. and J. Thomson (eds), *Beyond Farmer First*. Intermediate technology publications, London

Lundvall, B.A. (1992) *National Systems of Innovation: Towards a Theory of Innovation and Interactive Learning*. Pinter, London.

Maffie, J. (2009) 'In the end, we have the Gatling gun, and they do not': Future prospects of indigenous knowledge. *Futures*, 41, 53-65.

Mbile, P., DeGrande, A. and Okon, D. (2003) Integrating Participatory Resource Mapping (PRM) and Geographic Information Systems (GIS) in humid lowland sites of Cameroon, Central Africa: A methodological guide. *Electronic Journal on Information Systems in Developing Countries 14*.

McAdam, R. (2000) Knowledge management as a catalyst for innovation within organizations: A qualitative study. *Knowledge and Process Management*, 7(4), 233-241.

McAdam, R. & McCreedy, S. (2000) A critique of knowledge management: Using asocial constructivist model. *New Technology, Work and Employment*, 15 (2),155-168.

McConchie, J.A. and McKinnon, J.M. (2003) Using GIS to produce community based maps to promote collaborative natural resource management

Msuya, J. (2007) Challenges and opportunities in the protection and preservation of Indigenous Knowledge in Africa

Muchie, M. and Baskaran, A. (2006) Introduction: The Problem of Integrating ICT within National Systems of Innovation: Concepts, Taxonomies and Strategies, In Muchie, M. &Baskaran, A. (eds.) *Bridging the Digital Divide : Innovation Systems for ICT in Brazil, China, India, Thailand and Southern Africa*, Adonis& Abbey, London, pp. 23-50.

Murdoch, J. and Clark J. (2005) Sustainable Knowledge, *Geoforum,* New York, 25.2: (115-132)

Mytelka, L.K. & Smith, K. (2002) Policy learning and innovation theory: An interactive and co-evolving process. *Research Policy*, 31, 1467-1479.

Nakashima, D.J. and Reed, M.G. (2005) Application of Native Knowledge in EIA: Inuit. Eiders and Hudson Bay Oil, *Canadian Environmental Assessment Research Council: Hull*, Quebec

Nelson, R.R. & Winter, S.G. (1982) *An Evolutionary Theory of Economic Change*. Harvard University Press, Cambridge, MA.

Niosi, J. (2002) National systems of innovations are 'x-efficient' (and x-effective):Why some are slow learners. *Research Policy*, 31.

Nonaka, I. (1994) A Dynamic Theory of Organizational Knowledge Creation, *"Organization Science*(5:1)

Norfolk, S., Nhantumbo, I. and Pereira, J. (2003) Community based natural resources management in Mozambique: a theoretical or practical strategy for local sustainable development? The case study of Derre Forest Reserve,

Mozambique, Sustainable Livelihoods in Southern Africa Research Paper 10,Institute of Development Studies

Norgaard, R. (2003) Co-evolution of Economy, Society and Environment. Real Life Economics. *Routledge:* London

Nuwanyakpa, M. (2006) Ethnoveterinary healing practices of Fulani pastoralists in Cameroon: combining the natural and the supernatural. Serials collection: Indigenous Knowledge & Development Monitor

Pimental, D., McNair, L. and Buck (2002) The value of forests to world food security, *Human Ecology* 25, pages 91–120.

Polanyi, M. (1967) *The Tacit Dimension*. Anchor Books, New York.

Popadiuk, S. & Choo, C.W. (2006) Innovation and knowledge creation: How are these concepts related? *International Journal of Information Management*, 26,302-312.

Pritchard, L. and Sanderson, S.E. (2002) The dynamics of political discourse in seeking sustainability. Chapter 6 in L.H. Gunderson and C.S. Holling (eds),

Panarchy: understanding transformations in human and natural systems. *Island Press.* Washington, D.C.

Ramisch, J.J. (2002) Understanding Soil in its Social Context: Integrating Social and Natural Science Research within AfNet – Paper presented at the Social Research conference. Looking back, looking forward: Social Research in CGIAR System, hosted by CIAT, in Cali, Colombia].

Rhodes, R.E. and Bebbington, A. (1999) Farmers Who Experiment: An Untapped Resource for Agricultural Development. Lima, Peru: International PotatoCenter (CIP).

Richards, P. (2002) Culture and Community Values in the Selection and Maintenance of African Rice," in Stephen B. Brush and Doreen Stabinsky, eds., *Valuing Local Knowledge: Indigenous People and Intellectual Property Rights,* Washington, D.C.

Rogers, E. (1962) *Diffusion of Innovations*. The Free Press, New York.

Röling, N.G. (2000) Gateway to the global garden. Beta/gamma science for dealing with ecological rationality. 8th Annual Hopper Lecture, October 24,2000.Canada: University of Guelph.

Rosenberg, N. (1976) *Perspectives on Technology*. Cambridge University Press, Cambridge.

Rundstrom, R. A. (2006) Mapping, Postmodernism, Indigenous People and the Changing Direction of North American Cartography. *Cartographica* 28(2): 1-12.

Sahay, S. and Walsham, G. (1997). Using GIS in developing countries: Social and Management Issues. *UNIDO Publications*: Vienna.

Schoenhoff, D.M. (1999) The barefoot expert: The interface of computerised knowledge systems and indigenous knowledge systems. Westport, Connecticut: Greenwood Press.

Schultze, U. and Leidner, D. E. (2002) Studying Knowledge Management in Information Systems Research: Discourses and Theoretical Assumptions, "*MIS Quarterly* (26:3), 213-242.

Schumpeter, J. (1934) *The Theory of Economic Development*. Harvard University Press, Cambridge, MA.

Sherpa, N.S. (2005) *International Workshop on Traditional Knowledge*. Organized by United Nations held in Panama City, 21-23 Sep. 2005.

Sillitoe & Marzano (2009) Futures of Indigenous Knowledges, Department of Anthropology, University of Durham

Sorensen, C. & Lundh-Snis, U. (2001) Innovation through knowledge codification. *Journal of Information Technology*, 16, 83-97.

Spender, J.C. & Scherer, A.G. (2007) The philosophical foundations of knowledge management: Editors' introduction. *Organization*, 14 (1), 5-28.

Sutton, D.C. (2001) What is knowledge and can it be managed? *European Journal of Information Systems*, 10, 80-88.

Swan, J., Newell, S., Scarbrough, H. & Hislop, D. (1999) Knowledge management and innovation: Networks and networking. *Journal of Knowledge Management*, 3 (3), 262-275.

Tabor, J. A. and Hutchinson, C. F. (2004) Using Indigenous Knowledge, Remote Sensing and GIS for Sustainable Development in Indigenous Knowledge Monitor Vol. 2, (1) April 1994

Thompson, M.P.A. & Walsham, G. (2004) Placing knowledge management in context. *Journal of Management Studies*, 41 (5), 725-747.

Toffler, A. (1990) Power shift: Knowledge, Wealth and Violence at the Edge of 21$^{st}$Century. Bantam Books, New York.

Thrupp, L.A. (1989) Legitimizing local knowledge: from displacement to empowerment for Third World people. Agriculture and Human Values, 6(3),13–24.

Turnbull, D. (2009) Futures for indigenous knowledges. *Futures*, 41, 1-5.

Ulluwishewa, R. (1999) National Knowledge, National IK Resource Centers and Sustainable Development. Development Administration and Management, University of USAID Guinea

UN (2007) Traditional Knowledge and Folklore, *11th Session – Inter-governmental Committee on Intellectual Property and Genetic Resources*, World Intellectual Property Organization, Geneva,

Valsala, P.V. and Kutty, G. (2002) National Experiences with the protection of expressions of folklore/traditional cultural expressions; India and The Philippines. http://www.wipo.int/tk/en/studies/cultural/expressions/study/kutty.pdf

Warren, D. M. (1991) Using Indigenous Knowledge in Agricultural Development. *World Bank Discussion Paper* No.127. Washington, D.C.: The World Bank.

Warren, D.M. (2004) *Indigenous knowledge, biodiversity conservation and development.* Keynote address at the International Conference on Conservation of Biodiversity in Africa: Local initiatives and institutional roles. Nairobi, Kenya, 30 August-3 September.

Warren, D. M., and Rajasekaran B. (1993) Putting local knowledge to good use. *International Agricultural Development* 13 (4): 8-10.

Warren, D.M., L.J. Slikkerveer and D. Brokensha (eds) (1997) *The Cultural Dimension of Development: Indigenous Knowledge Systems*. London: Intermediate Technology Publications.

Walsham, G., & Sahay, R. (1999) GIS for District-Level Administration in India: Problems and Opportunities. *MIS Quarterly, 23*(1), 39-65.

Weick, K. (1995) *Sense making in Organizations*. Sage, Thousand Oaks, CA. World Bank (1997) Knowledge and Skills for the Information Age, The First Meeting of the Mediterranean Development Forum; *Mediterranean Development Forum*, World Bank, (1999) World Development Report 1998/1999:Knowledge for Development.

Zadeh, L.A. (1965) Fuzzy sets and systems. In: Fox J, editor. *System Theory*. Brooklyn, NY: Polytechnic Press, 29–39.

# CHAPTER 2

## The Role of Diaspora in Knowledge Exchange Network for National Growth and Social Change

*Tewabech Bishaw*

### Introduction

The outflow of educated and skilled individuals from developing to developed countries has remained a complex development challenge and is increasingly becoming a critical issue in the knowledge-based globalized world. It also continues to be extensively debated with varied interpretations (AHEAD 2004). Indeed the issue is of serious concern since it is estimated that there are about 200 million migrants globally (one in every 35 persons) some of whom are highly skilled 23 (IOM 2004).

The migration of skilled individuals from developing countries is often attributed to the complex and shifting interplay of the "push" and "pull" factors: from poor, underdeveloped, politically unstable and or/conflict prone countries to those that have stronger economies, are politically stable and offer good security and better socioeconomic and professional opportunities. In recent years, South –South migration has also emerged as a growing phenomenon the impact of which has yet to be well understood. For example, South Africa, Botswana and Namibia have attracted senior professionals from many African institutions from the rest of the continent. Although the source countries may be losing these experts, in the long term such moves may be the seed for broader integration in regional development.

In the present day global economy, knowledge has increasingly become a critical factor in production, distribution of goods and utilization of services. It is also essential in ensuring national competitiveness in the world market. As a result, for most developing countries, maintaining the appropriate human resource base in science, technology, engineering and innovations is a major policy challenge. These countries continue to suffer from extreme shortage of skilled human resources, often attributed to brain drain, which is emerging as a crucial factor that may constrain the development of knowledge-based competitive societies.

Rich and developed countries, such as those in Europe and North America, the Middle East and Australia, have, on the other hand, strategically opened their doors to attract highly skilled and talented human resources from developing countries to meet their socio-economic demands. Some of these countries even adjusted their national migration and education policies and designed special visa programs to selectively attract highly skilled persons from Africa and elsewhere (Maru 2005). Others actively recruit for specific skills that are in short supply created mainly by the increasing demand and an aging population in their country (Kapur & McHale 2005). Similarly, deliberate internationalization of higher education systems is another key strategy to attract foreign students who are a very good source of talent and labour for the host countries as they often continue to live there after they graduate. It is clear that the brain drain from nations in Africa and other developing countries around the world has not yet become a public policy concern of the advanced countries to stem the inflow of skilled labour.

David Shinn in his paper entitled "*African Migration and the Brain Drain*" (Shinn 2008) stated that an estimated 300,000 African scientists, engineers and professionals live and work outside the continent. Between 1987-2001, a total of about 50,000 South African professionals left for the UK, Australia, US, Canada and New Zealand. In 2000, approximately 65,000 African born physicians and 70,000 African born nurses were working in developed countries. These represent about one-fifth of all African-born physicians and one-tenth of the nurses. It is also estimated that about 30,000 African's holding

PhDs work outside the continent (Shinn 2008). In 2006, the World Health Organization (WHO) reported that out of fifty-seven countries worldwide suffering from a severe shortage of health workers, thirty-six were in Africa. Since the 1990s, for most developing countries, including Ethiopia, about a quarter of their citizens with tertiary education live in OECD countries (Kapur & McHale,2005). This proportion is expected to rise with increasing globalization and demographic changes in the developed world. According to IOM, since 1990, over 20,000 skilled professionals have left Africa every year, and available data also show that African countries spend more than USD$4 billion annually to employ over one hundred thousand expatriates to fill professional gaps created by migration. This represents about 35% of total official development aid to the continent. Furthermore, the unabated brain drain has diminished the ability of developing countries' to recoup their investment in higher education and training. Given this background, the much needed skilled intellectual brain power of professionals in the Diaspora can be considered a loss in billions of dollars.

However, in recent years, mounting evidences have brought about global optimism and a paradigm shift in migration and development thinking. There is increasing recognition on the contribution Diasporas could make in the development of their country of origin. With the appropriate policies and strategies in place, many developing countries could increasingly tap in to their Diaspora intellectual resource to support their country's development. The development contributions of the Diaspora not only include foreign currency remittances but also service as visiting scholars, creation of virtual networks, and generally contributing to shaping the direction of the scholarly environment in their country of origin. As the history of some newly industrialized countries like India, South Korea, Hong Kong, Taiwan, Malaysia, China and the Philippines indicate, members of the Diaspora are powerful social and economic forces, capable of successfully facilitating the improvement of the overall standard of living in their country of origin. The Diaspora can be positive driving forces for innovation in many sectors of society. Correspondingly, some African countries such as Ghana, Egypt,

Nigeria and South Africa have initiated strategies to mobilize and tap into their Diaspora resources for remittances, knowledge exchange and social, economic and cultural development initiatives (Shinn 2008). Accordingly, there has been a new initiative to tap African expatriates for the development of the continent with the African Union (AU) inviting Africa's Diaspora to actively take part in the region's development. For instance, the priority of the New Partnership for Africa's Development (NEPAD) is to develop Africa's human resources and reverse *brain drain*. African leaders often meet with their Diaspora professionals to discuss ways in which they could contribute to their respective country's development. Several consultative meetings involving regional and international organizations and institutions have come up with several resolutions and recommendations seeking to define a mix of modalities to effectively engage the Diaspora in the continent's development. The AU Commission has even resolved to classify the African Diaspora as the "Sixth Region" of the continent. However, actual tangible action to effectively engage the Diaspora has not been realized yet and progress seem to be slow.

## Ethiopia's Diaspora Situation

Ethiopia, with its plan to grow rapidly and the un-stemmed outflow of skilled professionals, has no choice but to develop and take action take decisive actions to implement effective enabling policies and strategies and engage the Diaspora's full and synergistic partnership to fully express its development potential.

The exact number of Ethiopian professionals in the Diaspora and their respective areas of discipline are currently not known. There is no census, documentation or database that is centrally available. Paucity of information on migration of professionals or their return is a serious setback to exactly quantifying the enormity of the problem of brain drain or gain in Ethiopia. Destination countries probably have more complete information but are not always presented in a way that reflects the impact on the source countries.

There is general agreement on the trend in migration. The exodus is unabated and it is still on the rise! In some areas, even a loss of a few individuals may be a significant percentage of the total manpower in that particular profession.

Ethiopia began and continues to lose its highly skilled professionals in large numbers since the mid-1970s. However, today, the exodus continues, further depleting the meager human capital available in the country. Based on anecdotal information, it is estimated that the country has lost over 30-35% of its skilled and educated professionals over the last four decades. For instance, David Shinn (Shinn 2008) reports that between 1968 and 2006, Ethiopia trained 3,728 doctors (excluding those who were trained externally). Of these, about 3,000 left the country, and in 2006, only less than 700 doctors were left to serve in the country. There were more Ethiopian doctors in the East Coast of the US than in Ethiopia and even more Ethiopian doctors in Chicago that in Ethiopia (Shinn, 2008). Regarding the characteristics of Ethiopians abroad, it is estimated that out of the more than 450,000 Ethiopians in the United States, over 30 percent have first degrees and above[1]. Also in 2007, 150 professionals left Ethiopian Airlines for jobs in the Middle East. Similarly, it is estimated that around the 4000 Ethiopians and Germans with Ethiopian background, about 20 percent hold a degree at the bachelor's level or above (Schlenzka 2009). According to a study by Adamnesh (2008) of the total number of Ethiopian Expatriates in the OECD, 31.2% are highly skilled. Such unabated exodus of professionals from Ethiopia has resulted in decline in the country's (a) capacity for competitiveness in the emerging knowledge-based global economy, (b) its capability in the production of science and technology, (c) in quality and access to basic services and, (d) in its ability to develop science, industry and academia.

These professionals in the Ethiopian Diaspora can be classified in to two main groups. The first group is made of individuals who completed their education in Ethiopia and then emigrated, mainly consisting of scientists, engineers, health professionals and entrepreneurs, etc. The second group involves those who studied abroad, found jobs, established families and be-

---

1 US Census, American Community Survey, 2006 and 2007.

came permanent residents or citizens of the host country. Additionally, there is also a third group that is emerging. These are second and third generation Ethiopians in the Diaspora, an energetic group that is a product of both cultures, often conferring them with special advantages from having derived the synergy of the two cultures. This dynamic group is open to change and also has the propensity of being innovative and productive. This writer believes that with proper mentoring and motivation, this group has a tremendous potential to contribute to science and technology development in Ethiopia. Meaningfully engaging these young professionals to partner and team up with their counterparts as well as with senior professionals and scientists for mentorship and knowledge transfer at home can help Ethiopia to leapfrog into higher technological transformation and development. (Bishaw, 2010)

## Promoting Diaspora Participation in Knowledge Exchange Network

There is a growing realization by the Ethiopian Government on the importance of the Diaspora as a development partner and professionals in the Diaspora also increasingly recognize the unique role they can play in this endeavor. In this regard policies and strategies are being developed to facilitate and encourage Diaspora participation. One such example is the introduction of the yellow card for citizens of other countries with Ethiopian Origin that would entitle them the benefit of a citizen in socio-economic and investment sectors. This has made an impact in attracting and facilitating the Diaspora participation. National policy and guidelines introduced to motivate the Diaspora in remittances and investment has been instrumental in attracting increasing numbers of the Diaspora to take part in investment in various areas and consequentially, cash flow into the country through remittances has grown steadily. For instance, according to a recent study sponsored by the Ministry of Foreign Affairs and IOM (MOFA &IOM, 2010), officially recorded private cash remittance flows to Ethiopia have increased from 53.9 million USD in 1997 to 779.6 million USD in 2006. According to the same source, the National Bank of Ethiopia's (NBE) estimation of non-official remittance since 2001 also indicates that it has grown from 81.4 million USD in 2001 to 951.5 million USD in 2008. This has resulted in significant improvement over the last several years in attracting more Diaspora finances into the country.

While the above data suggest the increasing interest among the Ethiopian Diaspora to contribute to the development of their country of origin, not much has been done nationally in terms of mobilizing and promoting Diaspora knowledge and technology transfer and exchange. The country's ability to tap into the rich intellectual resources that exists in its Diaspora population has been constrained due to the limitation of the national policy which is a major opportunity lost. The national policy is not very clear or specific on the strategies to encourage and attract professionals in the Diaspora to participate in knowledge exchange and transfer and it lacks clarity in actions to facilitate and promote knowledge exchange. However, this does not mean that there is no national interest to mobilize and engage professionals in the Diaspora to partner and contribute to national development. The new policy that is underway is expected to promote and support knowledge exchange. Especially, as the country has embarked on rapid knowledge-based development undertakings, there is big demand for highly skilled human resources to meet the expanding needs. The role that skilled professionals in the Diaspora could play in the country's development cannot, therefore, be over emphasized.

Some institutions of higher learning as well as sectors like health and education have information on their websites inviting professionals in the Diaspora to participate in their respective sectoral activities. However, progress in this regard has been very slow and uncoordinated. The Ministry of Foreign Affairs is the lead Ministry on Diaspora Affairs, The responsibilities of the Diaspora Affairs Office of the former Ministry of Capacity Building in Ethiopia that was mandated to facilitate Diaspora's contribution in knowledge and skills transfer has now been decentralized to individual sectors, institutions and NGOs. Unfortunately, most Diaspora knowledge exchange initiatives currently underway in the country are individually driven and self-motivated, emanating from personal and individual commitments. Few individuals in the Diaspora and in Diaspora networks have taken their own initiative to contribute their know-how using their own time and resources. This has resulted in uncoordinated, sporadic, haphazard and spontaneous engagements that have often proven unsustainable. Such uncoordinated arrangements have resulted in constraints and frustrations on both sides. In fact, sometimes, inputs and activities initiated by Diaspora professionals are perceived as a burden for institutions in Ethiopia as it would require planning and preparedness of these institutions to effectively and efficiently incorporate and utilize these inputs offered by the Diaspora. The absence of a national policy framework and implementation guidelines have led to the inefficient exchange and transfer of the rich intellectual resources that are available and so highly needed for institutions and human resources capacity building in Ethiopia. To facilitate and coordinate processes, serve as a link and to develop database on needs and resources for the purpose of matching, Alliance for Brain Gain and Innovative Development (ABIDE)[2] was established as an indigenous NGO with a sole purpose of facilitating processes and systems for Diaspora engagement.

There is a lack of information on existing systems, procedures and processes that would enable Diaspora professionals to establish linkages and

---

2 To partner with key players in the country and abroad and to contribute to improving coordination and networking, "Alliance for Brain Gain and innovative Development" (ABIDE) was established (the first of its kind in the country), in January 2006 as an indigenous NGO not-for-profit development organization to increase Ethiopia's opportunities for ***brain gain*** ( HLM strategic Planning Document March 2007). ABIDE endeavours to advocate for enabling policies and to facilitate institutionalized mechanisms, systems, processes for mainstreaming and to increase the ease through which professionals in and outside the country could link through networking and partnerships. It is expected to enhance the transfer of knowledge, and technology to contribute to the country's development through systematic and sustained engagement. ABIDE believes that the rich intellectual resources that exist within the Ethiopian Diaspora community could make significant contributions to Ethiopia's institutional capacity building if and when enabling systems and policies are in place. It also aspires to building synergistic and win-win partnerships, sustainable collaboration as well as mutually beneficial relationships among professionals in the Diaspora and those in the country thereby contributing to national capacity building and also offsetting the adverse effects of brain drain and in the long run contributes to attenuating brain drain.

effectively partner with professionals and institutions in the country. On the other hand, there is also a major information gap about professionals in the Diaspora by the institutions in the country that could have benefitted from their input. These institutions need to know: Who are the Diaspora professionals and how are they networked? Where do they live? What could be the extent of their participation in the development endeavors of the country? What are the possible institutional arrangements and the policies that would attract the Diaspora in knowledge exchange and transfer? What are the specific areas where Diaspora knowledge inputs are needed? What are the preferred modalities of collaboration? What about the question of sustainability? What would be the role of the current employer of the Diaspora professional? Because of the lack information, both parties are unable to establish effective relationships. As a result, knowledge exchange partnerships and collaborations between professionals and institutions in and outside the country have remained very limited. However, it is commendable that despite the many constraints and difficulties, some partnerships were still established between professionals in the country and those abroad. Regardless of how small or big they may be individuals have continued their efforts to maintain their relationship in knowledge exchange.

The large number of highly qualified and committed professionals in the Ethiopian Diaspora, if effectively engaged, could make significant contribution to the country's accelerated growth in science, technology and innovations to succeed in a knowledge-based competitive world economy. The role of knowledge exchange networks to facilitate brain drain into brain gain has been demonstrated in several countries that have embarked in effectively engaging their Diaspora professional communities.

## Strengthening Networks

Experiences from other countries who have immensely benefited from their Diaspora resources such as India, China, South Korea etc. confirm that knowledge exchange is facilitated when the willingness in the Diaspora was matched with enabling policy environment, coordinating mechanisms, preparedness of professionals and institutions in the country, and the availability of financial and related resources.

The Asian Development Bank Draft Manuscript (2006 ) on *"Converting Migration Drains into Gains: Harnessing the Resources of Overseas Professionals"* highlights the importance of knowledge networks in strengthening bonding social capital that encompasses solidarity and community identity. The study argues that with respect to knowledge transfer and exchange, it is first necessary to consider just what is to be mobilized and how these knowledge resources are created prior to considering the Diasporas' ability to mobilize and network. A clear understanding of who exactly the Diaspora members are from whom such contributions are sought is also equally important. While becoming member of a Diaspora professional network will not be sufficient, most critical is the members' common interest and commitment to give back to their country of origin. The centrality and commonality of goal to contribute to the development of the country of origin and its nationals is of paramount importance, and this need to be the focus of the partnership. The Bank study further argues about the important contribution of Diaspora professional networks' to migration, brain drain and brain gain and circulation, and it also states that these networks also facilitate knowledge development and transfer/exchange.

Forty-one knowledge based Diaspora networks tied to thirty different countries with specific efforts to link Diaspora professionals to the country of origin for the purpose of transferring knowledge were compiled and analyzed (Brown and Meyers 2006). These networks were categorized into four types: student/scholarly networks, local associations of skilled expatriates, expert pool assistance through the UNDP's Transfer of Knowledge Through Expatriate Nationals (TOKTEN) program, and intellectual/scientific Diaspora networks. The objective of these networks was primarily to create the links through which professionals in the Diaspora could effectively and productively be connected to the development of their country of origin without any physical return, temporary or permanent. Though the experiences differ from one country to the other, they all share the goal of systematically using their expatriate experts, scientists and engineers for development at home. Some lessons and policy implications may already be extracted from these recent attempts. They should pave the way for an efficient use of professional Diasporas. These are indeed strong potential resources for effective and mutually beneficial co-operation between developing and highly industrialized countries. Of the forty-one, fifteen were classified as intellectual/scientific Diaspora networks with an explicit purpose of promoting the scientific, economic and social development of the country of origin. The effectiveness of these networks are dependent on the interests and strengths of the network, the coordination and enabling environment that exist in the country of origin and the availability of the required resources to implement the transfer of knowledge.

Findings from in depth assessment and analysis of the 15 intellectual/scientific Diaspora networks show that good organization is required in a network of this kind in order to ensure communication, information-exchange and coordinated actions. This is where an interface or coordinating body appears necessary. The function of such a coordinating body would be to collect, organize and maintain the information needed for the systematic search of partnerships, but also to manage and promote the interests and actions of the multiple entities present in a network of this kind. The coordinating body would be responsible for opening up access to resources that can be used to generate action in the network. This body would consist of network members as well as interested parties from the national community. It would thus be a consortium of multiple partners. They incorporate actors beyond the skilled Diaspora to including governments, NGOs, donors, and other intermediaries, and would likely benefit where one or more of these actors assumed roles similar to that of a nodal firm. Alliance For Brain Gain and Innovative Development (ABIDE) is an example that can serve as a coordinating body and an intermediary that facilitates processes involved in effectively and efficiently facilitating the transfer of knowledge from the Diaspora.

Many associations of skilled expatriates are evolving into formal professional associations, which are becoming increasingly active in intellectual/scientific Diaspora networks. The highly skilled Diaspora may rely more on such professional networks, which are more likely to yield a match of skill levels to jobs. The most notable examples include, for example, the Silicon Valley Chinese Engineers Association, the Indus Entrepreneur and the Korean IT Forum. These associations fulfil a range of social network roles, including facilitating logistics, professional and technical advancement, identity formation and maintenance, and entrepreneurial investments particularly highlighting their role of bridging/linking the country of origin producers to the global economy.

## Enhancing the Role of the Ethiopian Diaspora in Knowledge Exchange Networks

In Ethiopia, most common modality of Diaspora's participation for knowledge transfer been on individual basis. These professionals contact counterpart professional colleague or directly with the departments and provides his/her services either through short visits, repeat visits, spending sabbatical and or research leave periods, short lectures, summer breaks, short seminars etc. Most benefiting institutions have been universities. Furthermore, institutional collaborations with research centers. Fulbright scholars, fellows of TWAS, etc. have spent their time in teaching, research, training etc. with their partner institutions. In the health sector, regular visits for specialized services have been utilized.

In addition, eestablishing and belonging to a functioning network for knowledge/exchange and transfer from Diaspora to country of origin will facilitate interaction between professionals in the Diaspora and at home. This could help in mobilizing a larger group and mix disciplines of the Diaspora and partners in the country for effective collaboration and win-win partnership. It can also facilitate coordination, resource mobilization and engagement on need-based interventions for sustainable development. In this regard Diaspora networks could be effective, provided there is commitment of the Diaspora to knowledge transfer to their country of origin, availability of a facilitated system within the country of origin and commonality of purpose and goal are ensured.

There are a number of formal and informal Ethiopian professional networks in the Diaspora that have and emerging to facilitate their contribution and partnership in the country's development endeavors. These networks help to establish links amongst professionals in and outside the country with varying degree of effectiveness. Following are a few examples of Diaspora networks and I may have excluded deserving groups and individuals, Some of these networks include Association of Ethiopian Health and Health Associated Professionals in Sweden (AEHHPS), Association for Higher Education and Development (AHEAD), People to People (P2P), Ethiopian Physicist Association in North America (EPA-NA), Ethiopian Professionals in Huston (EPH), Ethiopian North American Health Professionals Association (ENAHPA), Group Against Poverty and AIDS (GAP-A), Ethiopian Knowledge and Technology Transfer Society (EKTTS) , Ethiopian Tree Fund Foundation (ETFF), Books for Africa, Ethiopian Students' Association, international (ESAi), Network of Ethiopian Professionals In the Diaspora (NEPID), Forum International for Ethiopians Living in the Diaspora (FIELD), The Twinning Center etc. However, most of these inputs have been individually motivated, sporadic and unsustainable.

The above networks have been functioning with varying degree of efficacy depending on the strength and commitment of the members of the network. Their participation has been dependent on the objectives they have set and availability of the essentials required in performing their tasks effectively. The author recognizes the limitations of this paper since in depth assessment and analysis of each of the networks has not been systematically done. The paper, therefore, presents the suggestions based on personal experiences, lessons-learned from other country experiences to stimulate thinking and action to enhance the effectiveness of the knowledge exchange networks for Ethiopia. Following are, therefore, some suggestions to strengthen networks for effective knowledge exchange. The suggestions are divided in to three main groups, 1) the network, 2) country of origin and 3) management coordination.

## The Network

The network needs to be adequate and strong with members linked to the country of origin who are committed to the development of the same while also ensuring personal and professional fulfillment. Members of the network should be credible professionals interested in giving back with high standard performance and empathy for the challenges that their country of origin and its nationals face. Clarity of objectives is necessary to enable that the network's mission and activities are focused as well as relevant to the country of origin setting. The effectiveness of the network is dependent on the motivation of its members. The development of the country of origin and its people should be the prime goal of the network. The network should also be adequately resourced, and actively linked and networked with its own members and other networks with similar objectives and vision.

## Country of Origin

There needs to be clear policy strategy and an enabling environment to ensure the ease by which Diaspora professionals could engage in dialogue with partners in the country. National needs and priorities need to be clearly articulated and communicated. The partnership and engagement needs to be established in the context of trust and mutual respect. Policies, operating rules, regulations and guidelines need to facilitate the partnership. Clear monitoring mechanisms need to be identified to assess progress and to significantly contribute to mainstream development.

## Mechanism for Management and Coordination

To ensure effectiveness of the partnership, management and coordination mechanisms have to be in place and operational. The presence of intermediary organizations has proven to be effective to collect, organize and maintain the information needed for the systematic search of partnerships, but also to manage and promote the interests and actions of the multiple entities present in a network of this kind. Existing mechanisms have to be streamlined and others need to be put in place to ensure effective and coordinated collaboration and partnership. Such organizations need to be supported to optimally play their role towards the effective utilization of professionals in the Diaspora for the sustainable growth and development of the origin of country and its people.

## Conclusion and Recommendation

Diaspora knowledge exchange network is another effective way of engaging professionals in the Diaspora to support national development efforts. Through strengthened professional knowledge exchange and transfer networks, Ethiopian intellectuals, artists, academicians, scientists, social scientists practitioners, and young professionals and friends of Ethiopia in the Diaspora on the one hand and their respective counterpart professionals in the country on the other could establish partnerships to work together for the development and prosperity of the Ethiopian people and for social change. This would strengthen the synergistic collaboration and win-win partnership for sustainable cooperation, social, professional and economic fulfillment, and in the long run contribute to the countries scientific and technological development that could result in attenuating brain drain.

The motivation and decision of professionals, to stay abroad is a product of a complex mix of economic, political, social, cultural, and personal issues. It has become clear to most development thinkers that the incorporation of the brain power of the large Diaspora communities in national development is a crucial necessity for developing countries like Ethiopia. Although these communities have made significant contributions to their countries of origin, the contributions have generally been at micro, informal and individual levels. It is important that national, regional and global efforts need to be made commensurate with the enormity of the problem to tap in to the rich potential of the Diaspora intellectual and related resources including their second and third generation resources.

The paucity of data on the number of professionals in the Diaspora continues to be a major problem to have a clear understanding of the magnitude of the problem and existing variations. Guided by relevant Ministries, and organizations including NGO's such as ABIDE could be strengthened and mandated to undertake the task within specific time frame to have an up-to-date and user- friendly database on Ethiopian Diaspora professionals. Database that exists under different organizations can feed into the AU master database on African professionals.

It is important to take stock of the many decisions and resolutions of the AU and its Heads of States and Governments in regards to engaging the Diaspora, and (a) monitor the actions taken against those decisions and resolutions and assess the trend of implementation, (b) monitor the outcome and impact these actions have generated towards meeting the desired objectives.

Agreed upon monitoring indicators could be developed to measure the extent of implementation of the joint projects and undertaking. Country experiences and lessons learned could contribute to ongoing discussions and discussion to maximize the positive effects of Diaspora engagement in Ethiopia's capacity building and that of the Regions.

Several Diaspora professionals and networks have participated and contributed in their country of origin's development initiative. Undertaking systematic study at country levels to understand the issues and complexities involved to fully mobilize and engage Diaspora could reveal some of the challenges and opportunities experienced both by the Diaspora and host countries, institutions and professionals. Lessons learned can inform policy dialogue and new strategy development.

In the spirit of "Social Responsibility" rich and developed countries that greatly benefit from Diaspora intellectual resources could be obliged to engage in "Global Social Responsibility" activities and contribute to the development of the source countries. This will only be a fair action to compensate the source countries from the loss of their skilled professionals in whose education they had invested so heavily.

Developed countries in EU, North America, Australia, the Middle East are persuaded to ensure policy coherence and consider incorporating policies and strategies that would encourage and facilitate modalities that would allow professionals in the Diaspora to engage in the development efforts of their country of origin.

The Diaspora community is often individualistic. Many Diasporas are not organized making it difficult to effectively interface with them as a group. It would be useful if receiving countries in collaboration could support Diaspora communities to organize as professional bodies and participating in professional networks.

Mechanisms need to be in place to give full recognition to the achievements made by Diaspora professionals in their areas of specialization. Diaspora's contribution to the production of science, technology and other social and economic achievements both in the host and home country must be acknowledged.

## References

The Association for Higher Education and Development (AHEAD) 2004. Converting Migration Drains into Gains: Harnessing the Resources of Overseas Professionals, Draft manuscript, Asian Development Bank, Manila, Philippines, 2006

Kapur, D. and McHale, J. Give us your best and brightest: the global hunt for talent and its impact on the developing world. Washington, DC: Center for Global Development 2005.

Atnafu Adamnesh, (2008) Facets of Ethiopian out-migration and return: The case of Ethiopia, Unpublished, Envisioning Ethiopia , Addis Abeba

Maru M.T, The Commission of the African Union, Brain Drain and Development Support Policy Coherence, February 2008

Meyer, J-B. and Brown M. Scientific Diasporas: A New Approach to the Brain Drain, World Conference on Science , UNESCO - ICSU Budapest, Hungary, 26 June-1 July 1999

Shinn, D. H. Brain drain and migration in sub-Saharan Africa: making the best of a dilemma 2008. Unpublished.

Shinn, D. H. Reversing the brain drain in Ethiopia: a paper presented at the Ethiopian North American Health Association Conference, Nov. 23, 2002, Alexandria, Virginia, USA.

Bishaw Tewabech, (2010): The African Diaspora: A vital Resource for Science and Technology Transformation of Africa, Unpublished

Schlenzka, Nathalie, Potential of the Ethiopian Diaspora In Germany and the United States, June 2009. Unpublished GTZ

# CHAPTER 3

## Indigenous Knowledge Systems (IKS): Application and their impact on industrial activities of companies in Africa

*Medicine Magocha*

*To know where you are going, know where you are coming from.*
(Ziggy Marley-Reggae Star's Documentary "Africa Road Show")

### Introduction

Knowledge is a pivotal element in all facets of survival. In some sayings they asserted that "knowledge is power "This implies that without knowledge there is no production, no direction and no focus of life. Davenport and Prusak (1998:3) posit that

"Knowledge is a fluid mix of framed experiences, values, contextual information, and expert insight that provides a framework for evaluating and incorporating new experiences and information. It originates and is applied in the minds of knower. In organizations, it often becomes embedded not only in documents and repositories but also in organizational routines, processes, practices and norms"

This definition sounds enough, however it takes issues of ontology for granted. Knowledge is generated everywhere, every time and from everyone, Butler in Schwartz(2006:1) proposes that there are relationships between the individual and his social world; that is between the knowing social actor and the social groupings and contexts in which he or she participates and exists, and in which knowledge is socially constructed. This on a larger scope portrays that knowledge differs from time to time, place to place and from organization to organization. Knowledge is the conversion of sight into perception and from perception to application. Knowledge management is then described as an integrated, systematic approach to identifying, managing and sharing all of an enterprise's information assets like the databases, documents, policies, procedures and previously unarticulated expertise and experiences from individuals. When it comes to indigenous knowledge systems it would mean that the common phenomena from the people of certain citizenry must be shared and capitalized among the people for the generation of informed decisions to influence productivity.

Jordan the Minister of Arts and Culture of South Africa (2005) said;

IKS predates the imperial and colonial era. Despite the passage of time elements of it still are the lived experience of various communities. They have been preserved in the manner in which these communities deal with birth, the coming of age, maturity, marriage, old age and death. These ways of doing things have provided people with the survival strategies to deal with poverty and destitution, to create stability and how they co-exist with other communities. Much traditional knowledge could teach us how to interact with our natural environment and also to re-learn ways of healing that are efficacious and with few side effects. The performance of tales, song and dance so integral to traditional society still survives and has much to teach

And National Research Foundations (NRF) (2005) asserts that;

IKS refers to the complex set of knowledge and technologies existing and developed around specific conditions of populations and communities indigenous to a particular geographic area…These populations…retain some of, or their entire own social, economic, cultural and political institutions. … IK refers to the knowledge developed by these populations in themselves, as well as knowledge developed through interaction with other populations in South Africa

From both definitions it is noted that knowledge can be preserved, modified, altered and destroyed through research, education, training, coaching, mentoring and oral story telling (folklore) as well as reading. It will be of use if the radio and television broadcasting would be used to help disseminate indigenous knowledge systems for wide coverage. The paper aims to expose the set indigenous knowledge systems which are of great help to business activities. Justifiably this would assist in the formulation of trade policies; the paper would promote the development of products which matches the indigenous people. Sustainable development can be promoted by the interaction of the people and their knowledge sources. Knowledge is the gate way to informed decisions.

## Education in Africa Before Colonialism

History reflects that Africa had all forms of education during the pre-colonial era. The education system was mainly based in oratory approaches usually at early ages from (5-20 years); this age group was told folklore stories. The folktales were rich sources of the future indigenous knowledge systems depicting codes of moral conduct, policies of life, self-sustenance and ethical codes. If that were to be compared to present-day situation the folklore stage would be equivalent to primary education. From that stage they would graduate into adult world, where they would be exposed to village debates (politics of the day) that is especially for the boys. The boys would be instructed by the old men who would be associated with greater wisdom of life domains. The ladies would start learning issues of hospitality, housekeeping and household management, they will be doing this under the mentorship of old women who are experienced.

Senior boys would also get into practice doing things like iron and copper smelting producing malleable and ductile products (spears, axes, knives, hoes and arrows) though at a smaller scale it could suffice the needs of the population. This portrays that their knowledge sharing maintained them at a self-sustainable level. This knowledge was not tapped into modern day industrial activities.

People were educated on how to produce medicines from herbs, produce heat for smelting iron from metal smelters (*mvuto*). In Zimbabwe the existing government(2008) led by Zimbabwe African National- Patriotic Front (ZANU-PF) tried to revive the indigenous knowledge systems in medicinal areas by establishing the Zimbabwe National Traditional Healers' Association (ZINATHA) assisting in providing health services to the people of Zimbabwe. By so doing people had a choice between modernized medicines and traditionally extracted concoctions. These processes involved almost everyone and talents were identified at every stage of an individual growth and development and later on shared among the African people.

Knowledge was managed in a specific routine way in African communities which made people grow well and well informed. Most of the learned issues were derived from the life experiences which are really a tool of positive reinforcement (motivation) towards learning (change of behaviour) and acquisition of new knowledge.

The education of the current times when compared to that of the pre-colonial period it is not at par. The pre-colonial education was far better because it had practical which were much relevant than those of the present. Presently Zimbabwe have so many people who had graduated with degrees and diplomas in engineering but Zimbabwe Iron and Steel Company(ZISCO) is at halt none can sought out the blast furnace's technical faults or develop something else to substitute it. This case is not common to Zimbabwe alone but Africa over. There are a lot of academic doctors and professors in Africa but most of the African countries are still trailing the poverty line.

Bahra (2001:176) supports the pre-colonial education in Africa by exposing how story telling can be used in business concepts. The author give the story of Eskom where most of the personnel was illiterate but educated when it comes to African knowledge systems, so the Training and Development Head had to use Africa's tradition of verbal communication to involve personnel at all levels. This is the past success which is used to avert the present failure.

The indigenous knowledge systems used to maintain the discussion set up in a circular form, influenced most of the present lecture theatres to adopt a circular form as shown below

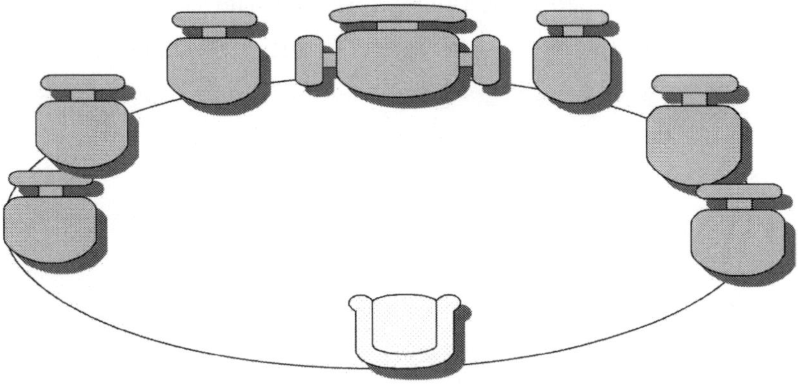

**Figure 1: The round-table set up derived from traditional folklore sitting format (From the research)**

This set up allows every member involved in the meeting, seminar or training to see face to face with the source of knowledge and get involved in the issue under discussion with total commitment.

The pre-colonial education equipped people with a number of skills which included the following;

| Oral education Folklore Stage(Primary) | Courtyard education (Secondary/ tertiary |
|---|---|
| Listening | Carpentry, Hunting |
| Singing | Black smithing and Iron or Copper smelting |
| Dancing | Curving, Weaving |
| Counting | Knitting, Cooking |
| Running | Budgeting |
| Socializing, Bargaining | Leading, Fighting (for protection soldiering) |
| Story telling | Entrepreneurial, Planning |
| Advertising and Public Relations | Architectural |

**Table 1: The Composition of indigenous knowledge systems skills sharing during the pre-colonial era (From the research)**

These skills forms part of the indigenous knowledge systems, without it was very difficult to meet the categories of being a person by standards (*Ubuntu*). These skills were imparted by experts who had the dexterities of doing that specific task.

During the pre-colonial era the indigenous knowledge was compulsory. Every child was supposed to go to the story teller's place to receive some education whilst the senior boys would go to the family court (dare) and ladies to the aunties. This was done to ensure that everyone is given a chance to discover what was expected at each stage of life. These usually took heuristic approach to life problems. Heuristics is a common present-day business approach which is generally cheap to spread valuable knowledge and information quickly.

People usually could learn by doing the thing thereby building experience for practical acquaintance with facts or events; knowledge resulting from the process (Swoden 2000). This process promotes collective experience which in turn brings diversity of technical expertise and consequently specialized work groups could be created for example those with hunting skills could collectively come together and form a team the same with those for curving and metal smelting.

## Indigenous Knowledge Systems and the Future Business Processes

Nel (2005) says;

"Indigenous knowledge systems (IKS) are about the knowledge, practices, values and ways of knowing and sharing in terms of which communities have survived for centuries. IKS is not only knowledge and practices of culture, custom, agriculture, medicine, biodiversity, ethno-numeracy, customary law… but also the rationality of these cultural practices and rites that effected social cohesion, creativity and artistry of dance and music, technology of fashioning, clothing and beadwork"

This definition seems to limit IKS to traditional systems of business which are archaic as if IKS are fixed and cannot change. However the thesis in the definition allows people to think in a positive way towards supplying people clients/customers or consumers with the right products, at the right time, with the right quality and for the right purpose. People had a substantial amount of information on IKS but they are not worried about applying it instead they keep it for history under the guise of heritage, sovereignty and nationalism. Mosimege (2005) postulates that, the recognition of the value of indigenous knowledge is clearly gaining ground world-wide, in South Africa it is ensconced in governmental policies relating to IKS in general, living heritage, trade and industry, biodiversity, science and technology, traditional health practitioners and traditional leadership, governance and sustainable development. Mosimege's expurgations reflect that in order for indigenous knowledge systems to survive there is need for the government to support, formulate and implement them in live activities. This can be modeled by a model below;

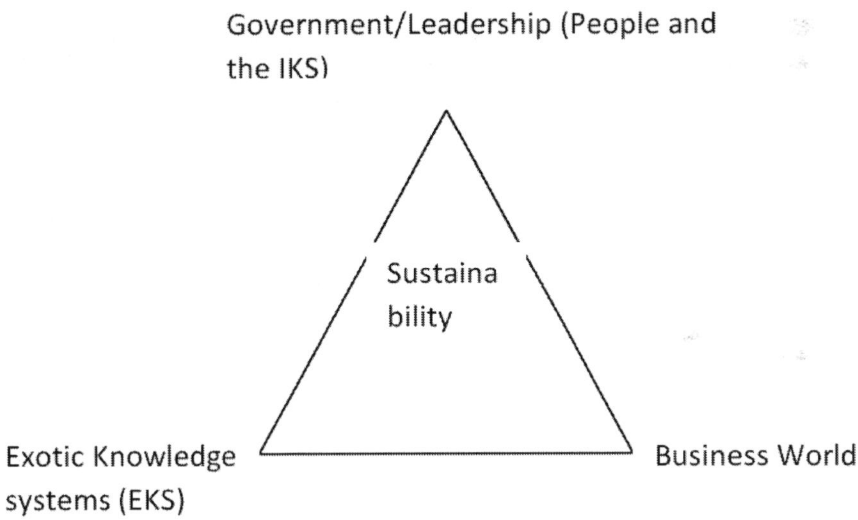


**Model 1: Forces surrounding sustainability (Derived from the research)**

The government has a role to play on regulating what the people (indigenous) would call knowledge. There is some type of thinking among the indigenous people which can destroy business which in turn would paralyze the sustainability of the economy. For example in Zimbabwe there is Chiadzwa Marange diamond deposit, the people in that place had their own knowledge, practices, values and ways on how to the products of the deposit could be benefited. Their perceptions and beliefs led to struggles of control between the Zimbabwean ministry of mines, the government leaders, the citizens in Chiadzwa area and other external interests to the extent that the diamonds from that area were condemned and labeled blood diamonds. Under such circumstances the state, people and corporate world must discuss then the generally acceptable agreement. Exotic knowledge systems must be mixed with indigenous knowledge systems within the rightful proportions to produce a good recipe for positive development.

The indigenous knowledge systems in Africa lost to exotic systems and they lost pride and popularity because they lacked contributions in terms of technological developments, product development and innovations. However despite the loss, by capitalizing on indigenous knowledge systems of Africa a new Africa can be generated. People must understand that most of the wars in the world, ethnic, civil or world wars are a result of undermining one's knowledge and understanding of life concepts. When other nations from the west, east and the north colonized Africa, they missed one thing—the IKS. They brought new life concepts familiar to themselves. That slowed the business processes and led to imbalances in the distribution of wealth, resources and knowledge.

For future developments and sustainable programmes in Africa the indigenous knowledge systems must be integrated into business. In order for both indigenous and foreign investors to serve Africa in the best way, they must understand the past, present and future methods, ways and approaches known by Africans. People must be ready to mix, socially, economically and politically. The interaction would allow people from Africa to heal from the legacy of colonialism and western knowledge imperialism. Due to colonialism and imperialism the African people had no opportunity to develop their countries in the way they felt plausible except to follow the protocols of the colonial masters. The idea of indigenous knowledge systems must not be centralized to African languages alone but must also be distributed to all faculties of learning in schools, colleges and universities. In fact people should learn IKS philosophies in all disciplines.

## IKS and the planning function in African Companies

To cut expenses from imports, African companies must plan for the Africans based on the values, knowledge, practices and ways of knowing and sharing. Brokensha, Warren and Werner (1980:1) emphasizes that development planners should consider accumulated knowledge and traditional skills and technology of the people among whom they work. The contributors in this issue are mostly social or cultural anthropologists but as of late other disciplines are also represented—geography, agricultural economics, political science, education and agronomy.

Before any business plan is approved it is better for the management to seek some knowledge from the indigenous people they are serving. In most cases people whom the business is serving have detailed knowledge of what they need not otherwise available to the management team. In most instances due to lack of IKS most football stadiums were built on very arable land suitable for farming and that one which was supposed for stadiums was left for farming. This affects industrial activities in Africa. In South Africa, there is Moloto road; rumor has it that, it was constructed without the consultation of the local people. That makes it a death trap; it has so many accidents each year. Planners have to do a lot of consultation before they get into decision-making. Company planners must work hand in glove with the education providers, the community and the state in order to avoid clash of interests and unnecessary criticisms from the indigenous knowledge managers.

Urban centres in Africa have a reflection of poor siting due to the fact that the town planners lacked foresight and insights of indigenous knowledge systems of the weather patterns and soil types. For example Gauteng in South Africa often receives a substantial amount of rains than Limpopo, yet very little farming activities are taking place in Gauteng but in Limpopo. This forces farmers in Limpopo to sacrifice effort and money through irrigation and intensive processes to meet the demand of food across the country. Under such circumstances planners have to consider soil type and rainfall patterns, given the case that Africa's economy is hinged on agriculture there is no reason to build town structures for residential on arable land when there are partially arid areas. James (1996:129) posits that, some believe that for economies to become sustainable and productive, the issues they confront can be understood, in part, through a historical approach to understanding how relationships with western economies affect and influence them. In this argument James portrays that the positive past should be ingrained and understood for improving the sustainability of African economies through agriculture and natural resources maintenance.

## Effects of Exotic Knowledge Systems on Planning of African Businesses

Exotic knowledge systems do not mix their values with the indigenous knowledge systems because of the capitalist mentality which concentrates on benefiting without considering the processes of benefiting. In most cases African indigenous knowledge systems are demonized and taken as backwards and dysfunctional. Semali and Kincheloe (1999:21) assert that western scientists maintain that much can be learned from a number of ethno-sciences which includes ethnobotany, ethno-astronomy, ethno-pharmacology, ethno-medicine and ethno-cosmology. Despite these benefits western science relegates indigenous knowledge systems to a lower order of knowledge production.

African businesses planning processes were based on the size of the moon and levels of growth of people. Western businesses planning were based on calendars. However all of them had an intention of taking action for the future, today. In African business context there was strategic planning, that is why the chiefs used to promote what is called *zunde* (a collective food warehouse for the whole kingdom). This was a strategic warehousing decision meant for the whole chiefdom to cater for shortages in times of poor harvest and drought. Similar to that are the Grain Marketing Boards (GMB) set by the colonial business and the silos. The differences between the GMB or silos and the *zunde* lies in fact for the *zunde* food was distributed free of charge to those in need but that from the GMB or silos is sold. Without money one cannot access it. The shocking thing with grain from GMB is when one is selling they are bought with the price for a lower grade when it comes to buying all will be in the same grade.

Exotic knowledge systems make the indigenous knowledge systems to be side lined. Semali and Kincheloe (ibid) expose that indigenous knowledge is tacitly de-contextualised severed of the cultural connections that grant it meaning to its indigenous producers, archived and classified in western databases and eventually used in scientific projects that may operate against the interests of the indigenous people. The major problem is the indigenous people themselves are not able to identify knowledge and keep that knowledge for competitive purposes, most of the products' raw materials are from Africa and by virtue, and the products are manufactured in the western countries and named after the producers in their language. When it comes to construction, the western architect only put the plan on paper and the African people do the donkey work. Whilst the African works through their knowledge, the western knowledge producer insists on testing the validity and reliability of the job via western scientific tests.

The planning of the African indigenes was not for mass production, but was for sustenance, the westerners' planning is for mass production and competition making business processes complicated. Those complexities in turn have large impact and pressure on the existing resources, contrary to that Jansky, Haigh and Prasad (2005:62)

Acquisition and evolution of IK is considered critical mechanism in achieving the environmental and social sustainability of resource use. Fielding and Kirsopp in Jansky, Haigh and Prasad (2005) suggest that IK has sustained local communities for generations, particularly those who have yet to meet any extension agents.

Despite the negative effects the exotic and the indigenous knowledge systems have to mix and produce a plausible and feasible business recipe to take African business activities on track. This would reduce the donor syndrome on African economies. The development of African business would mean a change at the world market business levels. If Africa also participates at the world market in the way Japan, Korea and other eastern and western countries participate the world would never experience serious business recessions and depressions. Above all democratic societies would be established, poverty disturbs social systems.

## Traditional Indigenous Knowledge Systems: Contributions to Present Business Situations

It is a common phenomenon that almost all societies through their developmental stages and business cycles, from developed to developing started as agrarian societies then get into green revolution and proceeded to industrial revolution up to information societies (where information and communication technologies are taking the toll) The existence of any society is anchored in food procurement as the first priority followed by other forms of leisure.

The following table, Table 2 demonstrates, indigenous knowledge systems that are applicable to present-day business

| Business | IKS applied |
|---|---|
| Agriculture | Cultural information on type of crops and |
| | the staple food, adaptive systems, soil |
| | fertility, society and environment, rainfall |
| | patterns, conservative nature, knowledge |
| | and technology, work patterns |
| | |
| Mining | Rock type, taboos and voodoos, previous |
| | processing methods, community |
| | requirements, mineral values, mining |
| | methods, reservation methods |
| | |
| Information | Word of mouth, blowing trumpets, |
| | appropriate register, cultural requirements |
| | beliefs, values and communication |
| | protocols |
| | |
| Tourism | Life histories, customs, traditional arts and |
| | structures, environmental resources, |
| | buildings, ruins, rivers, weather patterns |
| | sacred places |
| | |

| | |
|---|---|
| Health | Birth, growing, old age and death, diseases |
| | ethnomedicine, ethnopharmacology, food |
| | |
| Trade | Monetary values, other forms of exchange |
| | generally acceptable, product quality, |
| | product development stages, perception |
| | attached value |
| | |
| Education | Existing knowledge, language, literature, |
| | teaching methodology, curriculum, content, |
| | nature of students, learning history |
| | |
| Defense | Nature of battle formation, history of wars |

**Table 2: Indigenous Knowledge Systems Applicable to present-day Business Processes (Adapted from the research)**

In Africa tourism industry is one of the most lucrative businesses due to the fact that generally people are friendly and hospitable. Complementary to that people in Africa have kept their heritage for example Zimbabwe have maintained the ruins at Great Zimbabwe, Gokomere, Zihwa and Chibvumani; these have great reflection on the maintenance and reservations of indigenous knowledge systems. They are a source of knowledge to Zimbabweans and other people across the world. Zimbabwe is named after those stone structures the name reserves the knowledge behind Zimbabwean history, people of Zimbabwe have pride in these stones and the beauty associated with the natural resources of the country.

Indigenous knowledge affects the consumer behaviour of the people within a given society. Strydom (2004:53) identifies that cultural values, norms and symbols are created by people and are transmitted from one generation to another to ensure survival and also to facilitate adaptation to the circumstances of life. A consumer's culture further determines overall priorities in different activities and products. Thus, a product that provides benefits consistent with those of indigenous knowledge systems is much more desired by members of a certain culture and has better chances of being accepted in the market. Van der Walt (1998:85) asserts that South African society is fragmented into many cultural groups and subgroups. Marketing management must therefore be careful not to use symbols in advertisements that can be interpreted incorrectly or differently, and not portray unacceptable behaviour patterns. Effective communication can take place only if the theme of the advertisement reflects the cultural norms, values and symbols of the cultural group at which it is directed. This clearly spells out that indigenous knowledge is of significance towards the direction of business in any given community.

Brink (2000:35) alludes to indigenous knowledge systems in the following quotation;

"Some black and white culture groups, for example, interpret advertisements differently. For example flowers have special significance in the white culture, but in many black cultures, gifts in the form of flowers are frowned upon. The well-known *Five Roses* advertisement which said, "And the fifth rose is for you" was therefore not successful when used in media aimed at black consumers. Cross-cultural mistakes are also made in other countries. For example, in Germany, Pepsi's advertisement "Come alive with Pepsi" was presented as "Come alive out of grave with Pepsi".

Such presentations have adverse effects on business and product perceptions. These have to be rectified in a way which matches the indigenous knowledge systems in a continuous way. Knowledge is ever-changing so business should change on a regular basis.

When applying indigenous knowledge systems, companies must be careful of falling into the dungeons of patriotism, indigenization and nationalism. Knowledge systems are different from the mentioned principles.

## Benefits of Indigenous Knowledge Systems to African Business Thought

IKS closely links *Ubuntu* to business world. *Ubuntu* is a profound African philosophy based on the recognition of people's humanity through their participation as members of a community. A very simple translation of the idea is "people are people through other people" (Dugmore, Horner, Mooney, Nieftagodien and Lekgoathi 2007:245). Tutu stated in 1996:

"God has given us a great gift, *Ubuntu* .... *Ubuntu* says I am human only because you are human.... You must do what you can to maintain this great harmony, which is [constantly damaged] by resentment, anger, desire for vengeance [punishment for wrong doing]. That's why African (thinking) is restorative rather than retributive. (Wilson, Nesiah and van Zyl 2002).

In turn this would bring sovereignty to African business because they will be self-ruled and business people would be respected for their good, ethical and moral governance of their investments. They would be contributing greatly to the sustainability of their communities.

IKS promotes democratic business principles, such as the right to vote for a government, shareholders and business partners, equality, freedom of speech, movement, religion and thought. This allows the business world to take another turn. From indigenous knowledge, business can start expanding across African borders. Trade have been there in African continent since time immemorial, internationalization of any business need a clear description of the IK of that country in which the company need to start operations. Shepherd and Katz (2005:9) propose that internationalizing new firms face dual challenge of over-coming the liabilities of newness and liabilities of foreignness. On this instance they will be new and foreign in terms of indigenous knowledge systems. Hence, due to their newness, new firms are constrained in their ability to access external resources required for survival and growth. The foreignness relative to the foreign target market, internationalizing firms are disadvantaged relative to domestic firms in the issues of establishing business relationships. This act as some kind of protectionism to the indigenes but, however, the importance can be realized if the knowledge systems are opened to the world exposure. Knowledge systems act as a control measure, had it not been that effect Africa would have been robbed of all the resources by the western colonizing countries.

Indigenous knowledge systems control the insatiable desires of the two forces: socialism and capitalism. These are economic thoughts which can be destructive if they are not controlled. Socialists' theories argue that an economic system must be based on the ideal that those who produce the wealth must be owners. The capitalists' theories argue that an economic investment must be under the control of individuals who own the enterprises which produce wealth such as factories, farms and mines. This means that business ideology is to be under the control of certain categories of people but at this age of the knowledge societies wealth of any given society belongs to the people who have a common knowledge of the whole community. Capitalist mentality has high chance of creating a closed system. Socialist mentality can lead to the creation of an open system but the problem would lie in the question of who would take control of that system. Dugmore, Horner, Mooney, Nieftagodien and Lekgoathi (2007:81) purport that Mwalimu Julius Nyerere, former President of Tanzania, 1964-85 refined and implemented the policy called "African socialism". On this kind of knowledge he had a vision of a future Tanzania which is prosperous, self-reliant and classless society. Encompassed in his philosophy was Tanzanian socialism based on local resources rather than imported high-technology industrialization. Indigenous knowledge systems were highly promoted in Nyerere's policy; the emphasis of government and business was to be upon rural development and allowing rural areas to be self-reliant and self-sustained. This is a good approach to urban growth. Rural areas are reserves for the town activities. In Africa if there are no rural areas; heritage would have been no-more. In rural areas trees, wild life and cultural deposits are maintained better than they are maintained in urban technical archives, zoos and botanical gardens.

This case of Nyere's Tanzanian socialism is one of the knowledge systems which can be adapted to overhaul industrial activities of companies and governments in Africa. He suggested that Tanzania's mass of small remote rural areas must be gathered into larger, more effective villages, for government to provide better roads and rural markets combined with agricultural advice and improved technology. Better water, health and education facilities to be provided efficiently to larger centralized villages. The policy was known as *ujamaa* (Swahili word for extended family). A vital aspect of *ujamaa* was the promotion of the "African socialist" principles of communal labour for the benefit of the community towards improved and increased agricultural productivity, enabling communally cultivated fields to produce a surplus for sale to the towns or for export (Shillington 1995:417-419). This case is applicable to all African independent countries like South Africa, Zambia, Lesotho, Swaziland, Namibia, Botswana, Mozambique, Zimbabwe and Kenya to mention just a few. Zimbabwe attained land but they messed it up, because of inefficiency and poor service delivery. Nyerere behind this idea he was influenced by the way people used to live in historical times given the situation that some of the processes of evolution had started in Tanzania at Olduvai Gorge implying that all developmental processes also started there with the advent of tools. Above all the idea of collective settlements was well known in African Kingdoms, for there were the likes of Mwenemutapa, Luba-lunda and the Zulu state, people in these states stayed in clusters sharing common knowledge and resources like salt, gold, copper and iron. Many skills developed from these clusters in the Zulu state the developed "cow horn" formation type of formation on attacking the enemy in modern day war strategies they call it "surprise attack"

Egypt had a lot of indigenous knowledge; some of it is still applicable across the world in business sects today for example, the calendar, paper, pen and irrigation (*shaduff*). One can be justified to say indigenous knowledge systems are highly applicable to business processes towards sustainable development in Africa and the eradication or the alleviation of poverty in African states. (Ki-zerbo 2003:129-130) affirms that the gold of the Sudanic region and Zimbabwe became a source of prosperity for the Maghrib and the near east, while still strengthening the great empires of the western Sudan and Mwenemutatpa. Iron was traded in ancient times between North and South, while copper was the basis for transcontinental traffic towards the Indian Ocean. The salt mines on the edge of Sahara played an important role between the kingdoms south of the Sahara and the Arab-Berber people.

During these periods people were using their indigenous knowledge systems and were benefiting but later on they were dominated and exotic knowledge systems took over benefiting the colonialist- capitalist, who substituted the indigenes and overpowered the business processes to date.

Freire and Faundez (1989) find indigenous knowledge as a rich social source for any justice-related attempt to bring about social change. Indigenous knowledge systems are raised from the nature of our existence, our consciousness, our knowledge production and the "globalised" future. Furthermore indigenous knowledge counters western science's destruction of the earth. IKS has a tendency of focusing on relationships of human beings to one another and to the ecosystem. This finding concurs with that finding by Tutu of *Ubuntu* observed by (Wilson, Nesiah and van Zyl 2002). Western sciences have been reckless to both African people and their ecosystems. If one looks at apartheid in South Africa and colonialism in Zimbabwe one would see that the colonial masters were not bothered about issues of health and welfare of people in terms of human rights and the environmental pollution.

An appreciation of the IKS from business today comes from the activities like corporate social responsibility. This shows that companies are accountable and responsible to the probable damages that impacted on the people and the environment due to business processes. Some big companies are ploughing back the profits to the communities who were supporting them in business. They are giving donations and sponsorships sponsoring the HIV/AIDS victims and orphans, soccer, education and support programmes for the poor.

IKS gives the indigenous people an opportunity to give decisions on those aspects of business that affect their lives. It also promotes team work. In African economies people used to work in teams for example in *nhimbe* (beer parties) work was done collectively by a group of people sharing ideas and pooling their effort and resources together to promote productivity, efficiency and effectiveness. It was not easy for one person to build a house alone so had to call others for help the food and beer that were to be provided after the group task was not a payment but was just a token of appreciation. It is highly motivating, if African labourers are allowed to form work teams basing on their values and interests. In their groups at *nhimbe* African people used to sing at work to boost morale and spreading knowledge through songs. In like manner in modern industries there must be improvements in workplace morale by adjusting ecology and ergonomics to make work much more interesting, pleasurable and motivating by boosting the morale of employees.

Indigenous knowledge systems are beneficial to business applications and company activities in Africa. This paper is encouraging business practitioners in Africa to take cognisance of Africans and their values to meet the needs of Africa and its environment towards user-friendly production, ethical labour practices and user-friendly consumption.

## Factors Affecting the Applications of IKS in Industrial Activities in African Companies

There are a number of factors hindering the applicability of IKS in industrial activities in African companies. They range from;

Nature of education imparted to the people of Africa by the western scientists and philosophers.

Lack of terminology for technological developments, product developments and innovations

Cultural disorientation and language displacement

Divide and rule imposed by certain leaders

Neo-colonialism and cultural imperialism

Globalisation

Romanticisation

Demonisation of other business practices

Epistemological hegemony

Power and scientific colonialism

Lack of money and degradation in terms of economic levels and status

Fear of change – fear of the unknown

Subjugation- who is allowed to proclaim the truth.

## Conclusion

Applying indigenous knowledge systems in business scenario is a daunting process. However if it can be applied it can bring much success than failure; in fact it would bring harmonized business systems which would allow the people and ecosystem to interact in a friendly way. Every society has its indigenous knowledge systems, to avoid unnecessary business challenges those indigenous systems must be valued and given priority in business decisions-making processes. It is found that knowledge systems can be mixed to produce a hybrid of knowledge systems. For example the western (exotic) knowledge systems can be mixed with African indigenous knowledge systems. They should be mixed in equal proportions to avoid domination, oppression, inferiority, over-dilution and colonization of the other. IKS can be used in management functions and that knowledge can be derived from proverbs, stories, riddles, songs, food storage structures, buildings and nature reserves to mention a few.

IKS can change for the positive or negative from the old IKS to the new IKS. Presently given the fact that the African IKS have a strong bias towards the western knowledge systems, African business activities must manipulate and capitalize on the hybridism and reinforce their business systems which would be highly competitive on the world market. The major point is knowledge systems must be kept open otherwise they would become archaic and rendered obsolete and useless. On final note it should be however, noted that every person is an indigenous somewhere so people should not take pride in being indigenes of certain community and relax. It's encouraged that one should participate towards the sustainability of that indigene, complacence have destroyed many economies which were once first class countries for example Zimbabwe.

## References

Bahra, N.2001.Competitive Knowledge Management. New York: Palgrave.

Brink, A., Cant, M., Machado, R., Rudansky, S& Theron, D. 2000.Customer Behaviour. Pretoria: UNISA

Brokensha, D., Warren, DM, Werner, O. 1980.Indigenous Knowledge Systems and Development. Washington DC: University Press of America.

Davenport, T.H &Prusak, and L.1998.Working Knowledge: How Organisations manage what they know. Boston: Harvard Business School Press.

Dugmore, C, Horner, EA, Mooney, K, Nieftagodien, N, Lekgoathi, S.2007. Viva History. Florida Hills: Vivlia.

Freire, P., & Faundez, A.1989. Learning to Question: Pedagogy of Liberation. New York: Continuum

James, V.U.1996. Sustainable development in third world countries: applied and theoretical perspectives. Westport: Praeger

Jansky, L., Haigh, J. M., and Prasad, H. 2005. Sustainable Management of Head Water Resources: Research from Africa and India. Hong Kong: United Nations University.

Jordan, P.2005.Heritage Transformation Indaba. [Online] URL:http://www.nhc.org.za/index.php?pid=109&ct=1

Ki-zerbo, J.2003.General History of Africa I: Methodology and African Pre-history. South Africa: New Africa Books.

Mosimege, L.2005.The role of indigenous knowledge systems (IKS).South Africa: University of the Free State.

Nel, J. 2005.The role of indigenous knowledge systems [Online] http://www.uovs.ac.za/faculties/content.php?FCode=01&DCode=156&id=6336

National Research Foundations.2005.Indigenous Knowledge Systems [Online] http://www.nrf.ac.za/focusareas/iks/

Semali, M.L., and Kincheloe, J. L.1999. What is Indigenous Knowledge? London: Falmer Press.

Schwartz, D.G. 2006.Encyclopedia of Knowledge Management. London: McMillan.

Shepherd, AD., and Katz, J. A. 2005. International Entrepreneurship: Advances in entrepreneurship, firm emergence and growth volume 8. United Kingdom: Elsevier.

Shillington, K. 1995.History of Africa. New York: St Martin's Press.

Strydom, J. 2004.Introduction to marketing. Cape Town: Juta

Swoden, D. J. 2000. "The ASHEN Model: an enabler of action". Knowledge *Management* 3 (7)

Van der Walt, A., Strydom, J. W., Marx, S, & Jooste, C. J. 1998.Marketing Management. Kenwyn: Juta

Wilson, R. A., Nesiah, V., and van Zyl, P. 2002. "Challenging Restorative justice" *Human Rights Dialogue*, Series 2 No. 7 winter

# PART II:
# INDIGENOUS KNOWING AND EVERYDAY CHALLENGES

# CHAPTER 4

## Mixing Indigenous with Exogenous Knowledge in the Formulation of Agricultural Policy in Ethiopia: The Principle of "A Foot on the Ground"

*Demese Chanyalew*

### Introduction

Ethiopia is one of the few African countries which recorded an agriculture sector growth and budgetary appropriation more than the targets set by the African Head of States while launching the Comprehensive African Agricultural Development Programme (CAADP). Between 2004/05 and 2009/10 the average annual growth rate of Ethiopia's agriculture sector has been about 8.4%. Agriculture has been growing at the same time by paving the way for the growth and expansion of services and industry.

On the other hand, the recorded growth of the agriculture sector and the shift towards services and industry seem unpalatable to several opposition politicians and some academicians. There are those who dispute that EPRDF's[3] government agricultural development led industrialization (ADLI) strategy is wrong. Many politicians in the opposition camp believe the strategy should be industry development led. They also accuse the government being incapable of eliminating hunger and reduce poverty to the level of their expectations. They argue that government failed in this regard because it depends on agriculture which is an unpromising sector to

3 Ethiopian Peoples Revolutionary Democratic Front

depend on. In this connection they also blame EPRDF government's policy and strategy making principles, specifically "A Foot on the Ground" which promotes indigenized policy and strategy formulation.

This paper points out that the misconception that ADLI is the agriculture sector strategy should be corrected. ADLI anchored policies and strategies of the country have contributed to the recent positive economic growth, including the growth of the agriculture sector. Despite its high growth and significant positive contribution to Ethiopia's economy very little has been done to agriculture itself in terms of capital investment. There are still policy, strategy and institutional challenges and issues which the EPRDF government has to review or formulate in order to attain development explained not only by growth but also by changes in eliminating hunger, reducing food insecurity, poverty, and unemployment.

In order to address the above and other related suppositions this paper first establishes definitional and conceptual clarities on core terminologies and concepts. This is followed by a brief discussion of the recent performance of Ethiopia's agriculture based economy. Subsequently, the policies and strategies behind the observed performance are highlighted. Challenges and issues for policy, strategy, and institutions' review are discussed in section five. Finally concluding and way forward statements are presented.

## Definitions and Conceptual Underpinnings

In terms of recognizing the recent years growth of the agriculture sector of Ethiopia, one source of confusion among politicians of various domain and some academicians is the misconception of "agriculture." Hence, it is essential to have a clear definition of agriculture and conceptual underpinnings and definitions of development, growth, policy, strategy, programme and institution.

Agriculture by many urbanites and some academicians at a university level is perceived only as farming and mostly as crop farming. Some even fail to understand farming has crop and livestock components. Livestock is the major sources of livelihood in the lowlands, predominantly pastoral areas. Some also fail to comprehend the depth and width of agriculture's service sub-sector which includes policy, strategy development, and institutional issues contained in agricultural research, technology multiplication, extension, education and training. The agriculture sector by definition also includes agri-business (warehouse, transport, etc.) and agro-processing. Ethiopia's present manufacturing sector is laden with agro-processing industries that are related to or stand alone in the areas of textile, leather, and sugar industries. Agriculture also includes the input and output marketing and trade businesses.

In order to have an informed debate on the correctness of the recent economic performance of Ethiopia's agricultural economy it is important to highlight the theoretical knowledge underpinnings of economic development and growth in general and agricultural development and growth in particular.

The theoretical knowledge of development establishes two major definitional components: growth and change. These components have been used and sometimes abused by being considered in isolation. For example some discredit or reject the recent year's high economic growth reported by the EPRDF government arguing that problems of poverty and hunger in the country are still highly prominent, and hence the statistics of economic growth, including the high growth rate in the agriculture and allied sector is fictitious. This is despite the established knowledge of macroeconomics that not only in Ethiopia but in many other countries of the world, including developed economies, that growth does not necessarily result in the reduction of poverty to expectants level. It is possible that growth and distribution, which is core to the change component, can be inversely related. In the case of Ethiopia, studies indicate that a one percent economy growth leads to a 1.7% reduction of poverty level (MoFED[4]: GTP, 2010).

4 MoFED: Ministry of Finance and Economic Development, and GTP: Growth and Transformation Plan

Macroeconomics theory reveals that economy wide growth can be measured using gross domestic product (GDP) or sector specific measures derived from GDP, for example agricultural gross domestic product (AGDP). These are measured by monetary value of final goods and services produced in a given period of time, usually in a year. GDP or AGDP in a simplified model could be illustrated using a demand side or supply side expressions as follows:

| Demand side | Supply side |
|---|---|
| GDP => AGDP = f (C, I, G, (X-M))<br><br>C=Consumption,<br><br>I=Investment,<br><br>G= Government expenditure<br><br>X-M=Net Export | GDP=> AGDP = f ((L, N, K)/ T)<br><br>L=Land,<br><br>N=Labour,<br><br>K=Capital,<br><br>T=Technology |

The above specifications assist to explain the functional and structural relationships of various actors in the economy by recognizing the critical role of the product, input and financial markets. All these, however, are relatively more useful to define and explain the growth component of development than the change component.

Change often is measured by reduction of poverty, hunger, food insecurity, unemployment as well as improvements of inequality in terms of resources and wealth use and distribution. Indeed, the major problems Ethiopia is facing today are poverty and hunger. Though poverty does not necessarily entail hunger, the reverse is tautological. Hence tackling the problem of hunger and in its broader context food insecurity requires great attention from government, NGOs and other development partners.

In the present Ethiopia context, in order to understand the conceptual underpinning of poverty reduction and hunger elimination, it will be prudent to use a modified model expressed as follows:

GDP or NP = f (NR, HR, IF, PI)[5]

Where:

NP=National production

NR= Natural resources

HR= Human resources

IF= Infrastructure (economic and social)

PI= Policies, strategies and institutions

In this model, it is assumed that the influence of foreign economies on NP is evident. Growth is a positive change in NP due to changes in one or more of the factors on the right hand side. It magnifies that poverty and hunger reduction and elimination requires sound policy, strategy and institutions.

Here policy is conceived as a statement of course of action set by the Federal and Regional State governments of the Federal Democratic Republic of Ethiopia (FDRE) in the management of socio-economic development affairs. It could be formulated and implemented in the form of laws, rules, regulations, directives and broad goal oriented guiding declarations that affect different economic and social agents and institutions.

Institutional issues have organizational and relational dimensions. Both dimensions are footed on laws, regulations and directives too. The organizational framework covers the structure of the government and the place a given sector, say agriculture, is given in the government set up. Strategies serve to bridge policies and programmes. Programmes are assumed to emanate from the policy and strategy that has surpassed them by formation. A programme is a broader cost center of a public body or a broad objective of expenditure. Projects are expected to come under programmes and have entry and finish date, i.e., limited life time. Sectors programmes are expected to last long and usher several projects of varied project life.

---

5 This is a conceptual framework used in CAADP Ethiopia Study, 2009.

## The Performance of Ethiopia's Economy, Specifically the Agriculture Sector

*Trends in GDP and AGDP*

In Ethiopia when the agriculture sector performs well so does the overall economy. This is well captured by the GDP and AGDP positive correlation (Figure 1). A ten years trend, 1990s till 2006/07, shows that the agriculture sector had experienced an average of about 10% growth rate. As shown in Table 1, during the Plan for Accelerated and Sustainable Development to End Poverty (PASDEP) period, the average annual growth rate of the agriculture sector has been about 8.4%. During this period the overall economy growth was about 11%. This growth is attained concurrently with a shift from agriculture to services and industry. The percent share of agriculture from GDP declined from 53% to 42%between 1995/6 and 2009/10 (Figure 2). The share of the service sector has increased from 34% to 46% during the same period.

| Sector | Growth% | Share from GDP % |
|---|---|---|
| Economy wide (real GDP) | 11 | 100 |
| Agriculture and allied | 8.4 | 41.56 |
| Industry | 10 | 12.87 |
| Services | 14.3 | 45.57 |

**Table 1: Average economic growth (%) and sectoral percent share from GDP during the PASDEP Period[6].**

Source: MoFED, GTP 2010/11-2015/16, Vol.1, November 2010.

Within agriculture it is the crop subsector which is leading (Figure 3). As of 2006/07 crop agriculture share from GDP was 30% while its share from AGDP was 65%. For the same period the share of the livestock, and hunting subsector and that of forestry was 12.3% and 3.9% from GDP and 26.5% and 8.4% from AGDP respectively. Evidently, for a country with the largest livestock population in Africa and tenth in the world the performance of the

6 From 2004/05 to 2009/10

livestock sector is still very low. The pastoral system of production, which is mainly livestock, contributes 9% to the national GDP.

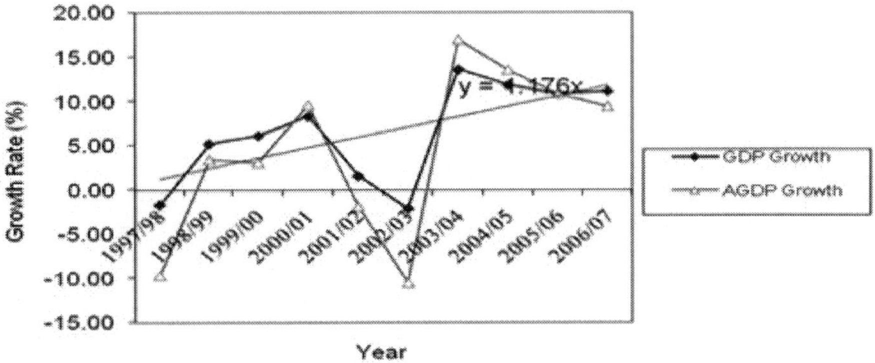


**Figure 1: Growth Trend: GDP Vs AGDP**

Source: MoFED, 2009

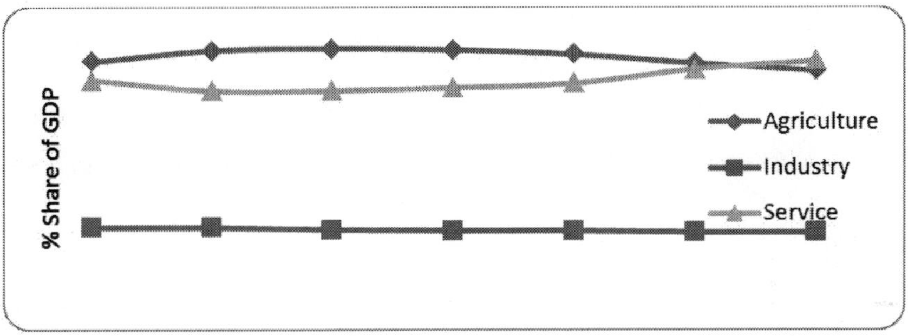


**Figure 2: Percent Share of GDP by Major Sector Classification**

Source: MoFED, 2009

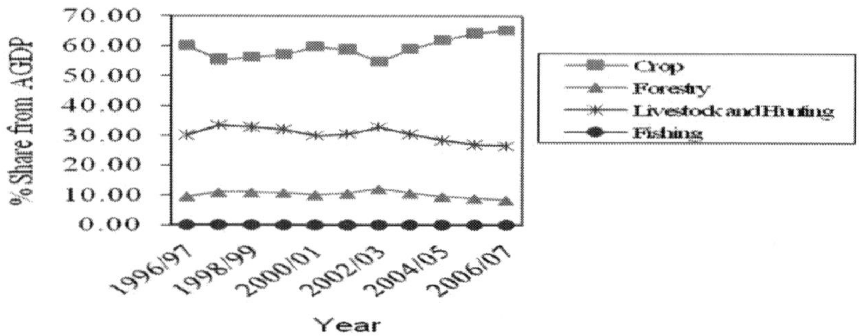


**Figure 3: Percent share of Crop, Livestock, Forestry and Fishing from AGDP**

Source: MoFED, 2009

*Trends in Food Security*

Food security is one area of focus to assess the performance of the agriculture sector specifically and the overall economy at large. On the production side[7], agriculture has contributed positively to make the nation food secure. There has been moderate increase in crop productivity but in order to reflect on total production level it has been challenged with the marginalization of previously high potential cultivated land due to acidity problems. For example about 30% of previously high potential cultivable land in Amhara region was reported to become marginal land as of 2009. It is also known that annually on average 1-1.2 million hectares of cultivated land remained unharnessed and this is estimated to be about 15-20 million quintals of crop pre-harvest loss. It is also documented that annually on average 15-25%, depending on crop types, produce is subjected to post harvest loss. Despite these, as Figure 4 shows grain equivalent food production has been on the rise. Milk production is on rise and per capita milk consumption which is still below Africa average is showing improvement and is closing to Sub-Saharan average (Figure 5).

7 To comprehend the achievements, it is essential to understand the three core elements of food security: production, income and entitlement.

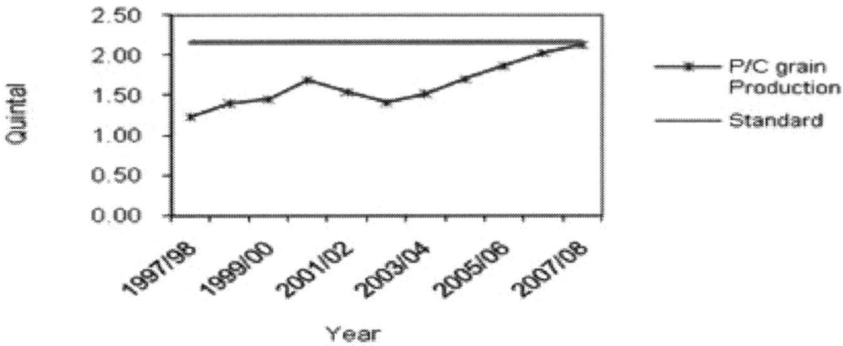


**Figure 4: Trends of Per Capita Grain Production**

Source: CAADP Ethiopia Study, 2009.

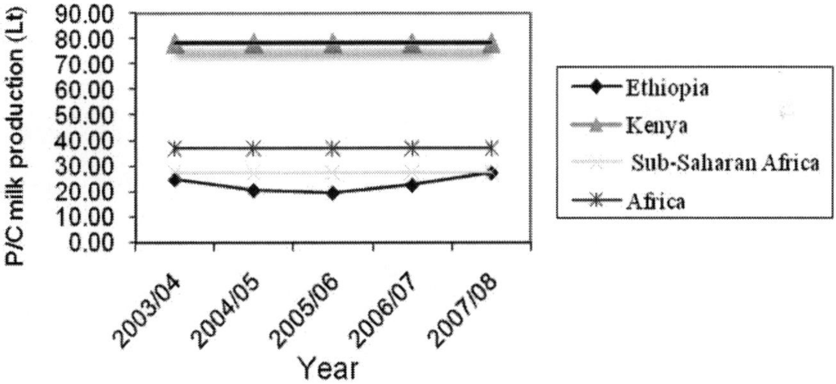


**Figure 5: Comparative Trend in P/C Milk Consumption: Ethiopia Vs Others**

Source: CAADP Ethiopia Study, 2009.

*Trends in agro-industry establishments*

The number of manufacturing establishments is increasing in the agro-industry sub-sector (Figure 6). The crop and livestock agriculture is the major supplier of raw materials for the emerging and expanding industrial sector which is being led by the textile, leather, and sugar industries.

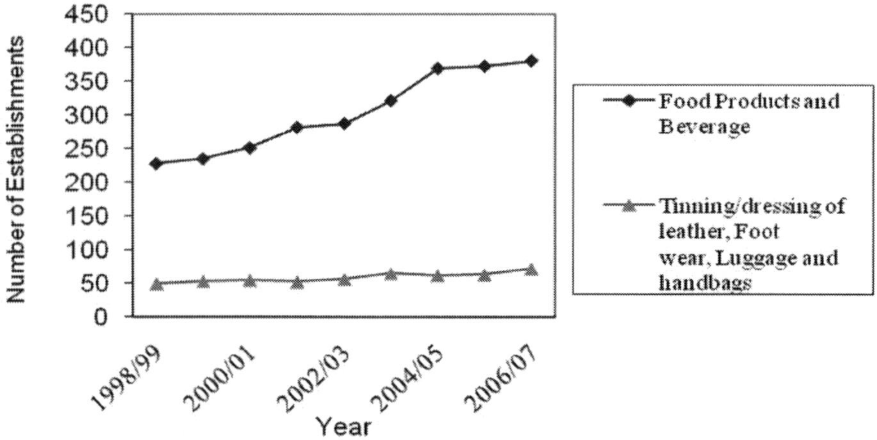


**Figure 6: Trend in Numbers of Agro-Industry Establishments**

Source: Central Statistics Agency (CSA).

*Trends in Agricultural Trade: Export and Imports*

Agriculture also remained the major source of export revenue. Available data and information reveal that exports increased, export diversifications taking place but imports do not prop up agriculture.

The value of total exports of Ethiopia increased from 602 Million US$ to 1466 US$ from 1997/98 to 2007/08 with an average growth rate of about 2% (Figures 7 and 8). This change is a result of quantitative, qualitative and price changes. The quantity of traditional export items has increased and also new export product mix is created including the export of flowers (Figure 9).

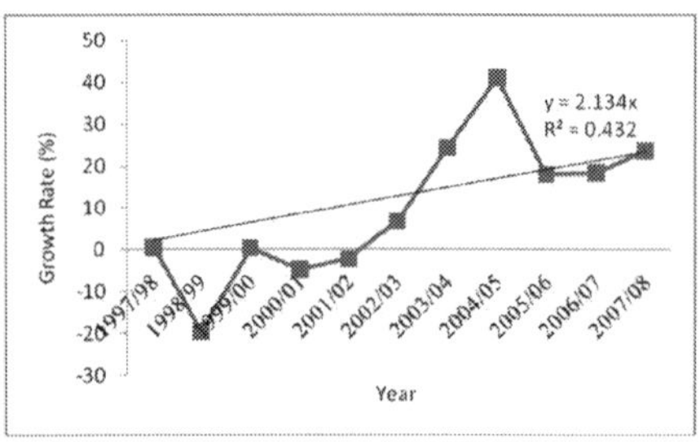


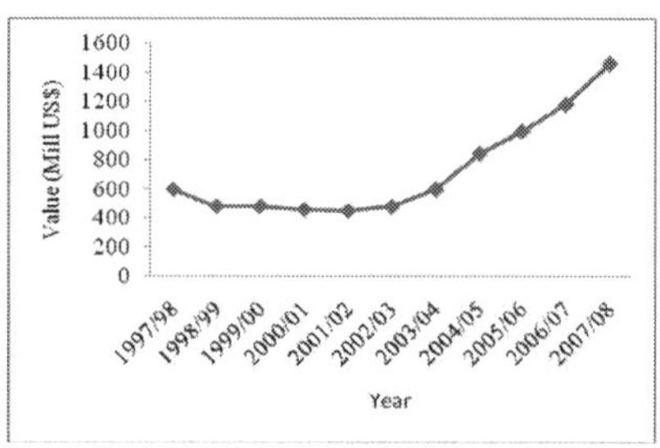


**Figure 7: Trend in Export Growth Rate**

**Figure 8: Trend in Total Export**

Source: CSA, National Bank of Ethiopia (NBE), and MOFED.

Total volume and value of agricultural commodities exports have shown positive trends. Coffee, pulse and oil seeds are the three main export agricultural commodities in volume terms. In terms of value although coffee is the highest generator it is with a decline percent share trend. This signals that export product diversification strategy is showing a positive result (Figure 10). Pulses export is showing an upward trend. Pulse export increased sharply from 2002/03 to 2007/08. Oilseeds export also showed a positive growth trend but the volume exported drops sharply in 2007/08. Live animals export have started to grow while the newly established floriculture industry is expanding with an increasing share of the export value.

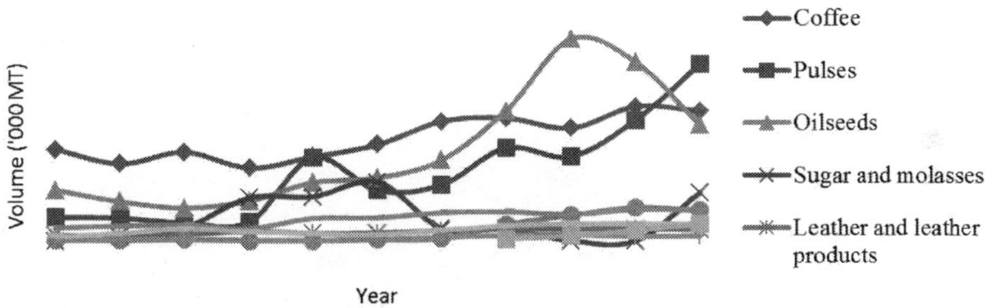


**Figure 9: Trend in Export by Commodity**
Source: CSA, NBE, and MOFED.

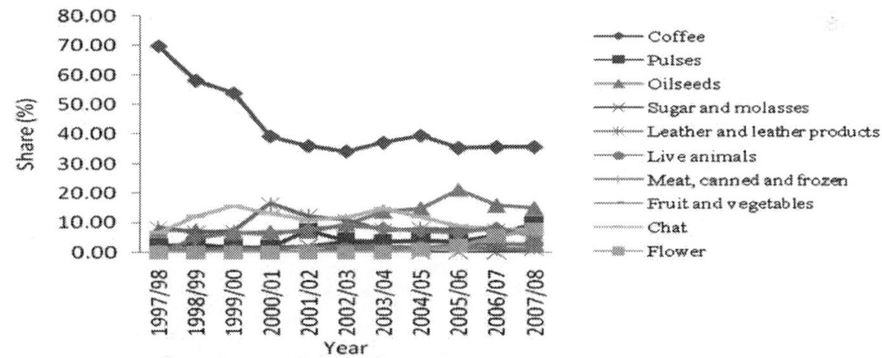


**Figure 10: Trends in Export Share by commodity**
Source: CSA, NBE, and MOFED.

In general, although the total value of exports is on rise since early 2000/01, the contribution of exports to GDP has not changed much (Figure 11). On the other hand the percent share of imports is on rise with an increasing deficit in trade. Implying the country is importing a lot with declining foreign exchange generation. This increasing import has not been in favor of supporting agriculture to modernize itself. The huge export revenue generated from agriculture has been used to make the industry and service sector, specifically transport to grow and expand.

On the average for the last ten years more than 75% of Ethiopia's import by end use is for raw materials, fuel, capital tools and consumer goods (Figure 12). The major imported raw materials are semi-finished goods, chemicals, fertilizers, and textile inputs. Semi-finished goods have been the leading imported items for the last 18 years. In recent years fertilizer importation follows this (Figure 13).

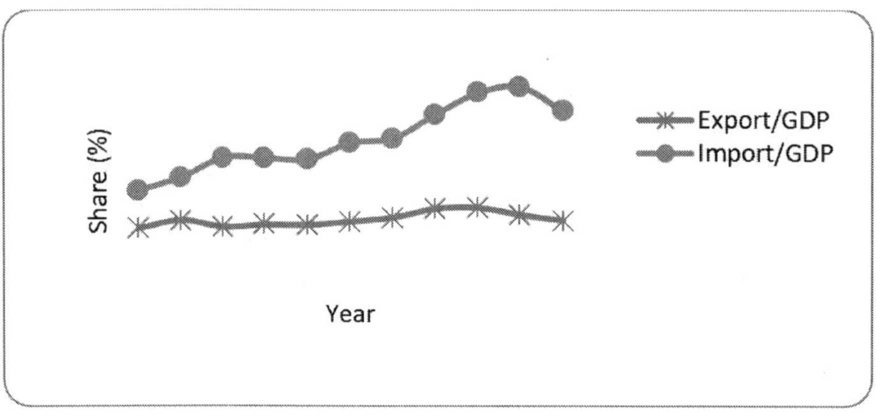


**Figure 11: Share of Export and Imports from GDP**
Source: CSA, NBE, and MOFED

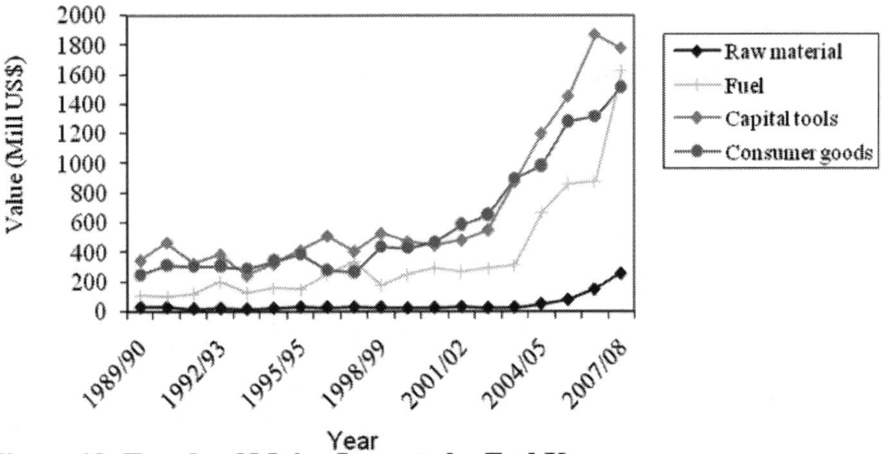


**Figure 12: Trends of Major Imports by End Use**

Source: CSA, NBE, and MOFED

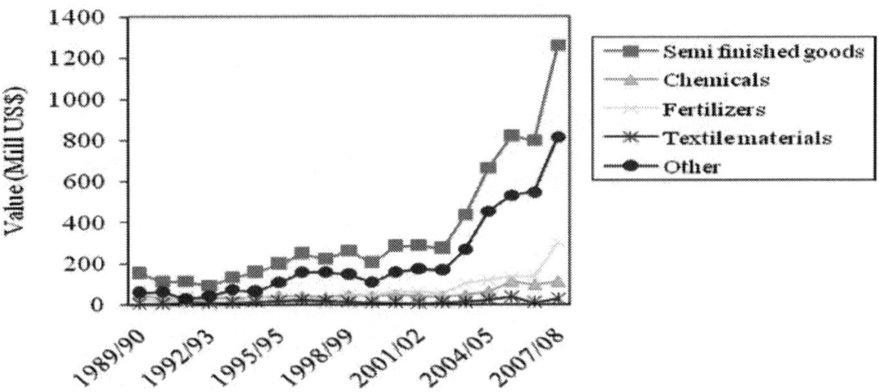


**Figure 13: Trends of Raw Material Import by Category**

Source: CSA, NBE, and MOFED

In the importation of capital tools[8], industrial goods have had the highest share followed by the transport sector (Figure 14). The share of agricultural tools import is negligible and remains constant. In other words, despite agriculture's contribution to the growth of the economy its share of capital (modern investment) is negligible. So far it has performed superbly with

8 Includes machinery and equipment

little capital investment. Of course recent studies indicate the government is determined to increase its investment in modern and commercial agriculture. This is reflected in the current five year development plan document, known as GTP (2009/10-2014/15).

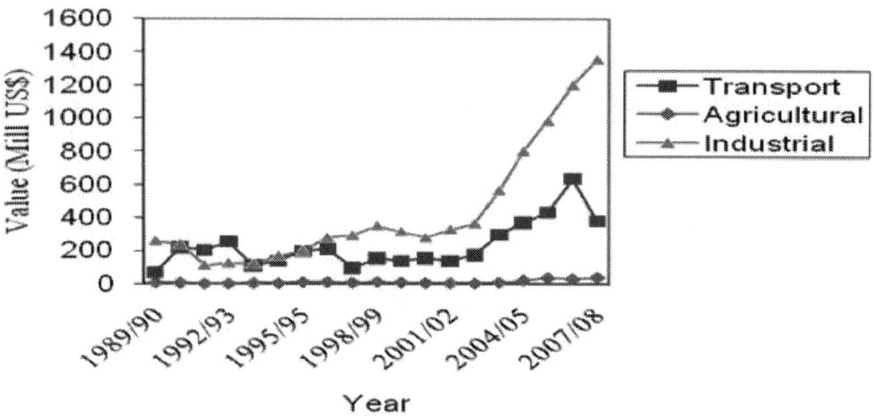


**Figure 14: Capital Tools Import by Component**
Source: CSA, NBE, and MOFED

## Policy, Strategy and the Principle of " A Foot on the Ground"

The above section indicates that agriculture in Ethiopia is the major source of income, producer and supplier of food, and raw material for the growing and expanding agro-processing based industry sector. It also generates export revenues that have been instrumental to the expansion of industry and service sectors. As indicated above since 2004/05 the average annual growth rate of the agriculture sector in Ethiopia in terms of AGDP exceeds the CAADP target of 6%. Of course this has happened by the government's heavily inward looking policies and strategies which have been also supported by increased government budget appropriation to the agriculture and natural resources sector. The budget exceeded the CAADP target of 10%. Generally the total Federal budget allocated to agriculture and natural resources has increased from 9.4% in 1993/94 to 15.5%in 2007/08 (MoARD: CAADP Ethiopia Study, 2009).

Simply the trend analysis indicates that the performance of Ethiopia's economy is inseparable from the performance of the agriculture sector. Given the definition accorded earlier, agriculture undeniably is an engine of growth and driver of economic transformation in Ethiopia. By implication this indicates that Ethiopia's agricultural policy was inseparable from the national economy policy and vice-versa. It also reflects that indigenous policy stances are rewarding particularly if the principle of incrementalism[9] and conditional policy making is adopted.

The above trend analysis results, however, are not palatable to some, particularly those in the opposition wings of Ethiopia's present day politics. Knowingly or unknowingly they either make an outright rejection of the growth statistics or formulate perplexing statements and explanations by mixing the macroeconomic knowledge of growth with the distribution knowledge of development.

Some academicians and politicians in the opposition argue that the economic policies and strategies of the EPRDF government have been erroneous. EPRDF's inward looking policies and strategies, according to them, did make the country remain in the vicious circles of hunger and poverty, hence the growth statistics is fictitious. Though the author of this paper thinks that poverty and hunger are the daunting challenges of the government and people of Ethiopia, on the other hand he concurs to the growth results. The author of this paper also believes that indigenous policy and strategy initiatives accompanied with selective external advices will make Ethiopia to attain its goal of a middle income country status by 2020[10].

AUC-NPCA (2010) CAADP Technical Review report indicate that the country is on track for halving its poverty rate by 2015. According to this report the country's $1.25 per day poverty rate fell from 71.3 percent in 1990 to 30.4 percent in 2009. The food poverty head count decreased from 44% in 1999/00 to 38% in 2005/06 and 28% by the end of the PASDEP period (2009/10).

---

9 This principle promotes that policy is built step by step, and wise decisions, as well as mistakes of the past are the foundations for current and future policies.

10 USD1000 per capita of the 2005 equivalent.

No country in the world is immune of policy deficiencies and hence should be on a standby to make policy reviews following the dynamic domestic and international politics and economic undertakings. Policy making is a dynamic process. Recognizing this and understanding the EPRDF's government readiness to review policies (MoFED: RDPS, 2003) and also recognizing the principles of policy conditionality, the agriculture sector policies and strategies at present are those which paved the way for the achievements made in the sector so far. In reality the last ten years high economic growth of Ethiopia would have not been obtained without introducing the policy, strategy, and institutional changes that were compartmentalized in the two successive five year development plans known as SDPRP and PASDEP[11].

First of all to discuss about a rightness or wrongness of any policy and strategy it is essential to identify and state those policies and strategies. In order to get a detailed understanding of the agricultural policies and strategies of the EPRDF government, specifically to relate them to the attained economic growth in the last two decades, the major documents one has to read are:

Revolutionary Democracy Development Directions and Strategies (text in Amharic, 1992 E.C.[12]),

Rural Development Policy, Strategies and Instruments (text in Amharic, 1994 E.C.), and

Related sectors strategies, successive five year development plans and related studies[13].

Careful examination of these documents enables to note the government's conditional policy setting strategy and the refinements and adjustments made by policy makers subsequently.

---

11 Sustainable Development and Poverty Reduction (SDPRP) is Ethiopia's first poverty reduction strategic plan and its successor development plan is known as PASDEP.

12 Ethiopian Calendar

13 The following are relevant to have a good grasp of the present EPRDF agriculture and pastoral development policies and strategies: *Industrial Development Strategy* (2002), Implementation Capacity Building Strategies and Programmes (2003); SDPRP, 2002 and PASDEP, 2006; the Millennium Development Goals Needs Assessment (MDGs-NA) for the Agriculture and Rural Sector (MoFED & UN 2004).

In Ethiopia, a comprehensive policy and strategy document that encompasses natural resources, agriculture, pastoral affairs, and rural development issues was in vain till 2001. In November 2001, the EPRDF government issued 'Rural Development Policies, Strategies and Instruments (RDPSI)'later on edited and issued in English as RDPS (2003). RDPS is the base document to know and analyze Ethiopia's agriculture, pastoral and rural development policies and strategies at present. According to this and those related documents listed as footnotes, the trust of the agriculture and rural development policy of Ethiopia is that the nation is set to build a market economy in which a broader spectrum of the Ethiopian people are beneficiaries, dependence on food aid is eliminated, and rapid economic growth is assured.

RDPS also reaffirms the government's strong belief in its core strategy of the national economic development that is ADLI. ADLI has been also explained as a pro-poor growth strategy of the economy (Mekonnen, 2002), and never was designed as a specific strategy of the agriculture sector. According to RDPS the strategy for the agriculture sector was the "agriculture-centred rural development" (ACRD) strategy.

The fundamentals of Ethiopia's agricultural development strategy, ACRD, as clearly put in PASDEP are:

Adequately strengthen human resources capacity and its effective utilization,

Ensuring prudent allocation and use of existing land,

Adaptation of development path compatible with different agro-ecological zones,

Specialization, diversification and commercialization of agricultural production,

Integrating development activities with other sectors, and

Establishment of effective agricultural marketing system

Based on these fundamentals several sub-sector and thematic strategies have been set.[14]As stated in the RDPS document the basic principles governing agriculture development *policy* in Ethiopia are

---

14 For details refer the CAADP Ethiopia Study, 2009.

*The labour intensive strategy*
*Proper utilization of agricultural land*
*A foot on the ground*
*Taking different agro-ecological zones into account*
*An integrated development path*

A close examination of the fundamentals of agricultural strategy formulation and the basic principles governing agriculture development *policy* in Ethiopia divulge that policy and strategy making in Ethiopia has been mixing both indigenous and exogenous knowledge. The former is enclosed in the principle of "A Foot on the Ground".

"A Foot on the Ground" principle recognizes that it is necessary to record and pass on to the next generation practical experiences and *indigenous knowledge*, and to make effective use of this knowledge in our development efforts. In other words the strategies for agricultural development will seek to draw opportunities for growth inherent within the available manpower and technology. This principle, indeed, has been central to the inward looking policies and strategies of the EPRDF government which also has attracted the attentions of opposition politicians and some academicians to make them throw critics on the governments overall economic policies and strategies. For example, in the academic circle, one can see the critique made on an article titled "Ethiopia: Agriculture Policy Review" (Dessalegn, 2008, pp.129-51). This review seem to suffer from the chronic problem of defining agriculture and agricultural policy narrowly from the point of view of land and agricultural extension services and secondly from a misconception of ADLI, which in this article is addressed by equating ADLI as a strategy of "Walking on One Leg", a prelude to explain EPRDF's "A Foot on the Ground" principle.

On the opposition politics side, for example, in the 2002 E.C election period, during a debate between political parties on agriculture, a failure to understand the philosophical underpinning of "A Foot on the Ground" principle led one opposition debater to condemn the strategy by taking a normative stance. According to him, the strategy is "wrong" because it advances "*Ye-ankassie*[15]" pace development or amputated development. Assuming that this was not a deliberate twisting by the debater, such stance, though may seem right to the individual, it may signify to the audience that the debater is either less knowledgeable about the subject matter or is confused to politicize the strategy for a wild political game irrespective of the relevance of the strategy to agriculture and rural development in Ethiopia.

Again, it is important to underline that the RDPS document states that "*a-foot on the ground*" is a principle which advances a proposition that if the country has to grow and develop then it should not be by neglecting its indigenous knowledge and technologies that are still relevant, *ceteris paribus.* This indeed is what we Ethiopians have to build on. We should use what we have first and improve on it when need arises by introducing from outside. The footing of this principle may even be traced back in the writings of Schumpeter (1949). Schumpeter once argued that *by "development," we shall understand only such changes in economic life as are not forced upon it from without but arise by its own initiative, from within.*

It is also important to note that in the policy and strategy formulation of the present Ethiopia, the principle of "A Foot on the Ground" cannot be easily comprehended unless one analyzes EPRDF's Revolutionary Democracy. For many the revolutionary democracy document at face value is a political document presenting mainly political issues pivoting on democracy. According to this author, this document is a base document, which all sectors, including agriculture, policies and strategies have been anchored on. This, however, does not mean all policy and strategy stances explained in the document are right and good. But, at least, the document give a footing that internally generated policies and strategies, whether they are right or wrong in the views of outsiders, have to be considered so long as the government believes it serves the interest of the nation. Besides, the revolutionary democracy document did contain paradigms that have challenged the traditional liberal or neo liberal free market economy thinking and theories.

---

15 Amharic phraseology to explain a person walking with one leg.

The revolutionary democracy document of EPRDF starts with outright rejections of liberal and neo-liberal free market economy and adopts the "Developmental State" paradigms. An in-depth analysis of these issues may be outside the realm of this paper. However, it is possible to concur that the EPRDF government aloofness from externally dominated policy and strategy formulation, and its commitment on the paradigm of developmental state formation seems to contribute to the positive achievements made in the agriculture sector and the overall economy thus far.

## Challenges and Issues for Policy and Strategy Review

Being cognizant of the registered positive economic growth and the presence of sound economic development policies and strategies, still there are questions and issues to raise and discuss if Ethiopia's endeavor to prosper has no limit. The following are few of the selected challenges and issues:

Are those progresses made in the agriculture sector and the economy at large, obtained with apt efficiency and efficacy of human, natural, and financial resources use?

Specifically, what has been the role of professionals in the country's policy, strategy formulation and implementation? Were they accepted by their knowledge and technical capacity or were they also subjected to become politically succumb?

What are the immediate issues that need to be addressed to make Ethiopia's land based resources, including forest, soil and water, conserved, i.e., for wise management and use that transcends the generations to come?

What specific policy and strategy changes are needed to use the countries abundant agricultural water resources to benefit the very many poor and smallholding farmers and pastoralists?

What policy, strategy or institutional changes are needed to supply and distribute critical modern agricultural inputs both to smallholder as well as emerging and expanding large-scale commercial agricultural enterprises?

What policy and strategy reviews are required to make the economy shift more to the private sector? Specifically to increase the nation staple food crop production? And garner an income from the livestock sub-sector that commensurate to the potential?

Which type and level of institutional capacity problems calls for immediate solution in order to implement existing sound policies and strategies?

The author realizes that each one of the above questions and the various issues embodied in them may require extensive explanation. For the sake of this paper, a highlight is presented in the context of earlier set conceptual framework, i.e., the nation's economic growth, hence its ability to eliminate hunger and reduce poverty, is a function of efficient and effective use and management of its human and natural resources, socio-economic infrastructure, and the policies and strategies that accompany these factors.

*Challenges and issues related to human resources*

Ethiopia's agriculture has achieved what it could mainly by making labour, specifically small farmers, to improve their work habits and traditional crop husbandry. Thus far the skilled and knowledge intensive professionals were not used and managed efficiently and effectively. Ethiopia's agriculture sector growth would have been more than reported above had agricultural professionals were used on the basis of their professional stands rather than their political submissiveness.

The author did make a remark on the role of agricultural professionals and their professional associations on the growth and transformation of Ethiopia's economy during his invited paper presentation at the third annual conference of Ethiopian Association of Agricultural professionals (Demese, 2007). In this conference it was pointed out that agricultural professionals and the government are not working together and concluded that if government, politicians (in the ruling or opposition camp), professionals and professional associations continue to operate in isolation then Ethiopia's agriculture and rural sector will remain, for several decades in future, without realizing meaningful growth and development. In the past for several decades what has come to Ethiopia in the form of economic aid and grant has made no significant change to solve the nation's food insecurity, hunger and poverty problems and again no matter the number of foreign experts imported and no matter the amount of finance in the form of development aid and grant comes from outside, Ethiopia will remain in the world's poor and food insecure, and perhaps hungry nations list for years to come unless it starts to look for the solution from within and make use of foreign aid and experts to work with development goals and instruments designed by Ethiopian professionals.

Put differently, the way the professional and skilled agricultural human resources is being used today still calls for concerted effort to minimize the wastage of domestically available resources and make a prudent use of the Diaspora, specifically the intellectual who can help in the transfer of technology and knowledge. The present biased and unknown motive of excluding domestically residing agricultural professionals from the growth and transformation arena, as some activities of the so called Agricultural Transformation Agency indicate, should be curtailed.

*Challenges and issues related to natural resources*

In the area of natural resources management land degradation and soil erosion are still issues that challenge Ethiopia's capacity to feed itself in the future. These issues require clear policy and strategy that accommodate all the agro-ecological zones of the country. Cognizant of the government's Forest Protection and Utilization Policy and Strategy (MoARD, 2007), still there is lack of clear strategy *to utilize forest and forest products* with an apt forest conservation practices that are well tuned to the interest of the local communities as well as to support the many poor and food insecure in rural agricultural areas.

Despite the fact that land policy is one area of policy which has been extensively covered through legislations and studies, there are still gaps in terms of adequate coverage of the *land tenure issues of Pastoral and Agro-pastoral areas*.

*Challenges and issues related to productivity and production augmenting inputs*

There are still strategic and institutional issues which should be addressed to enhance availability, access and use of productivity augmenting inputs, including improved seeds, improved breeds, fertilizer, pesticides, semen, vaccine, agricultural water, land, and labour.

The Agricultural Water Development and Management (AWDM) investment in Ethiopia should avoid size, scale and spatial and agro-ecology biases. In line with this there should be also a paradigm shift to water-led women-centred green revolution[16].Besides, sound policies and strategies should be crafted to halt the negative influence of HIV/AIDs on agriculture sector performance, reducing pre-and post-harvest losses[17].

So far the progress made in the agricultural sector is heavily dependent on non-capital goods and services reflecting the policy emphasis i.e., dependence on labour and land with the promotion and distribution of biological and chemical technologies listed above. Little is done in terms of making agriculture capital intensive. There is a lot to be done in terms of promoting private investment and commercialization of agriculture both in the crop and livestock sub-sectors. Again this does not mean that the small holder agriculture has been exhaustively exploited by using the relatively available abundant labour and land resources and employing improved biological and chemical technologies.

A pragmatic credit scheme for domestic investment on staple food crops production is highly required and essential to make the nation harness its potentials by its own investors. This could be for example by making domestic investors having access to capital goods by attaining up to 5% of the capital goods import of the country. Let the staple food crop sub-sector enjoy the 30/70 % own capital and investment loan arrangement the floriculture sub-sector has enjoyed.

---

16 For details refer to Demese, 2008.

17 Readers can read the recent extensive discussions on such issues contained in the CAADP Ethiopia Study, 2009 and the Agriculture Sector PIF: Ten Year Road Map (2010-2020) preparation report.

Agriculture should also get its own share to grow with modern practices both in terms of machinery and equipment and this can be started by encouraging commercial large scale farms to be capital intensive. In this connection, the urban peri-urban intensive high value cash crop and dairy activities expansion directive contained in the GTP (2010) of Ethiopia are commendable. But the expansion of such private agricultural investment should be done with proper compensation for land holders whose land is taken for this or related public investment purposes.

The policy and strategy focus on the livestock sub-sector is still inadequate. The Livestock sub-sector also lacks institutional focus. Because of these inadequacies in a country with the largest livestock population in Africa and tenth in the world the performance of the livestock sector is very low. As reported by Tesfaye et.al (2008) a lot has to be done in this sub-sector in the area of institutional and capacity building.

*Challenges and issues related to cross-cutting matters*

In addition to the above specific subject-matter issues, there are also policy, strategy and institutional challenges and issues revolving around climate change and disaster risk management, stakeholders participation, gender, spatial dimensions of budget allocations, use of sound Planning Monitoring and Evaluation (PM&E) and data system, as well as capacity development at grassroots level. Again the challenges and issues associated to these areas are extensively discussed in the recent CAADP and the agriculture sector ten year's policy and investment framework studies. Readers are advised to read these documents. Overall, it is important to note that recognizing and utilizing the strategic importance of engaging multi-stakeholders inclusive of a wide array of interest groups, national and regional governments, donor agencies, international bodies, the private sector, as well as local and national Civil Society Organizations starting from policy and strategy formulation up to implementation and M&E. Gender mainstreaming remains an issue to advocate for in the years ahead. It is the right of the food insecure and the poor who are in the so called food secure and high potential areas to get their share of the national resources allocation without being discriminated by project based fashionable classifications of intervention Woredas.

## Conclusion and the Way Forward

Aagriculture has been not only important but also an engine of economic growth and a sector leading Ethiopia to industrialize. Actors in the agriculture sector have to continue working hard and untiring in order to make agriculture's share of GDP decline at a faster rate than experienced during the SDPRP and PASDEP period. It is a very high growth rate, which is achieved through increased investment on modern and commercial agriculture that will hasten the rate of decline of agriculture's share from GDP and enable to achieve the average overall economic growth of 11.2% and basic industry growth rate of 20-21% set for GTP period (by 2015) and achieve the goal of a middle-income country status by 2020.

The principles contained in the revolutionary democracy document and the developmental state paradigms are inward looking, which promotes '*development from within*' policy and strategy in the agriculture sector, driven by agricultural land and labour utilization strategy, addressing the uniqueness of the country's diverse agro-ecologies via an integrated approach which maximizes on workable and useful indigenous skills and technologies. Hopefully the future will be better than the past because this inward looking policies and strategies that are based on Ethiopia's real economic conditions and situations continue to make Ethiopia remain on the path of economic growth but at the same time the government shall consider policies and strategies that narrow the rising disparity between the haves and have-nots in the country.

In general, in Ethiopia today economic history is in the making, a history that will become part of the future industrious Ethiopia universal history. As indicated in the previous sections, centuries old high dependence on agriculture is changing. In a period of a decade, since 1996, agriculture share of GDP declined by more than 10%. Studies indicate if agriculture is supported to grow faster than the rate it has been growing per annum in recent years, the share will reach to about 30% by 2020 (MoARD: PIF, 2010).

The future of Ethiopia is heavily dependent on today's Ethiopia renaissance and transition with integrated and all agro-ecologies tuned agriculture and natural resources use and management policy and strategy formulation and implementation. The renaissance era is hyped with the governments believe in Ethiopians and its policy and strategy formulation and leadership to construct the "***Hedasie Gedeb***" or the "***Renaissance Dam***" on the Abay (Internationally known as Blue Nile) river. The construction of this dam, and the many other huge national investment undertakings on social and economic infrastructure can be efficiently and effectively accomplished by maximizing on the use of available indigenous capacities, at least at the thinking and policy and strategy formation level, i.e., by continuing to take advantage of the principle of "***A foot on the ground***". This principle makes Ethiopian's to look inward and assess their past and present and believe that the problems of hunger, malnutrition, unemployment, inadequate and poor social and health services, overall abject poverty and underdevelopment are our own (Ethiopians) makings; they are the makings of neither foreigners nor unknown external forces. Put differently, the fate of Ethiopia rests in the hands of Ethiopians (Demese, 2004). No person or country from outside should be blamed. If this is transformed from a belief to actions and deed, and if the recent trend of reliance on domestically crafted policies and strategies continue, together with commitment to implement and good governance, then the vision that Ethiopia will be a middle income country by 2020 remains realistic.

## References

AUC-NPCA (African Union Commission, NEPAD Planning and Coordinating Agency). 2010. CAADP Technical Review-Summary of key Findings for Ethiopia, September 2010, Addis Ababa.

Chanyalew, D. (2008) A New Paradigm for Agricultural Water Development and Management: Analyzing Ethiopia's Position in Africa. In: Proceedings of the 10th Annual Conference of the Agricultural Economics Society of Ethiopia, Addis Ababa, May 2008.

Chanyalew, D. (2007) Transforming Ethiopian Agriculture and the Rural Sector: The Role of Agricultural Professionals and Professional Associations Including Academia as Drivers of Change in Growth and Development of Agricultural Innovations. In: Ethiopian Association of Agricultural Professionals, Promoting Value Chain for Transforming Ethiopian Agriculture and Rural Development. Proceedings of 3rd Annual Conference and Panel Discussion Sessions. Addis Ababa, 23-24, November, 2007.

Chanyalew, D. (2004) Agricultural Policy and Farm Price Support in Ethiopia. Commercial Printing Enterprise, Addis Ababa.

FDRE, MoARD, (2009) CAADP Ethiopia Study, Vol. I and II , Prepared by Demese Chanyalew, Getinet Gebeyehu, Goshu Meknonen, Yadddesa Dinssa and IFPRI, July 2009, Addis Ababa, Ethiopia.

FDRE MoARD, (2010) Ethiopia's Agricultural Sector Policy and Investment Framework (PIF): Ten Year Road Map (2010-2020). Prepared by Demese Chanyalew, Berhanu Adenew, and John Mellor. August 5, 2010, Addis Ababa, Ethiopia.

FDRE/MoFED, (2010) Growth and Transformation Plan (GTP) 2010/11-2014/15. November, 2010, Addis Ababa.

FDRE/MoFED, (2002) Sustainable Development and Poverty Reduction Program (SDPRP), Addis Ababa, Ethiopia.

FDRE/ MoFED, (2006) Ethiopia: Building on Progress- A Plan for Accelerated and Sustained Development to End Poverty (PASDEP) (2005/06-2009/10)." Addis Ababa.

FDRE/MoFED. Building on Progress: A Plan for Accelerated and Sustained Development to End Poverty Annual Progress report 2005/06

FDRE/MoFED, (2004) Millennium Development Goals Need Assessment: The Rural Development and Food Security Sector in Ethiopia, Addis Ababa, Ethiopia.

FDRE/ MoFED, (2003) Rural Development Policy and Strategies (RDPS). Addis Ababa.

Kumsa, T., Yigezu Z., and Ayana, I. (2008) Livestock Resource Potentials, Constraints and Opportunities for Intervention by Private Sector. Chamber of Commerce and Sida ,Addis Ababa, Ethiopia.

Manyazewal, M. (2002) An Overview of ADLI, Rural Development and Agricultural Performance in the 1990s. Presented at Ethiopia: Rural Development Workshop, Organized by FDRE/World Bank, November 13, Addis Ababa, Ethiopia.

MoARD, (2007) Ethiopian Forest Protection and Utilization Policy and Strategy. Addis Ababa.

MOFED and the United Nations Country Team (UNCT) (2004) Millennium Development Goals Report Challenges and prospects for Ethiopia. March 2004, Addis Ababa.

Rahmato, D. (2008) Ethiopia Agriculture Policy Review. In: Digest of Ethiopia's National Policies, Strategies and Programs, ed. By Taye Assefa. Forum for Social Studies (FSS), Addis Ababa.

Schumpeter, J. A. (1949) The Theory of Economic Development: An Inquiry into Profits, Capital, Credit, Interest and the Business Cycle. Harvard University Press, Massachusetts.

# CHAPTER 5

## Information for micro decisions and livelihood improvement: Outreach as a key challenge in Ethiopia

*Bedru B. Balana*

### Introduction

The World Bank's World Development Report (WDR) 1998/99, titled 'Knowledge for Development' asserts that "Knowledge is like light. Weightless and intangible, it can easily travel the world, enlightening the lives of people everywhere. Yet billions of people still live in the darkness of poverty – unnecessarily." (The World Bank, 1999, p.1). The Report focused on two important sorts of knowledge and two types of problems that are crucial for developing countries: 'Knowledge about technology' i.e., technical knowledge or know-how, for example, about farming or health and 'Knowledge about attributes', for example, about product quality, market information, or credibility of a borrower. The Report claims that developing countries have less technical knowledge than industrial countries, and the poor have less know-how than the non-poor. According to the Report, this unequal distribution of know-how across and within countries is termed as 'knowledge gaps'. Developing countries also face myriads of problems related to incomplete 'knowledge of attributes' which the Report termed as "information problems". Mechanisms to alleviate information problems, such as product standards, market information, and credit reports are fewer and weaker in developing countries than the developed ones. Information problems and the resulting market failures especially hurt the poor. This paper focuses mainly on the 'information problems' facing the developing countries and its impact on micro level decisions, choices and livelihoods with particular emphasis to Ethiopia.

'Choice' is the fundamental concept in economic science. 'Choices' can range from simple individual level daily routine decisions, for instance, one's choice to walk or use public transport or drive to a local supermarket; or a family's weekend meal choice between beef steak or chicken in a local restaurant to big national or global policy issues. Individuals, households, consumers, producers, governments and non-government agencies make dozens of choices or decisions routinely. Economic theory postulates that these actors are 'rational decision-makers' i.e., they strive to achieve certain objectives by allocating economic resources at their disposal (recognizing constraints or exogenous factors). The underlying assumption that underpins this postulate is that decision makers possess complete knowledge or information relevant to their decisions.

Neo-classical economic theory shows that decisions under incomplete information and imperfect competition result in sub-optimal allocation of resources (inefficiency) and social welfare loss. Producers' decisions depend, among other things, on the availability of information on input markets, output markets, and technology. Information on the qualities and prices of commodities are the key elements for consumer's purchase decision. Agricultural household's decisions, especially those in the subsistence farming as both a producer and a consumer, and the outcomes are highly dependent on the availability of and access to information. Missing or incomplete input and/or output markets and lack of information are endemic to agricultural producers in poor economies (Kay et al. 2000; Taylor and Adelman, 2002). This imposes significant hindrances to the economic welfare of rural households.

The World Bank WDR 1998/99 puts 'information problems' at the core of the difficulties that people, particularly the poor, in developing countries encounter in the struggle to improve their lives. Taking the case of 'information problems' in Ethiopia, the paper attempts to illustrate the effect of limited or no access to information on the social and economic development in the country. The aims of this paper are to: (1) highlight the key roles that access to information/knowledge play in advancing human wellbeing, and (2) outline the current situation and constraints to knowledge/information exchange in Ethiopia.

## Information/knowledge for human well-being

Information/knowledge advances economic and social development. It illuminates economic transactions, reveals preferences, gives clarity to exchanges, and informs markets. Lack of information causes markets to fail or never to come into being (The World Bank, 1999). The critical role that knowledge and information play in the development endeavours of countries has been well recognized in the 21$^{st}$ century. For example, the World Bank's WDR 1998/99 exclusively devoted to the impact of knowledge gaps and information problems on development and the ways these problems should be addressed. The Report further provides detailed analysis of the risks and opportunities that the global information revolution is creating for developing countries, and concludes that access to financial, technical, and medical knowledge is crucial to improving the health and living standards of the poor. Maxims such as 'Information is scarce resource!', 'Information is power!', 'Information is freedom!', 'Information is like light' and so on and so forth clearly show the acute demand of information for human wellbeing. From the development perspective both aspects of information i.e., knowledge gaps and information problems, are undeniably essential.

If so, what developing countries must do to address these issues? The 1998/99 WDR suggests three key ways developing countries need to focus in order to address the above two aspects of information: acquire, absorb, and communicate.

- Acquire knowledge: tap and adapt knowledge available elsewhere;
- Absorb knowledge: invest on education (universal basic education especially on women education, tertiary education such as science and engineering); and

- Communicate knowledge: ensure citizens especially the poor have access to information, e.g., by taking the advantage of Information and communication technology (ICT).

Knowledge creation is expensive. From the developing countries perspective; adapting, disseminating, and using effectively the stock of knowledge already created elsewhere in the world is extremely crucial because of the following major factors.

1. As knowledge creation requires huge investments on Research and Development (R&D), developing countries may not have sufficient resources.
2. Lots of knowledge is already available elsewhere, 'developing countries need not to reinvent the wheel.' So, if developing countries pursue appropriate strategy to acquire knowledge/information they can gain huge benefits. For instance, in the 1960's Ghana and the Republic of Korea had virtually at the same income per capita. By the early 1990s Korea's income per capita more than six fold compared to that of Ghana. It was claimed that about half of this difference is due to Korea's greater success in acquiring and using knowledge (The World Bank, 1999).
3. Huge stock of indigenous/local knowledge has not yet been fully exploited in the developing countries. By designing appropriate methods to tap this knowledge and establishing conducive environment for the traditional knowledge to flourish, developing countries can benefit immensely from the existing local knowledge.
4. The rapidly falling communication costs provide big opportunities for developing countries to transferring knowledge created elsewhere at low costs.

By putting appropriate policies in place to acquire and adapt knowledge; making use of the most of indigenous knowledge; and addressing information problems through institutional innovations, developing countries can enhance human well-being and livelihoods. The story of the green revolution and the Grameen Bank's Microfinance schemes are probably two of the best real world cases to illustrate how addressing knowledge gaps and information problems have benefited millions of poor people in developing countries.

## A. The green revolution (GR):

The creation and dissemination of new agricultural knowledge – the Green Revolution– resulted in a dramatic increase in crop yields. Yields have been more than doubled in much of the developing world (Asia and South America). Africa, however, has seen only modest increase in yield (Figure 1). As can be seen in Figure 1, over the period of four decades (1955-1997) agricultural yield have trebled in Asia and more than in South America.

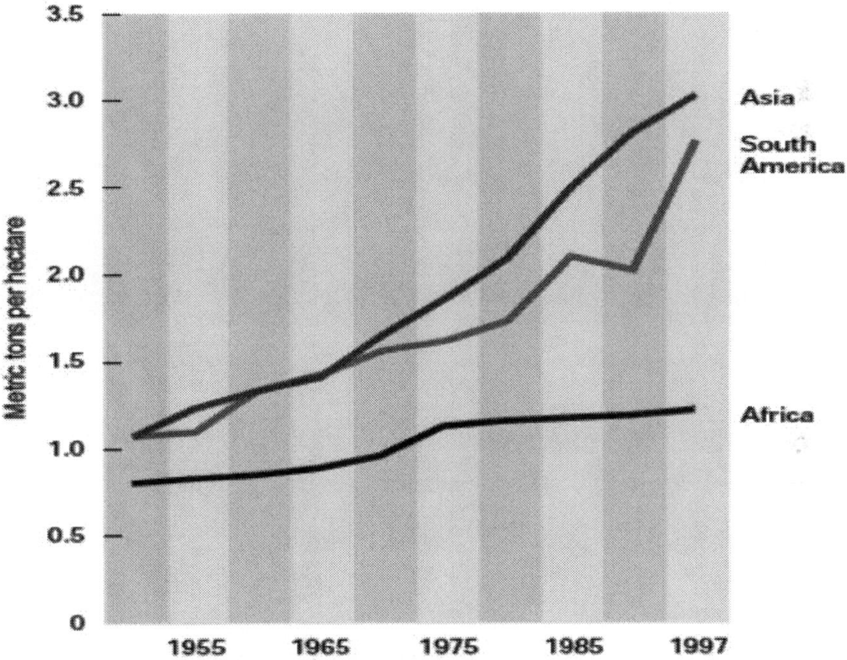


**Figure 1. The impact of green revolution on cereal yields by developing region**

Source: World Bank, WDR 1998/99, p.4

### B. Microfinance (Grameen Bank - Bangladesh):

Lack of access to credit and insurance is one of the biggest disadvantages that the poor face in developing countries. Lenders do not possess information on the creditworthiness of the poor and hence not willing to lend to the poor. Microfinance institutions are tackling the interrelated problems of credit and information problems to the poor. In this regard the Grameen Bank and Prof. Muhammad Yunus of Bangladesh pioneered the innovation of credits to the poor via microfinance institutions. The Grameen Bank not only lends to the poor but also promotes social developments through training, opening schools, and expanding communication networks. For his efforts to create economic and social development through microfinance and ultimately for lasting peace, Prof. Muhammad Yunus was awarded The Nobel Peace Prize in 2006.

### C. Other specific examples of success stories because of information and knowledge:

- Vietnam: Education has resulted in a massive reduction of poverty in Vietnam (poverty rate 68%; 54%; 41%; and 12% for households headed with no education, primary, secondary, and university education respectively).
- Brazil: Education and the mass media – a study found that parents who regularly made use of mass media have healthier children (a combined effect of education and mass media). Education is necessary for individuals to process information, but access to relevant information through the mass media is necessary for education to have an effect.
- Costa Rica: Government effort to provide people with health and sanitation information resulted in high life expectancy and low infant mortality. "With less than one-tenth the income per capita of the United States, Costa Rica boasts health indicators that compare favourably with those of many industrial countries. Costa Ricans live nine years

longer than their income per capita would predict, and infant mortality rates have fallen to industrial country levels. These impressive results are not by accident. Since the 1960s, Costa Rican governments have given high priority to the general dissemination of health and sanitation." (The World Bank, 1999).

- Rwanda: Government backed initiatives (training and awareness) on energy efficiency has achieved efficient charcoal burning stoves. "The women helped to design a stove that used one third less charcoal and produced less smoke, trained stove makers in producing the new design, and helped to market it. Three years later, one out of four city households were using the new stove-which was being produced and marketed without government subsidies." World Bank Press Release No: 99/1955WB news. http://web.worldbank.org/WBSITE/EXTERNAL/NEWS/0,,contentMDK:20017519~menuPK:34463~pagePK:34370~piPK:34424~theSitePK:4607,00.html (accessed on 01 April 2015).

Based on the preceding discussions, three lessons of particular importance to developing countries in relation to information/knowledge as a core element in the process of development are:

1. Institute good Policies: Examples – education policy; openness and transparency; competitive economy (particularly dismantling barriers to competition in the telecommunications sector);
2. Enhance collaboration (governments, donors, multilateral agencies etc.)
3. Recognize knowledge as the core of development and make all possible efforts to narrow knowledge gap and address information problems.

## Current situation in information flow in Ethiopia

Ethiopia is one of the most ill-fated countries in Sub-Saharan Africa (SSA) in terms of access to information (particularly in rural areas where supposedly 85% of the population reside). Information flow in Ethiopia is highly limited and controlled by the state. This has been revealed by dozens of evidences both from domestic and international sources (e.g., the reports of Committee to Protect Journalists (CPJ)[18]; Reporters without Borders[19]; Amnesty International[20]; Human Rights Watch (HRC)[21]; and US State Department[22]).

In its 2010 Human Rights Report on Ethiopia, the US State Department reports that:

> "While the constitution and law provide for freedom of speech and of the press, the government did not respect these rights in practice. The government continued to arrest, harass, and prosecute journalists, publishers, and editors. The government continued to control all broadcast media, including the sole television station, except three private FM radio stations. Private-sector and government journalists routinely practiced self-censorship. The broadcasting law prohibits political and religious organizations or foreigners from owning broadcast stations. Government-controlled media mostly reflected the views of the government and the ruling EPRDF coalition."

The Report adds:

> "In 2008 the parliament passed the Mass Media and Freedom of Information Proclamation. ..., the law ... grants the government the un-

18 See( http://www.cpj.org/africa/ethiopia/).

19 See (http://en.rsf.org/ethiopia.html). In 2010 Ethiopia received a rank of 139th in world press freedom out the 178 countries.

20 See (http://www.amnesty.org.uk/search.asp?q=+Ethiopia&submitted=-1)

21 See (http://www.hrw.org/search/apachesolr_search/Ethiopia)

22 For recent US State Department Report see human rights (including the status of mass media and flow of information in Ethiopia, see (http://www.state.gov/g/drl/rls/hrrpt/2010/af/154346.htm)

limited right to prosecute the media, criminalizes defamation of public officials, increases defamation fines to 100,000 Ethiopian Birr (USD 6,105), establishes "national security" as grounds for impounding materials prior to publication, provides government information officials exclusive discretion to withhold "sensitive" information without judicial review, and maintains the Communication Affairs Office's authority to regulate the media."

Human Rights Watch Jan. 2010 Report says:

> "Ethiopia passed a new media law in 2008 that improved upon several repressive aspects of the previous legal regime. The space for independent media activity in Ethiopia remains severely constrained, however. In August two journalists were jailed on charges derived partly from Ethiopia's old, and now defunct, press proclamation. Ethiopia's new anti-terror law contains provisions that will impact the media by making journalists and editors potential accomplices in acts of terrorism if they publish statements seen as encouraging or supporting terrorist acts, or even, simply, political protest."

A review of various reports such as those indicated above and other sources on the current status of information flow in Ethiopia show that:

- The government restricts access to the internet and blocks many web sites
- State monopoly of media and communication (TV, radio, telecommunication, and print media)
- The number of internet subscribers is very low even in SSA standard. Access remained extremely limited. For instance, according to the International Telecommunication Union (ITU) – the UN specialized agency for information and communication technologies – in the neighbouring Kenya as of June 2009 there are 3,359,600 internet users (8.6% of the population), whereas in Ethiopia only 360,000 Internet users (0.4% of the population).

- Mobile telephone network and text messaging are very poor. The number of mobile telephone subscribers was 6.1 million (about 7% of the population), according to the Ethiopian Ministry of Communication and Information.
- Call costs are among the highest even within the SSA rates. Ethiopia is one the few countries suffering from high call charges, text messages, and internet browsing. Using a 'Tele-discount' call system from UK to Ethiopia, the rate is >£0.20/minute, whereas to India or Nigeria the rate is less than £0.02/minute.
- Jamming/blocking broadcasting media is a wide spread practice in Ethiopia, e.g., it was reported that the Voice of America (VoA), Amharic service was jammed in 2010
- http://www.voanews.com/content/voa-amharic-broadcasts-jammed-in-ethiopia-86339587/153515.html (accessed 01 April 2015).
- Ethiopia has the second lowest Internet penetration rate (after Sierra Leone) in sub-Sahara Africa. Dozens of websites that happens to criticize government policies are blocked from access in Ethiopia (source: http://www.ethiomedia.com/andnen/2700.html)
- Freedom of speech, expression, and assembly are frequently restricted.
- Intimidation, harassment, extra-judicial prison, etc. of journalists are widely practiced, e.g., according to the Committee to Protect Journalists (CPJ), from Aug 2001-May 2011, 79 journalists fled Ethiopia which is the largest number of flee in Africa.

Apart from the wide range of institutional and governance constraints as exemplified above, the flow of knowledge/information in Ethiopia has also been impeded by a number of social and economic factors. As highlighted in Table 1, Ethiopia is much more behind in these and many related development indicators than most SSA countries. In a country where two-third of the population is illiterate; more than three-forth citizens below poverty line; and nearly one-half ages below 14 years, knowledge/information dissemination appears a practical challenge. Coupled with the institutional and governance constraints as indicated above; high level of illiteracy, wide spread poverty, lack of technical infrastructure, and cultural/religious barriers (including the diverse languages spoken in Ethiopia) portray the nature of challenges for the production and distribution of information and knowledge in Ethiopia.

| Indicator | value |
|---|---|
| Population (total) | 83 million |
| Average annual population growth rate | 2.6% |
| Population composition (age 0-14) | 44% |
| Per capita (GNI) (in US $) | 330 |
| Life expectancy at birth (in years) | M=54, F=57 |
| Adult literacy rate (ages 15 and above) | 36% |
| Population below $2 a day | 77.5% |
| Child mortality rate (under 5) | 104 per 1000 |

**Table 1. Key indicators**

Source: Extracted from The World Bank, WDR 2011.

## Conclusion

Knowledge, both technical know-how and information about attributes, makes a difference. Many developing countries lack the capability to create and adapt knowledge that has driven the success stories of many of the world's development. In the last 2 to 3 decades or so, a number of the world's poor countries, by adopting appropriate policies to increase knowledge transfer and exchange, have made a real difference in the livelihood of their citizens. Unfortunately, others such as Ethiopia have not exploited these opportunities. Many of Ethiopia's social and economic problems such as extensive poverty, low productivity, and illiteracy can be explained partly by the country's inability to acquire, absorb, and communicate knowledge. Narrowing 'knowledge gap' and addressing 'information problems' can improve human well-being in myriads of ways. Carl Dahlman, Director of the 1998/99 World Development Report says, "Knowledge can make the difference between sickness and health, between poverty and wealth. Governments that adopt policies to make the most of knowledge will have major advantages in improving the lives of their citizens." Of course, non-government agencies and individuals or groups having willingness, enthusiasm, capacity and ability can also make a difference through knowledge production, transfer and exchange. Let's do our best to 'exchange' whatever knowledge at our disposal and in whatever way we can to contribute our '*Dirisha*', to borrow this 'novel' word from Mr. Obang Metho (Ethiopian human rights activist), to alleviate the poverty of millions of our fellow country men and women through information/knowledge production and exchange.

## References

Amnesty International. See (http://www.amnesty.org.uk/search.asp?q=+Ethiopia&submitted=-1) (Accessed on 01 April 2015)

Committee to Protect Journalists (CPJ), 2010. Report. See http://www.cpj.org/africa/ethiopia/) (Accessed on 01 April 2015)

Human Rights Watch (HRC). See (http://www.hrw.org/search/apachesolr_search/Ethiopia) (accessed on 15.08.2011).

Kay, N., Sadoulet, E., de Janvry, A., 2000. Transactions costs and agricultural household supply response. American Journal of Agricultural Economics. 82: 245-259.

Reporters without Borders, 2010. Report. See (http://en.rsf.org/ethiopia.html) Accessed on 13.08.2011).

Taylor, J.E., Adelman I., 2002. Agricultural household models: Genesis, evolution, and extensions. Discussion paper. University of California, Berkeley.

The World Bank, 1999. Knowledge for development. World Development Report 1998/99. Oxford University Press, Inc., New York.

The World Bank, 2001. Building institutions for markets. World Development Report 2002. Oxford University Press, Inc., New York.

UNDP, 2001. Making new technologies work for human development.

Human Development Report 2001. Oxford University Press, Inc., New York.

US State Department, 2011. Human Rights report. See (http://www.state.gov/g/drl/rls/hrrpt/2010/af/154346.htm (Accessed on 14.08.2011)

# CHAPTER 6

## Increasing Water Availability, Quality and Quantity through Public Expenditure: An experience from the public service delivery in Uganda.

*Samuel Galiwango*

### 1.0 Introduction

The water sector is one of the priority sectors in Uganda, as it directly impacts on the quality of life of the people and overall productivity of the population. Water supply and sanitation are among the key issues emphasized under the Millennium Development Goals (MDGs, Goal 7 section c) and National Poverty Eradication Action Plan (PEAP), which is the key government framework for ensuring poverty eradication through creation of an enabling environment for rapid economic development and social transformation. Water is a key strategic resource, vital for sustaining life, promoting development and maintaining the environment. Access to clean and safe water and improved sanitation facilities and practices are pre-requisites to a health population and therefore have a direct impact on the quality of life and productivity of the population. Besides domestic water supply, water is also vital for: Livestock Water Supply, Industrial Water Supply, Hydropower generation, Agriculture, Marine Transport, Fisheries, Waste Discharge, Tourism, and Environmental Conservation. Water, therefore, significantly contributes to the national socio-economic development and thus poverty eradication.

Despite Uganda's being well endowed with significant freshwater resources, the challenges of rapid population growth, increased urbanization and industrialization, uncontrolled environmental degradation and pollution are leading to accelerated depletion and degradation of the available water resources. Uganda is also faced with the challenge of low safe water coverage (59% rural and 65% urban, as of December 2003). In order to meet the above challenges, government initiated reforms in the water sector in 1997 to ensure that water services are provided and managed with increased efficiency and cost effectiveness. Comprehensive sector reform studies have been going on since 1998 and are due to be completed by August 2004. During these studies, detailed situation analysis of the sector was carried out resulting in the preparation of a comprehensive water sector Strategy, investment plans and time bound national targets for the sector up to 2015. In demonstrating its commitment to the reform process, government has already embarked on the process of implementing some of the strategic recommendations from these studies.

One of the key strategic outcomes from the reform studies is the adoption of a 'Sector Wide Approach to Planning (SWAP)' for the sector. The SWAP framework, which has been embraced by both government and the water sector development partners, has already proved to be the most appropriate mechanism for resources mobilization and implementation of the action plans. The SWAP framework also guarantees the participation of all stakeholders in the planning and implementation of water sector activities. This openness has resulted in increased confidence from the development partners who have now agreed to finance water sector programs through the regular government budget, contrary to the project specific funding characteristic of the past.

Despite the significant progress highlighted above, the sector is still faced with a number of challenges. The major challenge is establishment of strong mechanisms for effective, efficient and sustainable delivery of water and sanitation services to the end users, on the basis of the strategies and funding mechanisms established under the SWAP framework. This challenge, coupled with the inherent financial and human resources constraints at both national and local levels pauses the greatest risk for the untimely achievement of the sector targets.

Improvement on the sector services will require enhanced capacity of the sector institutions, in particular the Directorate of Water Development (DWD), to assume its new role of planning, supporting and supervising water and sanitation programs through improvements in monitoring systems and procedures. Government is now focusing on building capacity of the water sector institutions and Local Governments as well as promoting increased private sector participation and effective community participation in all water sector activities. The policy of promoting community participation based on demand-driven approach is being emphasized. Hygiene and health education are actively being pursued as an integral part of the drive to improve household sanitation.

The following paper therefore seeks to answer the following questions:

- What percentage of Ugandans use protected water sources?
- How much time is spent while collecting water from the water source?
- What reasons do households give for not using protected water sources?

## 2.0 Background

### 2.1 Water Sector Goal, Objectives and Targets

#### 2.1.1 GOAL

The goal of the water sector in Uganda is: "To manage and develop the water resources of Uganda in an integrated and sustainable manner so as to secure and provide water of adequate quantity and quality for all social and economic needs for the present and future generations with the full participation of all stakeholders."

### 2.1.2 OBJECTIVES

The key water sector objectives include:

1. To promote coordinated, integrated and sustainable water resources management to ensure conservation of water resources and provision of water for all social and economic activities.
2. To achieve sustainable safe water supply and sanitation facilities, based on management responsibility and ownership by the users, to 100% of both the rural and urban population in Uganda.
3. To promote the development of water supply for agricultural production in order to modernize agriculture and mitigate effects of climatic variations on rain-fed agriculture.

### 2.1.3 TARGETS

The national targets for water supply and sanitation in both urban and rural areas are shown in Box 2.1 below:

| | **Box 2.1 -<br>National Targets for Water Supply and Sanitation** |
|---|---|
| Urban Areas | Achieve 100% safe water coverage and 100% sanitation coverage in urban areas<br><br>by 2015, with an 80%-90% effective use and functionality of facilities. |
| Rural Areas | Achieve 77% safe water coverage and 95% sanitation coverage in rural areas<br><br>by 2015, with an 80%-90% effective use and functionality of facilities. |

## 2.2 Definition of Coverage:

The definition of coverage relates to percentage of the population with access to an improved water source within a walking distance of 1.5 Km in a rural area and 0.2 Km in the urban area. For sanitation, coverage refers to the percentage of the population with sanitation facilities in their place of residence.

## 2.3 Water Sector Policy and Legal Framework

The government has put in place a comprehensive policy and legal framework for the management of the water sector. The framework comprises of a set of policies and laws the most notable of which include: The National Water Policy (1999); The Water Statute (1995); The National Water and Sewerage Corporation Statute (1995), and the Local Government Act (1997). Though most of the above policies and legislation have been in force for more than five years, a number of provisions are not yet fully operational, especially at the local government and local community levels. Furthermore, some of the legislation needs to be revised to address the emerging issues in the sector like Private Sector Participation, Decentralization, and the SWAP.

### 2.3.1 National Water Policy

The National Water Policy (NWP), adopted in 1999, provides the overall policy framework for the water sector. The National Water Policy promotes the principles of integrated water resources management as a means to ensuring sustainable management and utilization of Uganda's water resources. The policy also emphasizes the recognition of water as being both a social and economic good, whose allocation should give first priority to domestic use. The Policy is based on the principle of "some for all, rather than all for some" adopted from the 1990 "New Delhi Statement". It anchors operation and maintenance as an important and integral part of all water and sanitation programs to ensure their sustainability. The policy also highlights the key role played by women in all water management and development activities.

### 2.3.2 Guiding Principles

The National Water Policy adopts the guiding principles for water resources management emanating from the United Nations Conference on the Environment and Development (UNCED, Agenda 21, Chapter 18) as detailed in Box 2.2 below.

| BOX 2.2: National Water Policy Guiding Principles |
|---|
| Freshwater is a finite and vulnerable resource, essential to sustain life, development and the environment |
| Management of water resources at the lowest appropriate level. |
| Recognition of water as a social and economic good with a value reflecting its most valuable potential use. |
| Integration of water and land use management |
| Recognition of the central role played by women in the provision, management and safe-guarding of water. |
| The important role of the private sector in water management. |

The National Water Policy also re-emphasizes the water resources management strategy as stipulated in the Water Action Plan (WAP) whose preparation was based on the same objective and guiding principles. The key strategy elements are grouped into those supporting an enabling environment, those guiding institutional development and those supporting the upgrading of planning capacity and prioritization of water allocation and use.

## 2.4 Progress towards achieving MDGs

In September 2000 at the UN Millennium Summit, world leaders agreed to a set of time bound and measurable goals and targets for combating poverty, hunger disease, illiteracy, environmental degradation and discrimination against women. Placed at the heart of the global agenda, these goals (eight in number) are now called the Millennium Development Goals (MDGs). Progress towards attainment of the MDGs is measured with the aid of 18 quantitative targets and 48 performance indicators. One of the targets – target 10 – addresses water supply and sanitation and calls upon member states to “halve by 2015 the proportion of people without access to safe drinking water and basic sanitation”. Target 10 is monitored through indicators 30 and 31, which are the proportions of people with sustainable access to safe water sources and improved sanitation respectively. Using 1990 coverage levels, the MDGs for Uganda are 61.8% if the population with access to improved water, and 84% with access to improved sanitation by the year 2015.

## 2.5 Issues, Challenges and Opportunities

### 2.5.1 Issues

Key issues affecting the sector performance include:

a. Inadequate financial resources: The investment requirement for rural water supply and sanitation, to meet the 2015 target, is about US$ 951 million. The corresponding investment for the urban water supply and sanitation is estimated to be US$ 481 million. The Government cannot finance this level of investment from local limited resources. The issue here, therefore, is how to raise the required financial resources to meet the 2015 targets for the water sector.

b. Inadequate capacity at district and lower levels to plan and implement sector activities. This has consequently resulted in low absorption of funds by the local governments.

c. Limited Private sector capacity to cope with the increased water and sanitation activities.

d. Weak coordination and management at both national and local governments’ level.

e. Inadequate involvement of local communities in the planning, financing, implementation, monitoring and management of community based water and sanitation developments.

### 2.5.2 Challenges

Despite Uganda's being well endowed with significant freshwater resources, the challenges of rapid population growth, increased urbanization and industrialization, uncontrolled environmental degradation and pollution are leading to accelerated depletion and degradation of the available water resources. In addition, the major operational challenge for the water sector is establishment of strong mechanisms for effective, efficient and sustainable delivery of water and sanitation services to the end users, on the basis of the established strategies and funding mechanisms.

### 2.5.3 Opportunities

In order to cope with the above challenges, government has taken steps to revamp the water sector through comprehensive policy, legal and institutional reforms aimed at increasing the performance of the sector in terms of outputs but also the efficiency in service delivery.

The opportunities available to ensure this transformation of the water sector include:

a. Poverty Eradication Action Plan (PEAP) – Which gives high priority to water supply and sanitation as one of the key intervention areas for poverty eradication.

b. Water Sector Reform – whose objective is to ensure that services are provided and managed with increased performance, efficiency and effectiveness, while maintaining the government's commitment to equitable and sustainable provision of domestic water supply and sanitation services.

c. Comprehensive Policy and Legal Framework – which guides the planning, implementation and monitoring of water supply and sanitation activities.

d. Good governance in both the management of public resources (through multi-annual financial planning and budgeting under the MTEF) committed to water supply and sanitation activities.

e. Private Sector Participation (PSP) which has brought in expertise and financing from the private sector to support government in the planning and implementation of water supply and sanitation activities.

f. Commitment from government and development partners for continued support of water supply and sanitation activities.

## 2.6 Future Outlook for the Sector

Given the achievements made so far, the water sector is on the right track to meet both its medium term and long-term objectives and targets. With the Sector Reforms expected to be completed in December 2004, the sector is bracing itself for a period of intensive activity as the reform recommendations begin to be implemented.

Implementation of the reform recommendations will not only be challenging but will also inevitably have heavy financial, technical and legal implications all of which have to be addressed to ensure successful and timely realization of set targets. All these pre-requisites have been addressed in details in the different sub-sector strategies and investment plans. The biggest challenge to the future of the water sector is mobilizing the necessary financial resources as stipulated in the sub-sector investment plans and putting in place the necessary implementation and coordination mechanisms for the efficient delivery of services to the end users.

It is envisaged that the private sector and the Local Governments will play a more active role in the delivery of water and sanitation services in the new liberalized and decentralized water sector. The challenge in this case will be building the capacity of both the Local Governments and the private sector to enable them play their role effectively. Increased stakeholder participation in the sector activities will also call for more effective coordination mechanisms and a coherent ME&R framework to ensure transparency and accountability in the sector and to minimize duplication of efforts and wastage of resources.

## 3.0 METHODOLOGY AND RESULTS

A sample of 7864 households, both rural and urban from the 4 different regions of Uganda (southern, western, central and northern) was considered. It comprised of respondents between the ages of 15-49 years (National Service Delivery Survey, 2004). This paper is based on descriptive analysis using SPSS statistical package. The following results were obtained.

### 3.0.1 Use of protected water sources.

The survey asked households to indicate their main source of drinking water in both the dry and wet seasons. The responses for the two seasons were similar except that rain water was a significant source of drinking water in some areas in the wet season. Nationwide, 60.2% of rural households used unprotected sources in the dry season compared to 88% of the urban households.

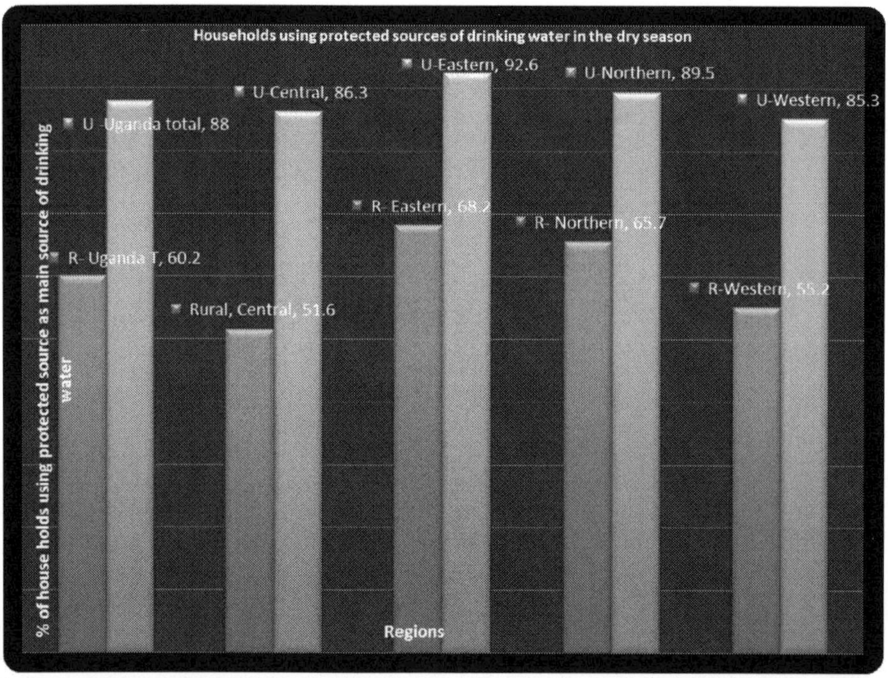


Source: NSDS 2004 Table 13.1.A. The following sources were considered protected; borehole, protected spring, gravity flow scheme, piped water in or outside dwelling or compound.

Access to protected sources of drinking water was closely correlated with economic status. In the poorest quintile, use of protected sources was 8% points below the national average of 69.4% while in the richest quintile (largely urban households), the use was 10% points above the national average.

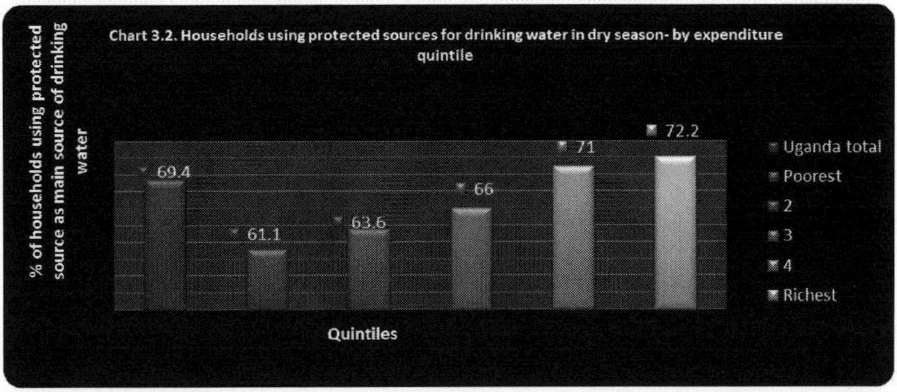


Source: NSDS 2004 Table 13.1.B. The following sources were considered protected; borehole, protected spring, gravity flow scheme, piped water in or outside dwelling or compound.

## 3.2 Time spent collecting water

Major disparities were found in the time households reported spending on collecting water. Average total collection times varied significantly among rural and urban areas, with the average rural household spending 60 minutes compared to 25 minutes for the average household.

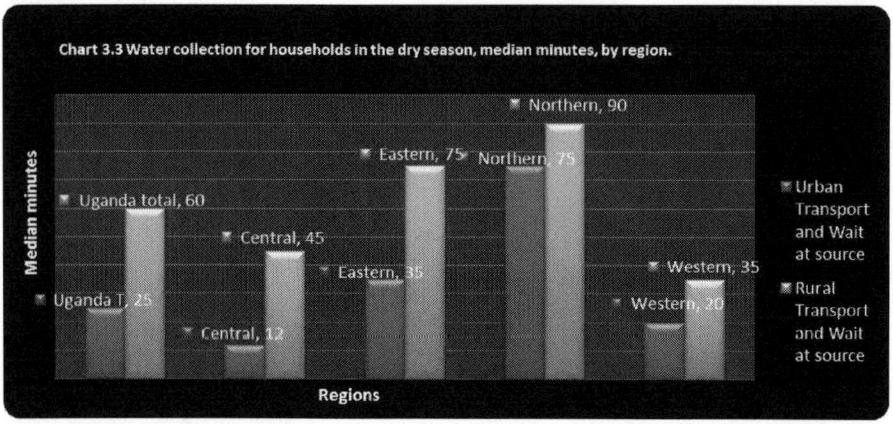


Source: NSDS 2004 Table 13.3.A. Total bar heights indicate total water collection time (wait and transport).

The above bar segments for the northern and eastern regions indicate a substantial burden on households from the waiting time reflecting crowded conditions at the source of the water.

Households appeared generally to use protected sources for drinking water where they were available. The chart below shows the main reasons given by 30% of households not using protected sources. Unavailability and long distance to a protected water source were the dominant reasons. This pattern was observed in both urban and rural areas in all regions.

Most households commented on whether the availability of safe water had changed during the prior 2 years. Nationwide, almost 44% of urban households indicated improvement, compared to 33% of rural households. Rural households in northern region gave the least favorable responses: 17.5% of them indicated improved access to safe water in the prior 2 years. Among the expenditure quintiles, the poorest quintile reported no improvement overall, while the richest quintile which was 57% urban, reported substantial improvements overall.

## Conclusions

Households' use of protected water sources for drinking was correlated with economic status. Households in the poorest quintile had less piped water and were more likely to use unprotected wells, lakes and river water than the national average. As in other quintiles, the dominant reasons for not using protected sources were "unavailable" and "long distance". The longer distances to water sources for the poorest households (an average of 0.8km in the dry season) and crowded conditions at water points are illustrated by the total collection time in the dry season; the average of 70 minutes for households in the poorest quintile as compared to 45 minutes nationwide.

## Recommendations

Provision of safe water sources should prioritize the poorest areas by way of allocating increasingly more funds. The districts that are least served with safe water sources are known by the Department for Water Department. These should be receiving increasingly more funding as compared to other districts.

## References

Brown, C. & Holcombe, A. (2004). In pursuit of the millennium development goals in water and sanitation. Water Policy, 6(3), 263–266.

Kiyonga, H. (1998). Health Policy Statement 1998/1999. Ministry of Health, Kampala, Uganda.

Knud-Eric Rosenkrantz (2006): "Equity, Access to public services in Uganda, Highlights of findings from the 2004 National Service Delivery Survey"

Lane, J. (2004). Perspective: positive experiences from Africa in water, sanitation and hygiene. Water Policy, 6(2), 153–158.

Rwabwoogo, M. O. (compiled by) (1997). Uganda Districts Information Handbook, 4th eds. Fountain Publisher, Kampala.

Statistics Department (MFEP), The 1991 population and Housing Census, Analytical report, 1995

United Nations, The Millennium Development Goals report, 2005 (2004): Poverty Eradication Action Plan2004/5-2007/8

Uganda Bureau Of Statistics 2006 Uganda Demographic and Health Survey, www.ubos.org 2002 Uganda Population and Housing Census, UBOS, www.ubos.org

White, G. F., Bradley, D. J. & White, A. U. (1972). Drawers of Water: domestic water use in east Africa. University of Chicago Press, Chicago.

WHO & UNICEF (2000). Global Water Supply and Sanitation Assessment 2000 Report, World Health Organization and United Nations Children's Fund.

WHO & UNICEF (2004). Meeting the MDG drinking water and sanitation target: a mid-term assessment of progress, World Health Organization and United Nations Children's Fund. E. W. Sugita / Water Policy 8 (2006) 529–537 537

# CHAPTER 7

## Putting Research Knowledge into Action: The Missing Link for Sustainability of Lake Tana Ecosystem, Ethiopia.

*Dessalegn Molla Ketema, Nick Chisholm and Pat Enright*

### Introduction

Lake *Tana,* located in north-west Ethiopia, is the second largest freshwater lake in Africa. It accounts for 50 per cent of the total inland water area of Ethiopia and feeds the Blue Nile River, which contributes about 85 per cent to the total flow of water in the Nile. It is important in ecological and economic terms, and it has both local and global significance. Lake *Tana* is selected by the Ethiopian government as one of the potential Growth corridor for development and reducing food insecurity problem. As a result, there are tremendous development activities that are taking place in and around the Lake Tana. Yet, despite the potential positive impact on the country's economic growth and alleviating poverty, these development activities are casting doubt on the sustainable use value of the Lake and its surrounding ecosystem.

Ecosystem sustainability is of growing concern to ecologists and other scientists who study the cumulative global impact of widespread environmental change and to social scientists attempting to develop policies to sustain the production of goods and services for a growing human population (Goodland, 1995). Therefore, understanding the big picture of the human environment nexus, with its complex interactions in and across ecosystems as well as in and across human systems, is essential if policy and action responses are to contribute to the goals of sustainable development and improved human well-being (UNEP, 2011). So far, long-term maintenance of ecosystem health is in conflict with the short-term interests of many stakeholders and policymakers (Bavinck. M. et al., 2005).

On the other hand, knowledge generation, synthesis and exchange are fundamental aspects of managing the future sustainability of natural resources, particularly healthy ecosystems (VCMC, 2002). On this regard, management of natural resources cannot afford to be the subject of just any single body knowledge such as the Western science, but it has to take into consideration the plurality of knowledge systems. Application of scientific research, local knowledge and incorporating the views of all relevant actors contributes both to the equity, opportunity, security and empowerment of local communities, as well as to the sustainability of the natural resources (Getz *et al.*, 1999).

The overall objective of this research is to throw light on and aims at providing a preliminary review of the major traits that cast doubts to the sustainability of the Lake Tana ecosystem and lesson learnt across the globe that witnessed unsustainable development activities leads to a complete or partial collapse of ecosystems. The study also highlights the importance of integrating different knowledge systems in an effective policy formulation and implementation process which ultimately leads to a sustainable ecosystems development and conservation.

## Research Methodology

### Description of the Study Area

Lake Tana is located in the country's north-west highlands at 12° 10′ 0″ N, and 37° 20′ 0″ E. The lake is natural and covers 3,000-3,600 $Km^2$ area at an elevation of 1,800m.a.s.l. It is approximately 84 km long, 66 km wide and has a volume of 28, 000 $Mm^3$, which makes it the largest in Ethiopia. The lake is a shallow, on average about 9 m and maximum 14 m. It is surrounded by densely inhabited area of 806, 545 residents (Tesfahun, Demissie 2004).

Lake *Tana* is an important source of fish both for the people immediately around the lake and elsewhere in the country (Berhanu et al, 2001). Its unique and isolated landscape includes forested islands, immense and varied wetlands and high mountain areas. The region is renowned for its biodiversity, and it is also the home of churches and monasteries dating from the 14th to the 16th century. Ninety per cent of the area's rapidly growing population of 2.5 million people depends on subsistence agriculture for their livelihoods. The productivity and sustainability of mixed farming practices depend on ecosystem goods and services, which rely on the functional integrity of the watershed's ecosystems- rivers, wetlands, lake, forests, pastures and soils.

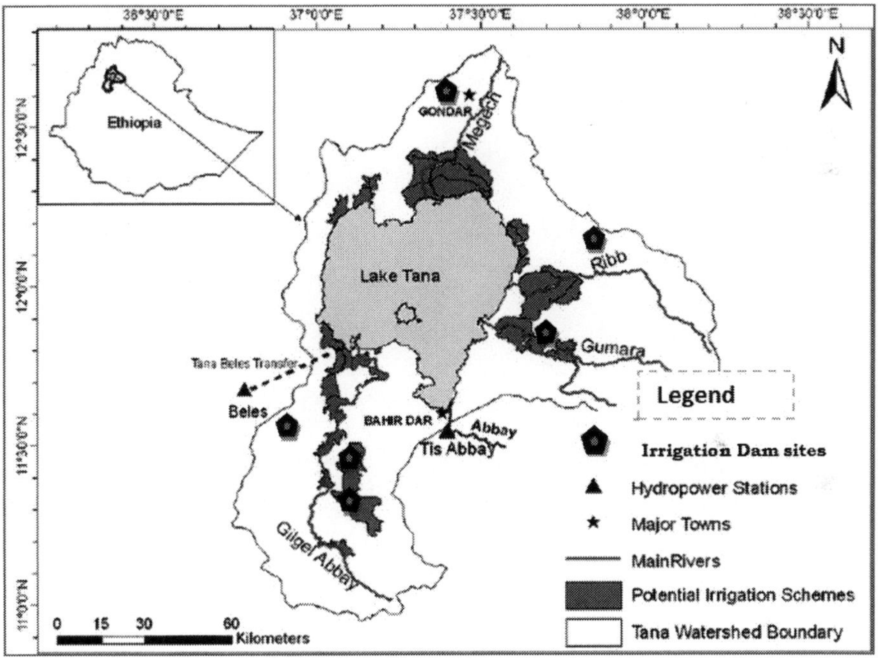


**Fig1: The Lake Tana watershed and development activities**

## Research Methods

In order to acquire the data with regard to the research objectives, primary as well as secondary data were employed. Data for this study was drawn from personal observation, key informant interview, a series of focused group discussions and reviewing of relevant policy documents.

## Result and Discussion

### The status of world's Ecosystems

The ecosystems on which we depend are not static; change is constant in nature. For millennia, the Earth has been transformed by advance of glaciers, volcanic eruptions, and the clash of continents. But today, human beings are the most powerful force of environmental change (MA, 2006). The Millennium Ecosystem Assessment (MA), an international synthesis that analyses the state of the Earth's ecosystems reveals that, over the past 50 years, humans have changed ecosystems more rapidly and extensively than in any comparable period of time in human history, largely to meet rapidly growing demands for food, fresh water, timber, fiber and fuel.

This has resulted in a substantial and largely irreversible loss in the diversity of life on Earth. The changes that have been made to ecosystems have contributed to substantial net gains in human well-being and economic development, but these gains have been achieved at growing costs in the form of: the degradation of many ecosystem services, increased risks of nonlinear changes, and the exacerbation of poverty for some groups of people. Some 60 percent of the ecosystem services examined in the Millennium Ecosystem Assessment-including fisheries and fresh water- are being degraded or used in ways that cannot be sustained. Global ecosystems conditions and trends show that, 50-80 per cent of original mangrove lost in costal ecosystem (Burke *et al.*, 2000), 50 per cent of world's wetlands lost during the 20th century has been recorded in freshwater ecosystem (Revenga *et al.*, 2000). Unless human attitudes and actions change, the degradation of ecosystem services could grow significantly worse during the first half of this century and is a barrier to reducing global poverty and achieving the Millennium Development Goals (MA, 2005; 2006).

## Learning from others as a development and policy guide: What happened to other lake ecosystems around the world? Lake Alemaya/Haromaya/ Ethiopia

The Alemaya lake in the Ethiopian highlands originally covered more than 175, 140 ha but had shrunk to 87, 910 ha (which is 50% of the original) in 1985 and to a mere 58, 600 ha (33.5% of the original) in 2002. It is now believed that the lakes have all but completely dried up. The loss of the lakes, which were a source of drinking water, irrigation and fisheries, has affected the livelihoods and well-being of more than 550, 000 people in the Ethiopian towns of Haromaya and Harar. Preliminary research seems to suggest that serious siltation has been a major factor in the destruction of the lakes. A dramatic increase in urban and rural settlements is also believed to have put tremendous pressure on natural resources in the area, including water resources (RCMRD, 2005). Increasing irrigation and domestic water use change in the local climate, and changes in the surrounding land cover are believed to be the causes of Alemaya's demise (Setegn et al., 2011).

## Lake Chad, Africa

Lake Chad, once one of Africa's largest freshwater lakes, has shrunk dramatically in the last 40 years. In 1963, the lake covered about 25,000 $km^2$. Today it is one-twentieth of that size. The size of Lake Chad has gone from 30,000 $km^2$ to 3,000 $km^2$ in 40 years, according to some sources - from 25,000 $km^2$ to less than 1,500 km2 between 1966 and 1997(Coe M.T et al.,2001). Major irrigation projects were built in the 1980s, which made use of the two main rivers, Chari and Longone, which supplement the lake. Increased irrigation demands are said to have significantly decreased the flow of water into the lake; some speculating that the irrigation demands, which increased a phenomenal four times between 1983 to 1994, have accounted for a 50% decrease in the lake's water level (Noury, 2009).

## Aral Sea, Asia

Formerly one of the four largest lakes in the world with an area of 68,000 $Km^2$, the Aral Sea has been steadily shrinking since the 1960s after the rivers that fed it were diverted by Soviet Union irrigation projects. By 2007 it had declined to 10% of its original size (Philip M et al., 2008). The region's once prosperous fishing industry has been essentially destroyed, bringing unemployment and economic hardship. The Aral Sea region is also heavily polluted, with consequent serious public health problems. The retreat of the sea has reportedly also caused local climate change, with summers becoming hotter and drier, and winters colder and longer (U.S. Geological Survey, 2007).The shrinking of the Aral Sea has been called "one of the planet's worst environmental disasters"(Daily Telegraph, 2010).

## The State of Lake *Tana* Ecosystem

Lake *Tana* Ecosystem has a global significance on which a huge livelihood, economic and socio-cultural groups are highly dependent. Lake *Tana* is also one of 250 lakes identified by Lake Net as having globally significant biodiversity. Lake *Tana* catchment is selected by the Ethiopian government as one of potential Growth corridor for development and reducing food insecurity problem. However, despite the positive impact on the country's drive towards achieving food security, Growth and Transformation Plan (GTP), these development activities that has been taking place in and around Lake Tana Ecosystem are casting doubt on the sustainability of the region.

Vijverberg, J., et al, 2009 summarizes the major treats as follows; despite the limited direct human influences on Lake Tana, both wetlands and surrounding catchment area have already been seriously damaged by human activities and most of the original forest has disappeared. The littoral region and wetlands of the lake are currently under severe degradation by the local inhabitants. Especially the area covered by papyrus has been decreasing recently (Teshale et al., 2001) by the ever growing human population. The local community is harvesting papyrus reed roots during low water level to use it as fuel wood. Farmers are cultivating the wetlands when the water is retreating. The abundance of endemic *Labeobarbus* species in Lake Tana decreased dramatically with ca 75% in 10 years' time (1991–2001). The destructive fishing operations during the spawning season (August–September) in river mouths and upstream on the spawning grounds in combination with alteration and destruction of spawning habitats may lead to extinction of 7 of the 15 endemic labeobarbs.

In the outflow of the Blue Nile from the lake, a large dam (Chere chera) has been constructed, resulting in increased silt load and turbidity of lake water and reduced water levels. The withdrawal of lake water for generating hydroelectric power and irrigation in combination with the predicted increased variations and progressive decrease of rainfall over the years is expected to cause severe water level fluctuations and reduced water levels in the coming decade. Conversely, Bahir Dar is a rapidly growing town, a six times population increase up to 1,800,000 inhabitants is expected in the next 50 years (Teshale, 2003). The current practice of discharging untreated industrial and domestic waste into the lake will have adverse effects on the quality of the lake water. Furthermore, pollution from agricultural sources as fertilizers, insecticides and herbicides are recently increasing.

Construction of four more major dams around the major tributaries of Lake Tana particularly at *Gumara*, *Ribb*, *Megech*, and *Gilgel Abay* are underway (see Fig 1). Tana-Beles power generation (estimated 460MW) by diverting water from the Western shores of Lake *Tana* through a canal (12km) in the western part around *Kunzila* and connected to *Beles* river is inaugurated in 2010. The project is expected to serve the country for the next 25 years (starting from March, 2010) without any major interruption.

A recent study shows that future development will exacerbate pressure on the lake. If all the planned development occurs, the mean water level of the lake Tana will drop by 0.44 meters (m), and the average surface area will decrease by 30 square kilometers ($km^2$) (i.e., 1%) and up to 81 $km^2$ (i.e., 2.6%) during some dry seasons. There will be prolonged periods of several years during which water levels will be much lower than they would be naturally. If environmental flow requirements (estimated to average 862 $Mm^3y^{-1}$) are maintained in the reach containing the Tis Issat Waterfall, the mean water level of the lake will reduce by a further 0.37 m and the average lake area will reduce by an additional 26 $km^2$. Without careful management these changes are likely to have severe ecological and social consequences (Mccartney, et al., 2010).

As a result of these treats and research evidences coupled with global climate change and environmental degradation, there is a speculation that Lake Tana could shrink drastically in size within two or three decades. Then the system may no longer able to support life and have an impact on the livelihood of the local people which ultimately affects the country's drive towards achieving the Millennium Development Goals (MDGs). Threats to Lake Tana and its catchment area have also serious implications for environmental and human security in Ethiopia and other riparian communities.

Therefore, keeping the health of Lake *Tana's* ecosystem for sustainable resource utilization and socioeconomic development of the local community in particular and the wider population in general, should be the priority agenda. However, the progression of degradation inside a lake often takes place on a wider and deeper scale than is apparent. If the seriousness of a lake problem is realized in time by scientific means, resource conservation and restoration measures may produce some promising results. Often, however, the symptoms of degradation remain unnoticed for a long period of time because of their incremental nature, and the introduced conservation and remedial measures may be too little too late. The level of ecological and water quality degradation may have already reached crisis proportions, suddenly leading to instant loss of ecosystem sustainability (ILEC, 2007).

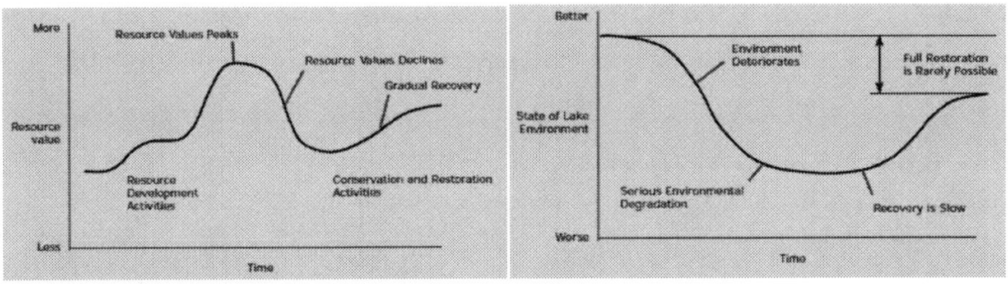


**Fig 2: Changing Resource Values and Resulting Environmental State of a Hypothetical Lake, (ILEC, 2007)**

As we see from the figure 2, once a lake ecosystem reached a serious environmental degradation, recovery is slow and full restoration is rarely possible. Especially, for the poor nations like Ethiopia where economical, institutional and technological capacity to reclaim the environment is untenable, it will be very difficult to preserve valuable environmental resources effectively. According to the Environmental Sustainability Index (ESI, 2005) which quantifies the likelihood that a country will be able to preserve valuable environmental resources effectively over the period of several decades, Ethiopia, ranks 135$^{th}$ out of 146 countries with a score of 37.9 which is the least score among the 40 NEPAD (New Partnership for Africa's Development) member countries just above Sudan (with ESI score of 35.9). Even when we compared with the 21 countries that are more than 50% Deseret, 15 countries scored from 39.8- 56.7. The ESI score also evaluates a country's potential to avoid major environmental deterioration (Esty et al., 2005).

## Understanding knowledge systems perspectives for the Lake Tana Ecosystem Management and governance

Natural resource governance cannot be dealt through the dichotomous division between scientific and indigenous knowledge systems; rather we need to explore diverse systems of knowledge triggered by specific configuration of political interests and cultural formations (Ojha et al., 2007). Knowledge related to natural resources and their management is stored in the mind and means of so many stakeholders and institutions, and in formal and informal ways that it is, ideally, hard to assess the legitimacy and dominancy of knowledge of one individual or knowledge communities to others. When stakeholders with diversity of interests, knowledge and power interact and collectively learn it is likely that there would be a synergy to develop social practices for the management and use of natural resources (Banjade et al., 2007).

Particularly transferring knowledge into action in ecosystem management appears to be a messy process which involves a complex series of interactions between the decision makers, producers and users of research. The problem of natural resource management and governance as perceived by government planners, policy makers, and researchers is quite different from the problems of primary concern to the users, who are trying to secure access to sustainable livelihood options by optimizing the use of their resources. This is because; decision makers, researchers and users inhabit different worlds (Caplan, 1979). Whilst researchers may revere theories and concepts, decision makers want concrete evidence which is relevant and easy to understand. And whilst researchers often take years to complete research studies, decision makers want answers quickly (Mitton, et al., 2007). Each side also speaks its own, highly technical language (Choi, et al., 2005).

Therefore, understanding different knowledge systems is imperative to manage and govern natural resources in a sustainable manner without compromising the needs of the future generation. According to Ojha, et al., 2007, practices of natural resource governance are shaped by knowledge systems of, and deliberative interfaces among, diverse groups of social agents that tend to vary both in terms of knowledge and other aspects of differentiation. Therefore, in terms of knowledge perspectives, five types of social actors were identified to better understand their perspective towards the Lake Tana Ecosystem (See Fig 3).

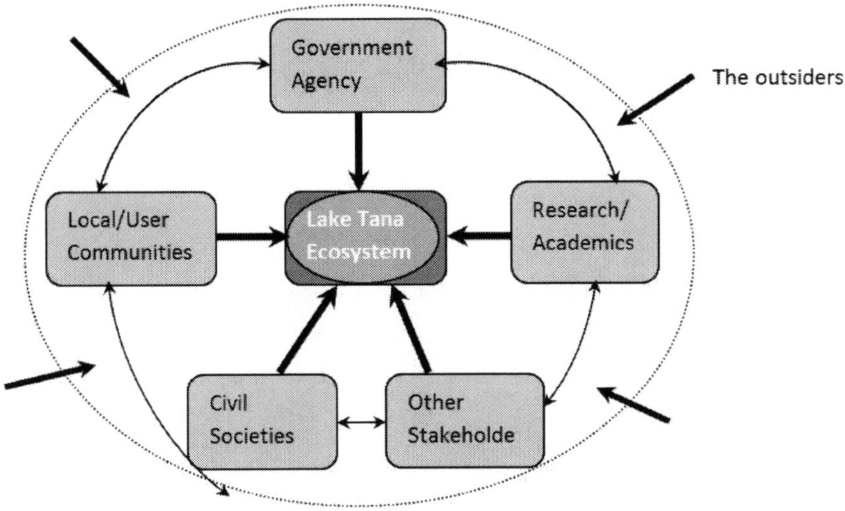


**Figure 3: Key components of Knowledge System clusters of the Lake Tana Ecosystem** (*Modified and adapted to the local context after, Ojha et al., 2007)*

Government agencies are ministries and administrative units at different level who are mainly involved in the formulation of policy, legislation, implementation, monitoring and evaluation of natural resource management and governance processes. Mostly these agencies are confronting with conflicting objectives towards natural resource management and utilization. Particularly, in the Lake Tana ecosystem, development goals are usually contradicting with the conservation goals. For example, Ministry of Agriculture and Rural Development with a broad objective of boosting production and marketing for food security and poverty alleviation, is trying to insist the farmers in the Lake Tana watershed to use improved agricultural technological packages particularly, fertilizer and pesticides. On the other hand, Environmental Protection Authority (EPA) and Institute of Biodiversity Conservation (IBC), with conservation and protective goals are striving to minimize the stress on the ecosystem through proposing the delineation of buffer zones, integrated watershed management approaches around the Lake Tana. Generally, these agencies have their own knowledge perspective to see and act accordingly towards managing and governing the Lake Tana Ecosystem. Government officials are typically unaware themselves of the conflicts and confusion caused by the contradictory goals, regulations, procedures, and plans of their various agencies to manage natural resources.

Local/user communities and research/academic institutions are believed to be the sources of indigenous and scientific knowledge systems respectively. Indigenous knowledge and resource use practice has been defined as a cumulative body of knowledge and beliefs handed down through generations by cultural transmission about the relationship of living beings, (including human) with one another and with their environment (Gadgil et al, 1993). Indigenous knowledge is also local knowledge unique to a given culture or society, mostly it contrasts with the international knowledge system generated by universities, research institutions and private firms. It is regarded as the basis for local level decision making in agriculture, pastoralism, natural resource management, and a host of other activities in rural communities staying very close to the nature (Farooquee et al., 2004). Local knowledge systems have been found to contribute to sustainability in diverse fields such as biodiversity conservation and maintenance of ecosystems services, tropical ecological and bio-cultural restoration, sustainable water management, genetic resource conservation and management of other natural resources. Local knowledge has also been found useful for ecosystem restoration and often has ingredients of adaptive management (Panday, 2002).

Special interest groups such as (civil societies and other stakeholders like UNESCO) range from those favoring development of resources to those favoring preservation as cultural heritage. Civil societies have the potential to promote equity, democracy in natural resource management and governance by facilitating and advocating discussions and negotiation among local users, decision/policy makers, and other stakeholders at different levels. Their responsibilities may include articulating the needs of the weak, working in remote areas, changing the attitudes and practices of local officials and nurturing the productive capacity of the most vulnerable groups of society (Chandhoke, 1995).

Very importantly, the outsiders views on Lake Tana Ecosystem is quite different from the insiders where the system encompasses the sources of Blue Nile, and one of the centers of global biodiversity. At international level for instance, the vision and aspiration of Egypt, Sudan and other riparian countries is always to see the constant inflow of Nile River throughout centuries. They wish not to see any development activities that threaten the hydrological cycle in the Nile Basin.

Though, understanding of these knowledge systems is not as such too difficult, accommodating the views, goals and interest of government agencies, researchers, local users and special interest groups is a challenge and complex in integration, because they have different perspectives and are operating in different contexts.

## Knowledge integration for policy formulation and implementation

The current un-sustainability state of the Lake Tana ecosystem might be either as a result of poor policy and management that could not integrate development and conservation in a sustainable manner or knowledge gap that integrate the policy makers, researchers and local users.

Basically, the integration of different knowledge systems into existing sustainable development goals, and natural resource management policies at international, national and local/user level will have a practical implication for sustainable use value of the lake ecosystem. This integration could occur in a variety of configurations: among science disciplines; across spatial and temporal scales; from science through policy, management, and education; and among knowledge providers, users. Effective management of the important environmental problems we face today requires integration of all of these interests (Bowden, 2002).

In political science or organizational analysis, decision-making processes are conceptualized as being informed by various types of information, including, in some cases, scientific evidence. Two general assumptions underlie policymaking. One is that governments, researchers or user groups possess, or can develop, sufficient knowledge on their own to form the basis for policy. The second assumption is that the world in which we live can be represented in simple models. However, it is found both assumptions untenable. Academics, policymakers and users have to interact 'to get the picture right'. In addition, they have to put the diversity, complexity and dynamics of governance issues right on the table (KOOIMAN, J. 2005).

Therefore, involving multi-stakeholders and incorporating their views, agendas and interest in policy framework will facilitate knowledge exchange and integration. Stakeholders often do not explicitly recognize the ways in which their knowledge and understanding frame their perspectives on common pool resource management policy. Knowledge about laws and institutions may be seen as providing both constraints and opportunities for common pool resource management, because this knowledge forces stakeholders to consider resource uses that are compatible with these wider policy processes (UNEP, 2011).

Conversely, involvement of an increasing array of stakeholders in natural resource management has created conditions for conflict and the processes of negotiations will inevitably lead to governance solutions. The emerging diversity of stakeholders has also given rise to the potential for pooling of diverse knowledge traditions in the practice of natural resource governance. But the actual deliberative interface has often been negatively affected by conflicts among different knowledge systems that have developed historically (Ojha et al., 2007).

As a matter of fact, knowledge has to be exchanged and translated among stakeholders in order to formulate better policy and implement successfully. Though, mutual understanding between researchers, experts and decision makers is often hindered by jargon, language, experiences, and presumptions about what constitutes persuasive argument, there should be a mechanism to link knowledge to action which requires open channels of communication between researchers, experts, decision makers and local communities but also requires that participants in the resulting conversation understand each other's context (Cash w., et al., 2003).

## Conclusion and the way forward

Generally the problem in the Lake Tana ecosystem is expected to worsen in the coming few years as population, hydropower and irrigation demands, and other development activities continue to increase. The analysis also reveals that there are global and regional research evidences that indicates the foresee treats for the Lake *Tana* Ecosystem possibly will have unprecedented impact on the region's social, economic and environmental conditions. Unless timely, protective policy intervention measures put in place, there might be a serious environmental degradation which ultimately ends up with a complete collapse of the whole ecosystem.

Development is a necessary condition for the survival of developing countries like Ethiopia. However, development should not be unsustainable to worsen the existing conditions. It has to be planned and implemented without affecting the sustainability of natural resource base, livelihood of the local community and the long-term economic development and the countries drive towards attaining food security and poverty alleviation. More importantly, the recent environmental catastrophe of the Aral Sea (Asia, by 2007 it had declined to 10% of its original size); Lake Chad (Africa) and Lake Haromaya (Ethiopia) should be a lesson and a guideline for any development activities in general water resources development in particular.

Conservation of natural resources through sustainable ecosystem management and planned development is the key to our secured future. There is a need to integrate research, ecosystems conservation and development to ensure sustainable resource use and livelihood security for the wider community in the Lake Tana catchment. Policy-makers need to develop policies and practices that distinguish between situations where conservation and development goals are compatible and situations where there may be conflicts.

Regarding Lake Tana, some researchers points out the need for more research to inform policy, I personally recommend more of policy intervention to protect the Lake Tana ecosystem from its foreseen treats and speculations, because the research knowledge that have been produced concerning Lake Tana coupled with the global and regional catastrophe that witnessed the disappearance of ecosystems are more than enough to inform policy. Therefore seeking an innovative ways of saving the Lake Tana ecosystem from a foreseeable treats and speculation is very imperative for those who have a stake in the Lake Tana catchment.

## References

Banjade, M. R., Luintel, H. & Neupane, H. R. (2007). Action Research Experience on Democratizing Knowledge in Community Forestry in Nepal. http://www.idrc.org/tehip/ev- 118402-201-1-DO_TOPIC.html

Bavinck. M, R. Chuenpagdee, M. Diallo, P. Van Der Heijden, J. Kooiman, R. Mahon and Williams, S. (2005) Interactive fisheries governance, Delft: Eburon Publishers. 72 pp.

Berhanu, T. Ralph, L. and Girma, Z. (2001) Development Initiatives and Challenges for Sustainable Resource Management and Livelihood in the Lake Tana Region of Northern Ethiopia. Proceedings of the Wetland Awareness Creation and Activity Identification Workshop in Amhara National Regional State. January 23rd 2001 Bahar Dar, Ethiopia.

Bowden, W. (2002) Integrated catchment management rediscovered: an essential tool for a new millennium. *In:* Manaaki Whenua Conference, 2002 Wellington, New Zealand.

Burke, L., Kura, Y. Kassem, K. Revenga, C. Spalding, M. and Mcallister, D. (2000) 'Pilot analysis of global ecosystems: Coastal ecosystems'. World Resources Institute, Washington, DC.

Caplan, N. (1979) The Two-Communities Theory and Knowledge Utilization. American

Behavioral Scientist; 22(3):459–70.

Cash, D. W., Clark, W. C., Alcock, F., Dickson, N. M., Eckley, N., Guston, D. H., Jager, J. & Mitchell, R. B. (2003) Knowledge systems for sustainable development. *Proceedings of the National Academy of Sciences of the United States of America,* 100, 8086.

Chandhoke, N. (1995) *State and Civil Society: Explorations in Political Theory.* Sage Publications: New Delhi.

Choi B. C.K., Pang, T., Lin, V. Puska, P. Sherman, G. (2005) Can scientists and policy makers work together? Journal of Epidemiology and Community Health; 59(8):632–637.

Coe, M.T. and Foley, J.A. (2001) Human and natural impacts on the water resources of the Lake Chad basin. Journal of Geophysical Research, 106:3349-3356.

Daily Telegraph (2010) "Aral Sea 'one of the planet's worst environmental disasters". *The Daily Telegraph* (London). (http://www.telegraph.co.uk/earth/earthnews/7554679/Aral-Sea- one-of-the-planets-worst-environmental-disasters.html. Retrieved 2010-05-01.

Daniel C., Marc, L. Tanja, S. and Alexander D. E.(2005) *2005 Environmental Sustainability Index: Benchmarking National Environmental Stewardship*. New Haven: Yale Center for Environmental Law & Policy.

Farooquee, N. A., Majila, B. & Kala, C. (2004) Indigenous knowledge systems and sustainable management of natural resources in a high altitude society in Kumaun Himalaya, India. *Journal of Human Ecology,* 16, 33-42.

Gadgil, M., Berkes, F. and Folke, C. (1993) Indigenous knowledge for Biodiversity Conservation. *AMBIO***22 (2-3):** 151-156 (1993).

Getz, W.M., Fortmann, L., Cumming, D., Toit, J. Hilty, J., Martin, R., MURPHREE, M., OWEN-SMITH, N., STARFIELD, A.M., WESTPHAL, M.I. (1999) Sustaining natural and human capital: villagers and scientists. *Science* 283: 1855-1856.

Goodland, R. (1995) The Concept of environmental sustainability. Annual Review of Ecology and systematic 26:1-24.

ILEC. (2007) Integrated Lake Basin Management: An Introduction. International Lake Environment Committee Foundation: Kusatsu, Japan.

Kooiman, J. (2005) *Fish for life: interactive governance for fisheries*, Amsterdam Univ Pr.

MA (Millennium Ecosystems Assessment). (2005) Ecosystems and Human Well-being: Biodiversity Synthesis. Washington, DC: World Resources Institute. MA (Millennium Ecosystems Assessment) (2006). Ecosystems and Human Well-being: Current State and Trends. Volume 1. Millennium Ecosystem Assessment. Island Press, Washington.

Mitton, C., Adair C. McKenzie, E., Patten, E., Perry B. W. (2007) Knowledge Transfer and Exchange: Review and Synthesis of the Literature. The Milbank Quarterly; 85(4):729–768.

McCartney, M.; Alemayehu, T.; Shiferaw, A.; Awulachew, S. B. (2010) Evaluation of current and future water resources development in the Lake

Tana Basin, Ethiopia. Colombo, Sri Lanka: International Water Management Institute. 39p. (IWMI Research Report 134). doi:10.3910/2010. 204

Noury, V. 2009. Lake Chad is dying. *New African*, 22-24.

Ojha, H. R., Chhetri, R. B., Timsina, N. P. & Paudel, K. P. (2007) Knowledge Systems and Deliberative Interface in Natural Resource Governance: An Overview. *Knowledge Systems and Natural Resources: Management, Policy and Institutions in Nepal*, 1.

Pandey, D. N. (2002) Indigenous sustainability science. *Human Ecology*.

Micklin, P. et al. (2008) "Reclaiming the Aral Sea". *Scientific American*. http://www.sciam.com/article.cfm?id=reclaiming-the-aral-sea&sc=rss. Retrieved 2008-05-17.

Revenga, C., Brunner, J. Henninger, N. Kassem, K. and Payne, R. (2000) 'Pilot analysis of global ecosystems: Freshwater systems'. World Resources Institute, Washington, DC.

RCMRD. (2005) Profile, Regional Centre for Mapping of Resources for Development, Nairobi, Kenya.

Setegn, S., Chowdary, V., Mal, B., Yohannes, F. & Kono, Y. (2011) Water Balance Study and Irrigation Strategies for Sustainable Management of a Tropical Ethiopian Lake: A Case Study of Lake Alemaya. *Water Resources Management*, 1-27

Tesfahun, G., Demissie, S. (2004) *Lake Tana: a socioeconomic synopsis.* Amhara Regional Agricultural Research Institute, Bahir Dar, Paper presented to The Lake Tana Resource Management Workshop, Bahirdar, Ethiopia.

Teshale, B., (2003) Influence of sediment on physico-chemical properties of Lake Tana. Workshop 'Fish and Fisheries of Lake Tana: Management and Conservation'. 6–8 October 2003, Bahir Dar, Ethiopia.

Teshale, B., Lee, R. & ZAWDIE, G. (2001) Development initiatives and challenges for sustainable resource management and livelihood in the Lake Tana region of Northern . In A. B. Dixon, A. Hailu & A. P. Wood (eds), Proceedings of the Wetland Awareness Creation and Activity Identification Workshop in Amhara National Regional State. January 23rd 2001, Bahar Dar, Ethiopia, 33–43.

UNEP , Peter Saundry "Inter-linkages: environment and policy web in Africa". In: Encyclopedia of Earth. Eds. Cutler J. Cleveland (Washington, D.C.: Environmental Information Coalition, National Council for Science and the Environment). [First published in the Encyclopedia of Earth April 13, 2007; Last revised Date April 13, 2007; Retrieved June 10, 2011 <http://www.eoearth.org/article/Interlinkages:_environment_and_policy_web_in_Africa>

U.S. Geological Survey .(2007) "Earthshots: Aral Sea" U.S. Department of the Interior. http://earthshots.usgs.gov/Aral/Aral. Retrieved 2008-05-17.

Victoria Catchment Management Council (VCMC), (2002). The *health of our catchments: a Victorian Report card 2002*. The State of Victoria.

Vijverberg, J., Sinning, F. & Dejen, E. 2009. Lake Tana: Source of the Blue Nile. *The Nile*, 163-192.

# CHAPTER 8

# Economic Contribution of Pastoral and Agro Pastoral Production to Food Security and Livelihoods Systems in Africa: The Case of Eastern Sudan, Eritrea and Western Ethiopia in the Horn of Africa.

*Zeremariam Fre and Bereket T. Tsegay*

## 1. Introduction

The pastoralists in the Horn of Africa survive in fragile ecosystems that are perpetually affected by drought and are continually threatened by desertification. Pastoralists and agro-pastoralists have therefore adjusted and adapted to these environmental challenges by evolving livelihoods mainly dependent on livestock and livestock-related activities. Because of their intimate understanding of these fragile ecosystems, with resources which are scarce and variable, the pastoralist inhabitants of the Arid and Semi-arid Lands (ASALs) have adopted a mobile and flexible pattern of resource use that has proven to be sustainable. Mobility is in itself an important ecological adaptation and may, in part, explain why some of these pastoral areas have a higher productivity in terms of protein per hectare than western and Australian ranches.[23] Mobility is key to the pastoral risk management strategy especially when they face acute shortage of water and pasture.[24]Moreover, Little and et al (2001) explain the risk of livestock loss during the erratic climatic changes is lesser in mobile than sedentary livelihoods. Inevitably, the ability of livestock to convert low quality biomass

23 Knips, V. (2004: 18) cited from Winrock International (2002)

24 Kaimba, G.K. et al (2011)

into high quality products such as meat and dairy products gives it a central position, particularly with regard to food security.

Over the last four decades there have been ongoing debates among natural and social scientists on the sustainability of pastoralism and whether or not pastoralism is a viable means to exploit the untapped resources of the arid and semi-arid environments. The trend in the past has been that natural scientists have viewed pastoralism as an outmoded and environmentally unsustainable system whereas social anthropologists and human ecologists have argued that pastoralism is a sustainable mode of production. Such contrasting or rather polarized perspectives have dominated the literature on pastoralism over the last half a century.

However, the current challenges and realities the pastoralist communities have been facing are forcing them to look for adaptive strategies and coping mechanisms to stand against all existing and new odds. The central theme of this paper is to reflect on the current experiences of pastoralist adaptation and their implications for their livelihoods and food security.

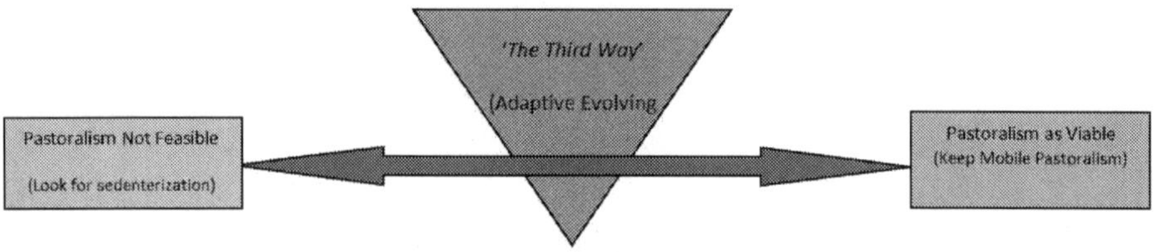


**Figure 1: Continuum on Understanding Pastoralism**
Source: Adopted from the arguments on understanding pastoralism

The aim of this paper is not to analyze the pros and cons of the natural and social science perspectives; but rather argue that traditional pastoralism like other rural livelihoods is under ecological and political stress which is forcing it to adapt to the new situation by adopting a semi-sedentary mode of livelihood. The authors draw arguments based on their experiences in the Horn of Africa demonstrating that while traditional pastoralism is adopting a more agro-pastoral mode of production it is nonetheless making substantial contribution to food security in the Horn Region. The authors reflect on whether this new pastoralist adaptation strategy guarantees better quality of life in which access to food is secured particularly among the vulnerable section of the population.

Livestock production systems are known to be predominant in the Horn of Africa with the livelihoods of millions depends on the rearing of animals. However despite the significant contribution pastoralism makes to local, national and regional economies, it lacks investment from government and private sector as compared to other sectors - which contribute less than the livestock sector to the regional GDP.[25]

The present case study is based on Western Eritrea, Western Ethiopia and the Eastern part of Sudan. The three regions share similar ecological, historical, cultural, and socio-economic ties, a situation that go beyond the colonial boundaries. The authors selected this region to be a case study as the livestock production makes significant contribution to the national economy and individual household's food security of the three countries. Despite the recurrent effects of climate change and environmental factors, pastoralists have shown tenacity and an ability to march beyond the challenges.

In the case of Eastern Sudan (Kassala State) there is ample evidence to show that pastoralist communities have opted for a semi-sedentary mode of livelihood where part of the family (wife, children and old folk) are permanently settled rearing small stock (sheep/goats) whereas men and older boys seasonally migrate with large stock (cattle/camels) to distant pastures during the dry season. During the rainy season the men and boys return to the homesteads where they can feed their animals and cultivate their land.

There is clearly an emerging semi-sedentary mode of production where pastoralism is facing different set of challenges and opportunities which have to be addressed from a different policy perspective which some experts call 'a third way'. The 'third way' is essentially based on the understanding that pastoralism is evolving and there is a need for more imaginative policies which take into account the dynamic changes in the semi-sedentary mode of

25 AU-IBAR (2010:4)

production which has become more dominant in the last two decades.

The paper uses primary data collected from the field by one of the authors in the last three years and also from related secondary sources including academic journals, policy brief documents, government statistics and other published articles. The paper applies conceptual framework and a broad definition of pastoralism in the Horn and its particular significance to the study area. It then elaborates the key challenges which are facing pastoralist communities, why such a situation call for 'new thinking' is necessary and therefore the "third way" approach to pastoral development. Furthermore, it links the challenge with the Social protection of the pastoralist communities. The last sections provide some analysis and some recommendations for a way forward.

## 2. Regional Context of Pastoralism

The Horn of Africa, comprising of Djibouti, Eritrea, Ethiopia, Kenya, Somalia, South Sudan, Sudan, and Uganda occupies an area of 5.2 million $km^2$ and supports a population of over 200 million people. The eight countries are under the regional umbrella body called the Inter-Governmental Authority for Development (IGAD). The arid and semi-arid lands (ASALs) and sub-humid lands constitute 80 percent of the total land area and contain more than 90 million tropical livestock units (TLU) that provide livelihoods for pastoralists.[26] With 60 percent of the land in IGAD countries being known as arid, the area of coverage in the countries differs from '100 percent in Somalia and Djibouti, over 73 percent in Kenya, 67 percent in Eritrea, 41 percent in Ethiopia and 1 percent in Uganda'.[27]

Getting an insight about the common features of pastoralism and agro-pastoralism of the Horn of Africa contributes to understanding the challenges these pastoral communities are facing in Eastern Sudan, Western Eritrea and Western Ethiopia.

---

26 IGAD (1990)

27 Knips, V. (2004:2)

According to Fre (1989) traditional pastoralism in the Horn Region in general and in Eritrea and Eastern Sudan in particular is predominantly semi-sedentary, and has the following characteristics:

- It lacks uniformity and specialization; there is almost no clan which can be referred to as 'purely' pastoralist;
- It cuts across ethnic, cultural and ecological boundaries; communities interact economically and share available range resources;
- Highland-lowland economic interactions among peasant (agro-pastoral groups) and pastoral communities is much more symbiotic than is often recognized;
- It is based on multi-species-based herding;
- It contributes significantly to the local economy through animal sales at local and regional markets including the Middle East.

As the *AU Policy Framework for Pastoralism in Africa of 2010* documented there is more positive trend of policy experience emerging among the East African countries, mainly in improving coordination and understanding pastoralism across national governments.[28] Though the awareness of the pastoral vulnerability is improving, there is still limited understanding and negative perceptions on the nature and magnitude of the cross-border trade from the side of decision-makers which leads to the lack of attention and backup to the sector.[29] As this sector has been a major contributor to the household and national economies, countries attempted to have relevant legal frameworks, policies and strategies. Given each country's particular context, policies towards pastoralism differ from one to another in the IGAD bloc. Some of these countries have clear legal frameworks whereas others do not have. According to Ibnoaf, the policy intervention constraints in these countries can be summed up as follows: [30]

28 AU (2010:15)

29 Pavanello, S. (2010:1)

30 Ibnoaf, M. (2008)

- The variable/unpredictable political environment.
- The weak capabilities of the private sector (in livestock trade and marketing).
- The disincentive to establishing and promoting functional pastoralist and other community based organizations.
- There is absence of active livestock policies at state and national levels.
- A low livestock profile in national economic planning, particularly observed in poverty reduction strategy paper (PRSP) which contains only general issues.
- Lack of understanding of the potential role of livestock sector in meeting the pro-poor development objectives.
- Low levels of data and information availability and the consequent poor analysis.
- Lack of participatory approaches in pastoral development and policy-making.

## 3. Pastoralist communities in transition and the underlying causes

There are a number of major factors that are playing key role in shaping traditional pastoralist livelihoods and pressurizing them to take alternative community-based adaptive strategies. The Horn of Africa region has been facing numerous socio-economic and environmental challenges due to both manmade and ecological challenges. The following issues are critical and it is worth highlighting them as they negatively impact pastoralist livelihoods and their production systems. Ultimately they undermine both the local and regional food security situations. The critical issues include:

- ***High and low intensity conflicts negatively impacting pastoral livelihoods:*** Many of the pastoral areas in Western Eritrea, Eastern Sudan and Western Ethiopia have experienced conflicts of various forms and intensities both intrastate and high intensity interstate wars over the last four decades. Proxy wars were also conducted as national regimes fought cross border wars in hot pursuit of opposition groups who used predominantly pastoral areas as their safe haven. Such conflicts have had a devastating impact on pastoralist production and livelihoods leading to the pauperization of pastoral communities, large-scale settlement and forced urbanization.

- The whole region, but Kassala State and Kassala city in particular have experienced massive population displacement both internal and across borders which included a large segment of pastoral and agro-pastoral population from both Eritrea and Western Ethiopia. The influx to the Kassala State created a pressure on the pasture resources and provision of social services.
- ***Land grabbing in high potential pastoral areas:*** This new phenomenon is leading to land encroachment onto pastoralist land and is fundamentally affecting pastoralist livelihoods, production and food security. As Fisher puts based on The World Bank's estimates, in 2009, alone, 'nearly 60 million hectares of land were purchased or leased in developing countries all of the world – an area the size of France'.[31] Looking at the magnitude of the Horn Region, he said that it is the most shocking example of the land grabbing. Land grabbers target high potential savannah and wetland (riverine area) and thus its availability to pastoralists has been rapidly diminishing because of agricultural expansion.[32] Under such circumstances, resource based conflicts among the various stakeholders seems inevitable. In some cases the land grabbing by local elites, commercial farmers, Transnational corporations (TNCs), foreign governments (mainly the Gulf and Asian countries) and the government itself is taking place in high potential areas which have already been denied to pastoralists in previous inter-

31 Fisher, S. (2011)

32 See Cotula, L. et al (2009) for more discussion of land grabs in Africa and in particular Ethiopia and Sudan.

ventions including in both colonial and post-colonial large scale commercial agriculture.

- ***Climate change and drought*** - The climate variability and change compounded with the above factors has resulted in creating recurrent droughts and famine as witnessed over the last many decades.[33] The source of food in these communities depends highly on the availability of the natural capital mainly grass and water resources. However, the decline in the amount of rainfall and vegetation, and ecosystem disturbances are creating a major challenge for them. This is contributing towards the diminishing livestock population and increasing food insecurity among the poor families.
- ***Revival and promotion of the Green Revolution (GR) inspired approach*:** The GR based policy approach which has been adopted by the national governments does not tend either to accommodate pastoralist livelihoods and production priorities or to recognize the significance of indigenous knowledge systems or sustainable biodiversity. Yet, pastoralists in the Horn Region still own and manage some of 'the most productive breeds of animals'[34] which are highly adaptable to their environment.
- ***The Prosopis Juliflora challenge***: Vast and high potential rangelands have been invaded by Prosopis at the expense of the indigenous vegetation which is vital to pastoralist survival as well as livestock productivity. Prosopis was introduced to the Region from South America more than 50 years ago as part of a desert encroachment control measure especially in Northern Sudan. Prosopis which is highly drought resistant with very deep roots has expanded to many riverine and savannah suppressing the growth of the palatable indigenous vegetation upon which livestock depend. For instance, in the Afar region of Ethiopia about 30 percent of the region, equivalent to 1 million hectares,

33 Omolo, N. A. (2010:98)

34 Fre, Z. (1989)

is swallowed by Prosopis.[35] Prosopis is partly toxic to livestock, less palatable and yet abundantly available in the rangelands. The alternative uses of Prosopis have not been properly considered, researched or supported by national governments, UN bodies or NGOs and Prosopis therefore remains as the most underutilized resource.

The desire to adopt changes emanates from the pastoralists themselves which challenges the commonly held view by some policy makers in the Region that 'pastoralists are resistant to change'. Indeed, the contexts provided above are the pushing forces that are encouraging the pastoral communities to accept change in their livelihood. Furthermore, the role of federal and local government and other stakeholders in providing the necessary social services can be highlighted as a pulling force that is attracting the pastoralists to adopt semi-sedentary livelihoods in Kassala State and possibly other parts of the Horn Region.

## 3.1 The emerging livelihood systems and opportunities

Because of the various challenges pronounced above, many former pastoral communities are adjusting to the new situation by adopting multiple survival strategies. The strategies of this 'third way' which are creating new livelihood opportunities as well as enhancing household food security include:

*i.* ***Creating stronger rural-urban socio-economic linkages***

The adaptation mechanism taken by these communities increased the rural-urban interaction enabling former pastoralists to have access to goods and services in nearby towns as the Kassala case of this study demonstrates. For instance, involving themselves in urban and peri-urban agricultural activities, adopting a semi-sedentary form of pastoralism with fixed homesteads but allowing mobility for larger livestock (camels, cattle), and commercialization of livestock production. Pas-

35 Headey, D. et al (2012:17)

toralists bring livestock, dairy products, hides/skins, artifacts, charcoal, wood produces etc. for sale and in return they buy a variety of food stuffs, domestic utensils, school materials for their children, farm implements, animal fodder and other related household items.

Improved road situation and better communication facilities allowed more access among pastoralist settlements. These infrastructural improvements have also created greater opportunities for rural pastoral community to increase their economic interaction with the urban centers. In the case of Kassala city fresh milk is supplied daily by small milking associations/groups of pastoralists who jointly hire trucks to carry milk from remote grazing areas to Kassala city.

From a food security perspective the milk is fairly cheap and this enables the urban dwellers to have adequate supply of dairy products in Kassala city, especially poor families who lack enough protein in-take. As the African Union (AU) in its "*Framework for Mainstreaming Livestock*" mentioned 'animal source foods are the best source of essential micro-nutrients like vitamin B12 that are essential for children physical and cognitive development.'[36] The adequate supply these pastoralists are making to the Kassala State is contributing to the price stabilization and food security of the communities.

***ii. Income source diversification***

As part of their adaptive strategy, they are gaining new skills to generate alternative income which compliments livestock production. Such widened economic opportunities include: rearing more small ruminants around the homesteads, poultry production for women and wage labor (farming or urban employment), petty trading in various commodities and livestock marketing.

***iii. Involvement in local government processes***

Pastoralists are asserting their rights by actively involving themselves in civic and political rights by asking for more representation in state

36 AU-IBAR (2010:2)

affairs. Such opportunities include participation in local elections to represent their communities at various levels, forming their own CBOs.

*iv.* ***Owning land in semi-sedenterized system***

By adopting a semi-sedentary mode of production, pastoralists are now able to secure land ownership while at the same time maintaining some form of seasonal mobility to distant grazing areas for larger stock (camels and cattle). Land ownership within their homestead not only secures tenure but widens their opportunities for small scale farming and keeps away land grabbers who traditionally occupied pastoral land under the pretext of such lands being unoccupied.

*v.* ***Ensuring adequate social services***

Many thousands of boys and girls from pastoralist areas now have access to formal education. However, some older boys who have to seasonally migrate during the dry season with their fathers and have to opt out of schooling. This has increased ratio of girls over boys in the nomadic schools. The semi-sedentary communities are also now more accessible and have easier access to human health and veterinary services provided by local authorities and NGOs.

*vi.* ***Improved communication facilities***

In the case of Eastern Sudan mobile telephone and broadband coverage even in the distant grazing areas is very good. The improved communication facilities serve multiple purposes among which better communication between the pastoralist families in the settlements and herders in remote grazing areas; communication with towns regarding livestock markets and commodity prices; and in case of emergencies (droughts, epidemics, communal clashes, etc.) contacting local authorities who are mostly based in towns.

*vii.* ***Change in gender roles***

Pastoralist communities in the study area have been traditionally male dominated with very limited political, social and economic roles for women beyond the domestic chores. As mentioned previously pastoralists in the sedentary context have more access to health, education and other services. Women being in a position to sell small livestock and its products; set up and manage micro-businesses manifest women's role in managing small stock in the homestead and cooperatives as well as women led CBOs. The women's role is further enhanced by their role as primary and secondary teachers in their villages. Women of a pastoralist origin are now able to join higher educational institutions and such transformation through widened opportunities for women is a growing phenomenon which is worth studying. Pastoralist girls are going to school in larger numbers (see v above).

## 4. Implications of Social Protection Policy on Pastoralism

Social investments promote societal justice and contribute to poverty reduction.[37] This was drawn upon Africa's experience during the 1960s and 1970s. Though social protection is a fundamental human right, pastoralists have been neglected in securing this right. They usually do not get the necessary facilities that help their survival in the drought-affected areas of the Horn. The problem of drought in the Horn of Africa is systemic and has led to a vicious cycle of famine. Over the last forty years, the Horn of Africa witnessed a number of development interventions which one after another failed to address the rights issues and ensure social security. Among pastoral communities, middlemen manipulation, unfair government taxation on livestock and ecological decline mainly cause inequalities due to the absence of the sharing of power, unfair trade and exploitation. The above problems have been exacerbated as most government policies focus more on urban and semi-urban areas rather than the rural areas which are predominantly occupied by pastoralists with their unique livelihood. Basic social services

37 Bangura, Y. (2011)

(health, education, hygiene and sanitation) are urgently needed to be developed in the remote pastoralist but in a manner which is responsive to the special circumstances, culture and way of life of the pastoral communities.

The pastoralists' key assets are livestock where there is a need for livestock based insurance or other alternative mechanisms to be introduced in order to support their self-adaptive strategies during the financial and environmental crisis. The high degree of market volatility and herders assumption about the expected risks of chronic drought pushes them to sell animals at lower prices. Such situation, force the herders to pay more for the cereals which is not comparable to livestock sales.[38] Such unfavorable terms of trade negatively affects the overall pastoralist household financial position, and they are thus exposed to social insecurity. Currently, they depend almost entirely on their own social, cultural and traditional means for their social protection. Designing long-term development interventions and policies that make social protection a backbone should be devised in order to reduce poverty and improve their wellbeing.

## 5. Case study: The significance of pastoral economy in Kassala Regional market

### 5.1 Overview: Livestock Production in Kassala

Livestock contribution in terms of draught power, crop manure and fuel (dung) will remain to be a major source of livelihood for the rural population and also a source of cash and social prestige. The most interesting part of the current livestock management change is the development of commercial-mindedness of the pastoralists.[39] This process-based attitudinal change is greatly contributing to the commercialization of livestock sector, food price stability and food supply of the Kassala communities and beyond. Livestock will continue to play a significant role in the national economy contributing well beyond the estimated 25 percent of agricultural GDP.

38 Little, P. et al (2001:151)

39 For more discussion see Livingstone, I. (1977:221)

In the case of Western Eritrea, it has to be stated that reliable figures on livestock and the extent of their contribution to domestic and international markets are unavailable. However, Nauheimer (1995) stated that in 1994 that 35,000 heads of sheep and goats were exported to Saudi Arabia from the Gash-Setit[40] alone earning for the country at the time an estimated $100,000. The export trend has been on the increase since liberation of Eritrea in 1991.

Pastoralists in the Kassala State (including some from Western Eritrea and Western Ethiopia) make significant contributions to the State and Regional economies. Generally, reliable data is difficult to get as the major livestock trade happens unofficially, but the following data gives us a general overview the sectors' economic and food supply contribution. Total numbers and value of livestock sold for local consumption or export is shown on Table 1 below. A total of 76 million Sudanese Pounds (SDGs) (equivalent to 38 million dollars) is a result of pastoralist production.

- Cattle (which are not exported) contributed 23 million SDG (30%)
- Camels (local consumption and export) contributed 32 million SDG (42%)
- Sheep contribution was estimated at 20 million SDG (26%)
- Goats share in local consumption and export was 1.1 million SDG (2%)

Revenues from the camel trade came first followed by cattle, sheep and goat respectively. The goat contribution to trade is underestimated because most of goats are slaughtered for home use and therefore remained not reported.

40 One of the then ten provinces of Eritrea (before the regional restructuring of 1996)

| Species | Local consumption recorded slaughter (Number) | Value at retail prices (1,000 SDG) | Export (Number) | Value at retail local prices (1,000 SDG) | Total (Number) | Value (1,000 SDG) |
|---|---|---|---|---|---|---|
| Cattle | 37,962 | 22,777 | - | - | 37,962 | 22,777 |
| Sheep | 79,115 | 14,241 | 31,783 | 5,720 | 110,898 | 19,961 |
| Goats | 7,129 | 428 | 11,234 | 674 | 18,363 | 1,102 |
| Camels | 1,114 | 1,337 | 25,717 | 30,860 | 26,831 | 32,197 |
| Total | - | 38,783 | - | 37,254 | - | 76,037 |

**Table 1: Livestock market output – Kassala State (2007)**
Source: Fre and et al (2009) mentioning the State Ministry of Animal Resources, Animals Slaughtered and Animals Exported report. (Note: Non-market local consumption is not included.)

## 5.2 Why is the Kassala market locally and regionally significant?

The Kassala city with a population of over 150,000 people is located 60 Kilometers from the Eritrean Sudanese border in the East and 180 kilometers from the Ethiopian-Sudanese border to the South[41]. It also located on the main tarmac road which connects the main sea gate of Port Sudan and the capital city Khartoum (See map below). The Ottoman Turks chose Kassala as the Eastern administrative center in 1840 and since then it remained to be the most important administrative and commercial center for Eastern Sudan, the neighboring countries and the Middle East. It was then the capital of Kassala region (1950), capital of the Eastern region (1980) and capital of Kassala State (1994).

The demography of Kassala is very diversified and mosaic in its nature – ranging from the Fulani of West Africa to the Rashaida Arabs from Yemen to several indigenous tribes whose livelihood occupations equally range from camel herding to small scale farming, urban agriculture, petty trade and commerce.

41 See more on the socio-economic situation of Kassala on FAO (2010:7)

Kassala is also well connected to the neighboring countries with whom it shares: natural resources (rivers), ethnicities, trade, marriage affinities and wage labor. Sudan has been also a major provider of petrol and grain to both Ethiopia and Eritrea which use the Kassala main road while at the same time remaining a major market for livestock owners from Eritrea and Ethiopia.

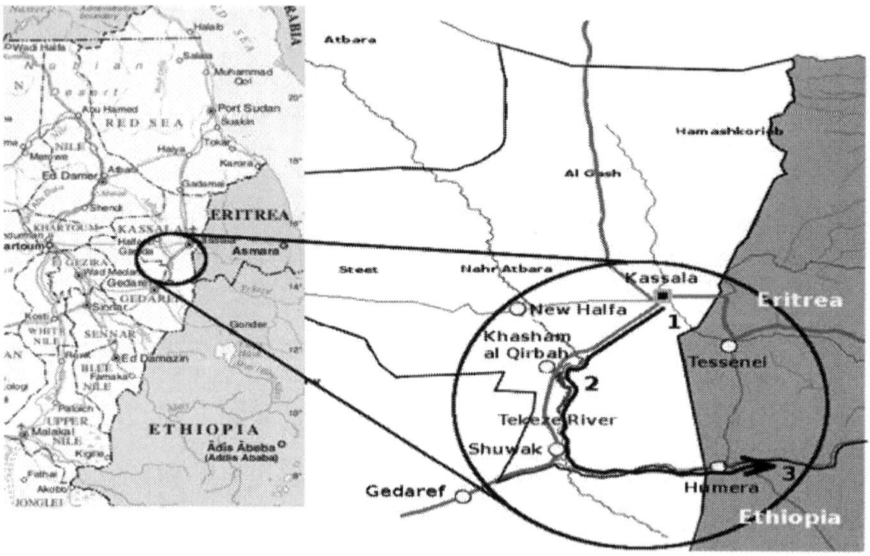


**Map 1: Positioning of Kassala city and pastoral mobility in Regional context**

Source: Fre, Z. and et al (2009)

Note: Black arrow on the map shows their annual movement from Kassala State area to Ethiopia and Eritrea. Numbers (1, 2, and 3) shows approximate places where livestock owners are in each season of the year. 1. Rainy season, 2. Harvest season and 3. Dry season.

The following are the main reasons why the Kassala regional livestock market is so important to pastoral communities of the three countries, namely Eritrea, Ethiopia and Sudan:

- ***Relative safety and security:*** Kassala is one of the safer border cities for both Eritreans and Ethiopians to exchange goods and services. The three countries have had their share of civil wars and interstate conflicts over the last three decades which at times led to border closures and thus limiting cross border trade. The Ethio-Eritrea border has been closed following the 1998-2000 border conflicts which means direct cross-border trade between the two countries has become impossible, thus making Kassala as the most viable market for cross-border communities. So the absence of formal cross-border trade agreements between Ethiopia and Eritrea resulted in to an increase in unofficial trade exchanges among Ethiopian, Sudanese and Eritrean pastoralists using Kassala livestock market as the most viable option.

- ***Proximity of the area:*** Kassala city is close to the Eritrean border and not far away from the Ethiopian border (see map 1) and this allows pastoralists from both countries to trade freely with pastoral communities of Kassala city in the Eastern Sudan. For Eritrean and Ethiopian pastoralists to reach their own capital cities hundreds of kilometers away would have added high transportation costs.
- ***Wider market options and competitive prices*:** The Kassala market attracts both local as well as global livestock dealers as there is a wider opportunity for better prices and exchange of other commodities given the strength of the Sudanese currency compared to the Eritrea and Ethiopian currencies. The Kassala market is also a major source of live animals for major centers like the capital Khartoum, Port Sudan and other intermediary cities. Kassala market also supplies livestock for the North African, Gulf and Arab countries consistently but the researchers could not obtain data on such an important market outlet.

## 5.3 Some major market drawbacks in Kassala

There are a number of drawbacks many of which are beyond the control of the pastoralist communities but such drawbacks clearly undermine livestock production, access to food and household food security. These include the following:

- ***Taxation:*** The problems of multiple and excessive taxation of the livestock sector is discouraging many poor pastoralists and investors.[42] As Aklilu revealed in his study 'Sudan probably applies the most excessive and complex fees and taxation system on livestock trading in the region.' Taxes are levied at various levels and in various amounts from the livestock production areas to the consumption areas. Government regularly collects taxes from the livestock sales in Kassala ranging from USD 1 to USD 3 per head of the animal depending on the type of the animal. Such tax is a major burden for the majority of low-income pastoralists who do not get any services from central government in-terms of animal health or other benefits for their livestock. Furthermore, the pastoralists (those who are Moslems) are also supposed to pay an annual religious tax (*Zakat*) to the religious leaders.
- ***Transport costs:*** Animals are trekked on hoof from the production areas to the main market in Kassala. Such distances range between 20 to 150 kilometers in one direction and animals lose a lot of body weight fetching lower prices once they arrive to Kassala market. If animals are not sold for any reason they have to be trekked all the way back to their settlements and the producers end up empty handed if not impoverished.
- ***Shortage of animal feed during the dry season (hunger gap):*** During the long dry season (January-June) large herds migrate seasonally to potentially high grazing areas far away from Kassala market, while small stock remain around the homesteads. This is the period when there is high demand for livestock products both in urban and rural areas. However, the major livestock are not accessible, being in far grazing areas. Intervention by NGOs and the government to improve the animal fodder supply would enable migrant herds to stay for longer periods around Kassala thus supplying the market with livestock produces during the dry season.

---

42 Desta, M. (2007:8)

- ***Exploitation by middlemen:*** The pastoralists who produce the livestock get far less profit from livestock sales compared to merchants and trans-continental market dealers. Pastoralist bargaining power is so low because of their lack of understanding of how the regional and global livestock markets operate. Also they do not have agents or their own middlemen who would promote their interests and put it as part of government policy agenda.
- ***Reduced pastoral labor force and abuse of child rights:*** The Region is a post conflict zone and is affected by human displacement which includes a large segment of the pastoralist population. Some of the younger men of pastoralist origin have abandoned pastoralism and taken up other occupations as traders, teachers, the army and seasonal employment as wage earners. This in turn has meant fewer people are now available for tending livestock and hired herders have to be remunerated. Poor families cannot hire herders and their kids end up looking after the animals thus depriving them from education. A large number of nomadic children from poor families in Eastern Sudan are illegally trafficked by Arab merchants as camel jockeys and taken to Gulf countries to train camels for rich Arab families. All these alternatives have reduced family labor within the pastoralist households increasing their vulnerability and food insecurity.

## 6. Conclusion

The Horn Region still faces major environmental, food security and political problems which are impacting on the pastoralist livelihoods and their economy. The paper argues that pastoralists have shown high degree of resilience and adaptation as demonstrated in the case study. The central argument in the paper is that pastoralist livestock production system is significant to food security and to reduce poverty. It provides nutritionally rich foodstuffs and gears the market dynamism that creates positive spill over on other sectors of the economy.

Based on the analysis, this paper concludes that given the contemporary challenges which are affecting the livestock system production (including climate change, intra and interstate wars, policy deficiencies), pastoralists have managed to develop multiple self-initiated coping mechanisms. Indeed, there is a need for a new paradigm shift in our understanding of the emerging pastoralist livelihood systems in the Horn of Africa. The new circumstances are pushing pastoralist families to a semi-sedenterized but partly mobile way of life is helping them to use both opportunities which exist in permanent settlements and mobile livelihoods. The adaptation strategy is being used by the communities of the Horn of Africa in general and Kassala in particular helps the pastoralists to remain pastoralist while gaining access to social services, increasing their bargaining power and owning permanent residential land which they might otherwise lose in the future due to the land grabbing and other forms of encroachments. Moreover, the new system is also allowing strong rural-urban livelihoods linkages and a change in the role of women in their families and communities as they started to hold more responsibility and authority in the absence of their husbands.

The paper concludes by stating that when governments design their Poverty Reduction Strategies, they do not take into account the contribution of pastoralism to their economies and in the next section the authors make specific recommendation for action.

## 7. Way forward:

To boost household food security status and the right to food, the following actions would be worth considering by all the stakeholders concerned about pastoralist wellbeing:

- Understand the evolving processes, current trends and pastoralists' adaptive strategies. This would lead to responsive policies and appropriate interventions.
- Recognize the contribution of pastoralist livelihoods as a means towards food security and economy of communities.

- Support for animal health services particularly in bordering areas to minimize the cross-border transmission of animal diseases and increase productivity.
- Engage in strong lobbying to curtail land grabbing in pastoral areas by the States and international companies by creating stronger alliances and synergies among pastoral peoples.
- Help maintain the biodiversity of the range lands by controlling Prosopis Juliflora - and finding alternative uses for Prosopis, including animal feed, fire wood, construction materials
- Conduct proper studies on livestock market chains in the border areas of Eritrea, Ethiopia and Sudan in order to understand the opportunities for trade available for pastoralists locally, nationally and internationally thus enhancing the pastoralist bargaining power.

## References

Aklilu, Y. (2002) An Audit of the Livestock Marketing Status in Kenya, Ethiopia and Sudan, OAU/InterAfrican Bureau for Animal Resources, Nairobi, Kenya

AU (2010) AU Policy Framework for Pastoralism in Africa: Securing, Protecting and Improving the Lives, Livelihoods and Rights of Pastoralist Communities, Department of Rural Economy and Agriculture, Addis Ababa, Ethiopia

AU-IBAR (2010) Framework for Mainstreaming Livestock in the CAADP Pillars, Interafrican Bureau for Animal Resources Nairobi, Kenya

Bangura, Y. (2011) Jobs and equity key to Africa's poverty fight: Progress on MDGs requires more than social safety nets, 19 Jan 2011, www.unrisd.org/unrisd/website/newsview.nsf/(httpNews)/073CF249ECD5C583C125781D0045128D?OpenDocument, Accessed on 22/10/2011

Bokre, H., Fre, Z. and Jardine, A. (2006) Pastoralism and Natural Resource Management in the Gash-Barka Region of Eritrea: A Literature Review, PENHA-NPA Publication

Desta, M. (2007) The Regulatory Framework for Trade in IGAD Livestock Products, IGAD LPI Working Paper No. 07 – 08.

FAO (2011) World Livestock 2011: Livestock in Food Security, Rome, FAO.

Fre, Z. (1989)Pastoral development in Eritrea and Eastern Sudan: Implications for livestock extension programmes, PhD thesis, University of Reading, Faculty of Agriculture, Reading, UK

Fre, Z. and et al (2009) The impact of increased food prices on rural-urban-rural relationships: The Case of Eastern Sudan, Study commissioned by IIED/UNFPA, PENHA, Khartoum, Sudan

Ibnoaf, M. (2008) A Policy Research on Livestock Sector in Kassala, Practical Action, Kassala, Sudan

IGAD (1990) Inter- Governmental Authority in Development (IGAD) Forum on Environmental Protection and Development of Sub-Regional Strategy to Combat Desertification, 1990, Vol. II, Country Reports-IGAD-NORAGRIC University of Oslo, Norway

Kaimba and et al (2011) Effects of cattle rustling and household characteristics on migration decisions and herd size amongst pastoralists in Baringo District, Kenya, in Pastoralism: Research, Policy and Practice 1:18.

Knips, V. (2004) Review of the Livestock Sector in the Horn of Africa (IGAD Countries), Livestock Sector Report Horn of Africa (FAO), September 2004

Little, P. and et al (2001) When deserts flood: risk management and climatic processes among East African pastoralists, in Climate Research, Vol. 19: 149–159, 2001

Livingstone, Ian (1977) Economic Irrationality Among Pastoral Peoples: Myth or Reality? In *Development and Change, 8 (1977), 209-230*

Nauheimer, H. (1995) Animal Health and Livestock in Gash Setit: Assessment of the Sector and Proposal for Improvement Report, June 1995 GTZ-MoA, Eritrea

Omolo, N.A. (2010) Gender and climate change-induced conflict in pastoral communities: Case study of Turkana in North-western Kenya

Pavanello, S. (2010) Working across Borders: Harnessing the potential of cross-border activities to improve livelihood security in the Horn of Africa drylands, Humanitarian Policy Group, HPG Policy Brief 41

# CHAPTER 9

## Comprehending Globalization and Local Development

*Amare Desta and Mammo Muchie*

There are increasing national and international concerns and uncertainties caused by the forces of globalization on local communities in the developing world. Although these are valid concerns, this paper will focus on related and pressing issues on how to identify, accommodate, modify or fully accept practices associated with globalized and localized knowledge. The authors of this article therefore hoped that the issues they are proposing within the paper can contribute to the ongoing debate on globalization vs. localization in both economically developed and the developing world.

### Introduction

Globalization is a word which is becoming too common in the new millennium, mainly due to the rapid growth of information and communication technology (ICT) in the last three decades. To many authors, different types of globalization can be identified even though most of the attention is in the areas of economy, technology, and culture (Brown and Lauder, 1996; Waters, 1995). Other authors such as Cheng (2000, 2001) also described varying type of globalizations, including technological, economic, social, political, cultural and learning globalization.

Authors such as Pratt and Poole, 2000; Green, 1999; Henry *et al.*, 1999; Jones, 1999; Ayyar, 1996; Brown and Lauder, 1996; Little, 1996; Fowler,

1994; McGinn, 1996 Council, 1999) outlined, how education for example, should be responsive to the trends and challenges of globalization and they further stated that this issue should be a major concern for academics, researchers and policy-makers. Furthermore there are also increasing demands for recognizing the values and contributions of local knowledge - widely known as indigenous knowledge (IK) for the wider societal development. It is believed that this recognition needs to be accompanied by appropriate support to maximize the effectiveness of education that is based on national needs which could reflect societal and cultural values and norms as this should be a prerequisite for ongoing human capital development. Therefore there is a need for a paradigm shift first towards localization and then towards globalization to address the wider societal issues in today's developing world.

Various other authors including Cheng, 2000; Waters, 1995 also argued that globalization is creating various opportunities for sharing knowledge, technology, social values and norms promoting development at different levels to individuals, organizations and societies across the globe.

A brief literature review on the promises of globalizations reveals the advantages of globalization to include:

to global sharing of knowledge, skills, and intellectual assets that are necessary for multiple development at different levels,

to mutual support, supplement and benefit to produce synergy for the development of countries, communities, and individuals.

to create values and enhancing efficiency through the above global sharing and mutual support to serving local needs and growth.

to promote international understanding, collaboration, harmony and acceptance of cultural diversity across countries and regions.

to facilitate multi-way communications and interactions, and encouraging multi-cultural contributions at different levels among countries.

As the above brief reviews suggest, globalization seems to be the mantra of the day by many countries as the only way to overcome and advance the ailing economies of the developing countries. The advocates of globalization further argue that the low-income countries should open up their doors and embrace the forces of globalization in order to "develop their societies and people". However there is equally a valued concern about the negative aspects of globalization on local communities – especially in the developing countries. The negative impacts of globalization could be classified as direct or indirect types of political, economic, and cultural colonization and the influences of economically developed countries and the rapidly-increasing gaps between the rich and poor and between the information have and have-nots areas in different parts of the world.

In particular, the negative aspects of globalization could include the following but not limited to:

exploiting local resources and mimicking and destroying indigenous cultures of the developing countries to only benefit a few advanced countries;

widening the technological gaps and digital divides between economically developed and developing nations that are hindering equal opportunities for fair global resource sharing;

facilitating legal/illegal avenues for a few economically developed or rapidly developing countries to economically and politically colonize/influence other countries;

maximizing imbalance and conflicts between areas and cultures;

increasing inequalities and conflicts between areas and cultures;

promoting the dominant cultures and values of the "developed world" and accelerating cultural transplants from advanced areas to less-developed areas.

This direct or indirect impacts of globalization on developing nations are complex issues and beyond the scope of this paper. However the authors of this paper believe that education and human resources development could be the key to redirect the impacts of globalization to the well-being of local communities. Therefore minimizing the unintended negative consequences of globalization requires immediate attention by the academics, researchers, practitioners and other development workers and this should include the educational and pedagogic reform agenda for national development in the developing world. It is therefore paramount to specifically assert and clarify the relationship between the locally immersed and widely practiced indigenous knowledge systems and the western "scientific" knowledge system in the context of globalization

## Local vs. global "scientific" knowledge

As various authors such as Desta (2009) argued local/indigenous knowledge and the innovations of local communities are crucial for low income economies in order to achieve cumulative growth as the knowledge of local communities has been tested and found to be valid in a local context and which are accumulated by their respective communities over long period of time.

In ancient and history rich African country like Ethiopia for example, the knowledge base and the existing social context, cultural assets and historical backgrounds is unique and the knowledge and the wisdom of the local people is unquestionably useful and valid despite its potential not being fully recognized and put to good use. As argued by Desta (2010), in contrary however the dominant "global knowledge" is vague as this concept is defined and expressed by different scholars or researchers with uncertainty. For example, for some global knowledge roughly defined and understood as the knowledge that is "scientific" and common to many areas of our world. If we accept this assertion, it is easy to understand that local knowledge certainly contributed or is a part the pool of global knowledge.

Furthermore due to the rapid advancement of ICT and the wide networking, the influence of some knowledge systems, particularly of the economically developed countries on developing nations is undeniably huge. It is therefore paramount for the main stakeholders (i.e. government and its development agencies; agricultural & environmental/education institutions including indigenous communities) in the developing world to find ways how local communities and people can share and exchange their know-hows' on equal basis with the dominant global knowledge system in order to bring a win-win situation to any area of global concerns. Otherwise the inevitable is going to happen (i.e. the local knowledge remain local and inaccessible to the development workers or worst the local/indigenous knowledge is going to irreversibly disappear altogether).

## Fostering knowledge in globalized world

As various authors such as (Yu, 2001) & Wada (1993) argued, for example many Asian countries including Japan and China have been interacting with their Western counterparts for more than a century. Many other Asian countries such as India, Thailand, and Korea also experienced a long history of building a knowledge base with the dominant "global force" while at the same time making an effort to maintain their cultural assets and local identities in order to face the challenges exerted by the external forces. As a sequence they largely managed to retain their local/indigenous knowledge base and absorbed external knowledge and relevant resources from the global knowledge system to simultaneously growing their local knowledge system.

The Asian countries experiences demonstrates that fostering local knowledge in globalized education system requires local identity and cultural roots and the curriculum and education system should therefore be based on local values and cultural assets but absorb only suitable global knowledge and technology to support the development of local communities. The selection of global knowledge therefore should depend on the needs of the local communities. The expected outcome of the national education system ought to develop a strong national community capable of addressing local demands and expectations with international outlook to contribute to the common cause of global concerns.

The benefit of the above approach is that the local/indigenous communities can maintain and develop further their traditional and cultural values while interacting with the input of external forces to appropriately utilize local human resources for the wider national and global development. Since the process is embedded within the local culture, it can be sustained by homegrown resources with the minimal or no outside support. Ultimately the successful growth of a local communities and the local knowledge system will contribute to the growth of the global communities.

In the other hand if the cultural roots and bases are not fully acknowledged and not valued, the growth of local communities will remain local and the conversion of global knowledge into local knowledge will be very limited. Furthermore, any development which may come along with the concept of globalization bound to remain superficial and unsustainable. As we argued elsewhere, one of the greatest consequences of the under-utilization of local systems is the loss of indigenous system which will result in the inefficient allocation of resources and manpower to inappropriate planning strategies which have done little to alleviate local or national poverty.

We therefore propose the following ***FIVE scenarios*** in order to provoke further thoughts and initiate debates on how best to link global and local knowledge.

## Scenario No 1: *"planting local seeds"*

The best scenario to counter balance the unintended consequences of globalization is to "*plant local seeds*" and only then accumulate the global knowledge along a given local expectation and demand. Therefore, the aim of fostering local knowledge is to accumulate global knowledge around "local seeds" and this could be the existing local demands and values to be fulfilled. This means one have to identify and maintain the core local needs and values as the fundamental seeds to accumulate the relevant global knowledge and resources accordingly. The authors of this paper believe that in this era of globalization, understanding and valuing the wisdom of local knowledge structure should be the basic criteria before considering to entertain or accumulate the dominant global knowledge and the education sector could be the driving force for such endeavor.

The expected educational outcome could therefore be to develop local communities who remains local with global knowledge and can act nationally with increasing global techniques. With local seeds to crystallize the global knowledge, there will be no conflict between local needs and the external knowledge to be absorbed and accumulated in the development of the strong and economically viable communities. In practice, it is easy to identify what the local communities wants or do not want from the overwhelming global inputs.

However one must not underestimate the influence of external forces that led people in developing countries to become increasingly dependent on outside expertise in the first place. It is equally important to acknowledge the fact that it may also not be easy to find a set of good *local seeds* or values that can be used to crystallize and localize the powerful global knowledge. Since the original nature and type of local seeds themselves may crystallize only those similar to themselves but not others, the knowledge to be fostered may be limited. However given the nature of one-way crystallization, there will be no clear contribution to the growth of global knowledge and communities.

## Scenario No 2: *Minimizing the domination of global forces*

How to avoid or minimize the total domination of global forces on their opposite local communities is often one of the key issues to many people in the developing world. Perhaps the best way how to resist or counter balance this threat is to continuously foster local knowledge in the globalized education system. This scenario (i.e. *minimizing the domination of global forces*) can be open for incoming global knowledge and resources but at the same time efforts should be made to limit or converge the local development and related interactions with the outside world to a manageable number or fixed homegrown/local framework. It means that fostering local knowledge in globalized education requires a "local framework" for filtering the incoming external knowledge and safeguarding the local development from the unintended consequences of globalization.

It is therefore important to identify homegrown "local framework" with clear ideological boundaries and social norms for education/curriculum design such that for example, all educational activities can have a clear local focus when benefiting from the exposure of wide global knowledge and inputs. This means local wisdom and concerns should be a core part of any educational activity. The expected educational outcome should therefore be to develop local communities with bounded global outlook, who can act locally with filtered global knowledge.

In so doing, one can ensure local relevance in globalizing education and avoid any loss of local identity while interacting with global forces. The authors of this paper firmly believe that the "local framework" should aim to protect the local interests from over-globalization. However in practice, it may be difficult to build up appropriate cultural or social boundaries to filter the global impacts and ensure local relevance because the boundaries may be too tight and closed to stop any necessary and meaningful interactions with the outside world and limit the growth of local knowledge. However regardless of whether the local framework is accepted or not, one have to reject the temptation to the idea of knowledge transfer (KT) but instead embrace knowledge exchange (KE) when appropriate.

## Scenario No 3: *Transplanting global and viable knowledge*

This approach places emphasis on identifying and transplanting the better key elements from the global knowledge to replace some "potentially weaker" knowledge and possibly reject some harmful cultural practices. This means the purpose of fostering local knowledge will be to mainly replace the unproductive local knowledge with the vital global knowledge through globalization or globalized education.

In globalizing education, the curriculum design should be very selective to both local and global knowledge with aims to choose the best elements from them. Therefore, the understanding of weak and strong elements in both local and global knowledge is necessary in education. Learning institutions for example, may be strongly encouraged to be open for transplanting any "good & viable elements" into local contexts. The expected educational outcome is to develop local and national communities with locally and globally mixed elements, that can act and think with mixed local and global knowledge.

The strength of this approach is its accommodativeness to transplant valid knowledge without exerting further pressure or burden on local culture. This approach can provide an alternative to learn and improve the existing local practices and developments. However in practice, it may not be simple to easily identify a weak or strong elements and equally it may be too mechanical to assume that the transplant and replacement with goodwill can be carried out easily without any cultural imposition on smaller or large scale.

## Scenario No 4: *Embracing global knowledge*

This approach assumes that it is faster and easier to digest and absorb certain relevant types of global knowledge for the growth of national development than to expand the local knowledge base. Therefore, the process of fostering local knowledge is to digest the types of global knowledge which are significant and relevant to local development and convert it as local lubricant for the development of the local communities.

This scenario suggests that the education system for example, could aim at enabling local communities to identify and justify for themselves why global knowledge is valuable and articulate how it can be utilized in order to support the national and local development. In globalizing national education system, the design of education activities should aim at digesting the complex global knowledge into appropriate forms that can feed the needs of local communities and their aspiration to growth. The expected educational outcome is to develop communities equipped with certain types of global knowledge, who can act based on relevant global knowledge system.

The main point of this approach is to digest and absorb the "useful elements of global knowledge" than to strengthen the local knowledge system. The roots for growth and development are based on the global knowledge instead of local innovations whether it is indigenous or not. Clearly, the process will be mainly one-way digestion and absorption of external knowledge and thus, its contribution of the local system to the growth of the global knowledge and resources is very limited. There is no clear local identity and vision for its growth but it is bound to depend on the knowledge and resource base of external sources and this may lead to one way of knowledge diffusion which is Knowledge Transfer (i.e. from economically developed Western nations to the developing nations).

## Scenario No 5: *Total globalization*

To smaller or newer countries without strong ties to specific homegrown culture, this approach could be an option as they may not be concerned as such with conservation of local values and cultural norms. As a consequence, they may be most concerned about the adaptation to the fast-changing global environment and economic survival. Therefore, they may consider fully embracing ***total globalization*** as they may be able to exploit the full use of the global knowledge with minimal local constraints. This approach considers that fostering local knowledge is only a process to fully use and accumulate global knowledge in the local context. Whether the accumulated knowledge is really local or the local values can be preserved may not be a major concern.

Accordingly the curriculum design for example, could include the full range of global perspectives and knowledge to totally globalize education in order to maximize the benefit from global knowledge and become more adaptive to changing environments. Therefore, to achieve a broad international outlook and apply global knowledge either locally be globally may be crucial to the education system. Cultural burdens and local values could be minimized in the design of curriculum and instruction in order to let local communities be totally open for global learning. The expected educational outcome is for example "*to develop flexible and open communities*" who can act and think globally.

The *total globalization* approach directly or indirectly force the local communities to embrace to be completely flexible and open to all global exposure such that it will either have none or bare minimum constraints in accumulating global knowledge and resources. Then, the development of local communities will be exposed to the "advanced communities experience" via the global forces. Of course, the strengths of this scenario are also its limitations, particularly in some culturally strong and pivotal nations like Ethiopia, China, India and others. Certainly there will be potential loss of local values and cultural identity in ancient countries; as a result local communities will potentially lose its direction and social cohesion due to the unintended consequences of the overwhelming forces of globalization. However it is very difficult to ascertain how the local communities can sustain their own livelihood and it is also difficult to predict the contribution of local communities to the national and international development.

## Summarizing the FIVE scenarios

Each of the above five scenarios have their own strengths and limitations to implement in different countries across different continent as each communities and the context they live in are different. It is therefore difficult to pick one or two scenarios or approaches and to say these are better than the others. However, the aforementioned FIVE scenarios can provide an opportunity for policy-makers, educators and researchers to consider, conceptualize and implement a viable human resources strategy by weighing the advantages and disadvantages of localization and globalization in order to ultimately strengthen their local knowledge base and achieve the desired development. Given the diversity and complexity of each country or local communities, it may be appropriate to consider combining multiple of scenarios to foster local knowledge and local know-hows. From the above scenarios therefore we can ascertain that appropriately fostering local knowledge and human development in the global era could contribute to the pool of global knowledge system.

## Conclusion

Given the increasing international concerns about the impacts of globalization on local communities, how to manage the realities and practices of globalization and localization in education for maximizing the benefits whilst minimizing the disadvantages for the development local communities inevitably becomes a major concern – especially for developing countries. After digesting the related scenarios of localized and globalized knowledge system in a context of globalization, in this paper we attempted to address these key concerns as FIVE different scenarios.

These scenarios have varied emphasis on global dependence and local orientation and therefore they may have their own strengths, and limitations in conceptualizing and managing the process of fostering local knowledge. Clearly, their implications for design of curriculum and instruction and their expected educational outcomes in globalization of education are correspondingly different. We believe that the above FIVE scenarios provide different approaches how best to foster local or global knowledge in the context of developing nations.

It may be obvious that each countries or local communities may have their own unique social, economic and cultural contexts and therefore, its tendency to use one scenario or a combination of scenarios in globalizing education may be different from that of others. It is therefore difficult to say one is scenario or better than the other. We may for example consider that *Scenarios No. 1, 2 and 3* may be more preferred in some culturally rich and ancient countries like Ethiopia, China, India and Japan.

At the same time, for those countries with less cultural ties or local values, one may suggest that the *Scenarios No. 4 and 5* may be more appropriate choices for development. However, we hope the above FIVE scenarios can provide a wide spectrum of alternatives for policy-makers and educators to conceptualize and formulate appropriate strategies which are suitable for each countries concerned in order to foster local knowledge and human resources for local, national and global development.

## References

Argyris, C., Scho¨n, D.A. (1974), Theory in Practice: Increasing Professional Effectiveness, Jossey-Bass Classics, San Francisco, CA., .

Argyris, C., Putnam, R., Smith, D.M. (1985), Action Science, Jossey-Bass, San Francisco, CA., .

Ayyar, R.V.V. (1996), "Educational policy planning and globalisation", International Journal of Educational Development, Vol. 16 No.4, pp.347-54.

Brown, P., Lauder, H. (1996), "Education, globalization and economic development", Journal of Education Policy, Vol. 11 No.1, pp.1-25.

Brown, T. (1999), "Challenging globalization as discourse and phenomenon", International Journal of Lifelong Education, Vol. 18 No.1, pp.3-17.

Cheng, Y.C. (1996), School Effectiveness and School-based Management: A Mechanism for Development, Falmer Press, London, .

Cheng, Y.C. (2002), "A CMI-striplization paradigm for reforming education in the new millennium", International Journal of Educational Management, Vol. 14 No.4, pp.156-

Desta, A., & Smithson, S. (2010) Indigenous knowledge in the context of Natural Resources Management: an information systems perspective, paper presented to UKAIS Annual Conference 2010, Oriel College, University of Oxford

Desta, A. (2010) Exploring the extent of ICT in Developing Countries pedagogical practices. Ethiopian electronic Journal on Research and Innovation Foresight(Ee-JRIF) 2010.

Desta, A. (2009) Reviewing the phenomenon of IK within NRM (Natural Resources Management) paper submitted to the inaugural ceremony of Ethiopian electronic Journal on Research and Innovation Foresight(Ee-JRIF). Vol 1, No 1 (2009)

Curriculum Development Council (1999), A Holistic Review of the Hong Kong School Curriculum Proposed Reforms, Government Printer, Hong Kong, .

Fowler, F.C. (1994), "The international arena: the global village", Journal of Education Policy, Vol. 9 No.5-6, pp.89-102.

Green, A. (1999), "Education and globalization in Europe and East Asia: convergent and divergent trends", Journal of Education Policy, Vol. 14 No.1, pp.55-71.

Henry, M., Lingard, B., Rizvi, F., Taylor, S. (1999), "Working with/against globalization in education", Journal of Education Policy, Vol. 14 No.1, pp.85-97.

Jones, P.W. (1999), "Globalisation and the UNESCO mandate: multilateral prospects for educational development", International Journal of Educational Development, Vol. 19 No.1, pp.17-25.

Little, A.W. (1996), "Globalization and educational research: whose context counts?", International Journal of Educational Development, Vol. 16 No.4, pp.427-38.

McGinn, N.F. (1996), "Education, democratization, and globalization: a challenge for comparative education", Comparative Education Review, Vol. 40 No.4, pp.341-57.

Mok, M.M.C., Cheng, Y.C. (2001), "A theory of self-learning in a human and technological environment: implications for education reforms", International Journal of Education Management, Vol. 15 No.4, pp.172-86.

Pratt, G., Poole, D. (2000), "Global corporations 'R' us? The impacts of globalisation on Australian universities", Australian Universities' Review, Vol. 42 No.2, pp.16-23.

Wada, S. (1993), "Reinspiring Japanese educational objectives", Peabody Journal of Education, Vol. 68 No.4, pp.76-87.

Waters, M. (1995), Globalization, Routledge, London, (2001), in Yu, H.P. (Eds),The Course of Modernization in China, Jiang Xu People Education Publishing House, Jiang Xu, .

Yuen, P.Y., Cheng, Y.C. (1997), "The action learning leadership for pursuing education quality in the twenty-first century", paper presented at the Fifth International Conference on Chinese Education Towards the 21st Century: Key Issues on the Research Agenda, Hong Kong, .

Yuen, P.Y., Cheng, Y.C. (2000), "Leadership for teachers' action learning", International Journal of Educational Management, Vol. 14 No.5, pp.198-209.

# PART III
# INDIGENOUS KNOWLEDGE IN COMPATIBLE WITH SCIENTIFIC KNOWLEDGE

# CHAPTER 10

## The Nano-Revolution and Economic Development: Is Africa Ready?

*Hailemichael Teshome and Mammo Muchie*

### Introduction

The nanotechnology revolution is set to affect everything from production systems to global power relations, from the way we lead our daily life to a future life without death. Once left to science fiction writers and eccentric futurists, nanotechnology is fast gaining ground as the orthodoxy of present-day science and technology (hereinafter S&T). If we are to follow some predictions, by 2020 most technology will have become nanotechnology or will comprise nanotechnology (Kurzweil, 2003; Davies, 2008). It has already triggered the rejuvenation of S&T with a redrawing of the contours between the various specialisations and their role in the economic development. In this sense, nanotechnology epitomises the whole issue of the governance of S&T proffering a chance to renegotiate the stakes of the developing world.

With the current precariousness of the comparative advantages in natural resources and cheap labour, innovation and S&T turn out to be the only choice for economic development. In the developed world economic reinvigoration and growth is entirely tied to S&T. This front has been neglected and ignored for far too long in the developing world especially in sub-Saharan Africa. Africa missed out previous technological revolutions like the Green and ICT revolutions. It cannot afford to miss out on the nanotech revolution for that would be a final blow hurling its people and economy into a state of irrelevance. The advent of nanotechnology will change this for the better if Africa engages with the phenomenon in a timely fashion.

## Nanotechnology – the science of things at the bottom

After more than two decades of active engagement and mammoth investment, there remains a

great deal of the disambiguation exercise on the 'ontological status' of nanotechnology. While a consensual definition has yet to be worked out, working definitions are widely adopted by various jurisdictions. Nanotechnology is conventionally defined as science and technology operating at the nanoscale and that scale is confined to the range of 1-100 nanometres. One nanometre (abbreviated as nm) is a billionth of a metre. To help with the imagination of the nanoscale, comparisons with the more familiar length scales are needed. A helpful comparison is made using the human head: one nanometre is to a human head what a human head is to the planet Earth (Gimzweski and Vesna, 2003, p. 3). A dollar bill is 100, 000nanometres thick while the human hair is 80,000 nm wide. The smallest of human cells, the red blood cell is 1000 nm wide.

The US National Nanotechnology Initiative (NNI) definition adds one more element on top of the size scale: the creation and use of novel properties and applications at the 1-100nm range.

## The official NNI definition reads as follows:

Nanotechnology is the understanding and control of matter at dimensions between approximately 1 and 100 nanometres, where unique phenomena enable novel applications. Encompassing nanoscale science, engineering, and technology, nanotechnology involves imaging, measuring, modelling, and manipulating matter at this length scale (NNI, 2010).

Indeed, the novel property rather than the scale range of 1-100nm tends to be the central defining concept of nanotechnology (Loder, 2005, p.3).

It is the strange properties of matter at the nanoscale and the possible applications thereof that made nanotechnology a subject of enormous interest. At the nanoscale, the classical laws of physics governing the macro world cease to operate and the laws of quantum physics take over and the strange properties of matter unknown at the macro level begin to dominate. At the nanoscale silver turns into a bioactive antimicrobial substance; gold melts at a much lower temperature than it does at the micro or macro scale; copper strangely becomes a poor conductor; aluminium behaves like chlorine; the soft carbon in the form of graphite becomes a hundred times stronger than steel when manipulated at the nanoscale turning into a much sought-after material with incredibly high strength-to-weight ratio. It is, thus, understandable why the exploitation, actual or potential, real or purported, of these strange properties has become the crucial concept in the definition of nanotechnology.

Though scripted in various official documentations across the globe, the US formulation in terms of length scale and novel properties is not meant to serve as the last word; that it is merely 'a working definition' needs to be stressed (Lee, 2010, p.800). Wood et al (2007, p.17) rightly suggest a detour in the course of engaging with the phenomenon: 'strict definitions maybe irrelevant as perspectives on how it is best pursued and what it can achieve become more important.' They point out that '[r]ather than seeing the issue of the field as a matter of definition or at least as defining it once and for all, it may be more helpful to approach it as a sociological issue' (id, p.12).

Hence, nanotechnology is understood here loosely as 'an umbrella term' for the scientific and technological activities where the conditions in the conventional definition prevail without excluding the features in the more ambitious and perhaps long-term aspects of the technology.

The accommodation of the interests of developing countries is dependent on the kind of definition that will be adopted. The adoption of a broad open-ended definition that would eschew the risk of fragmentation or under-inclusion of certain areas could be of greater crucial importance from developing countries perspective than it is in the developed countries.1

Accordingly, it is preferred to risk an over-inclusion with such broad understanding than risking under-inclusion and fragmentation (Wood et al 2003, pp.26, 29). Thus, no distinction is made here between nanotechnology as material, process, device or system; nor between the modest mid-term 'generations' or the more ambitious long-term generations of nanotechnology.

1 McHale (2009, p.71) discusses the problem of fragmentation in the context of Nano medicine research in the EU.

See Niosi and Reid (2007, p.436) discussing the same problem with emerging nations developing nanotechnology.

## The Nanotechnology Revolution and the Development Agenda c.2015

The year 2015 is a key date for endeavours both in nanotechnology and economic development. 2015 is the cut-off date for halving extreme global poverty as set out in the UN Millennium Development Goals (MDGs). This is also the year that nanotechnology is set to beat cancer according to the US National Cancer Institute (Cameron, 2007, p.289). In that same year nanotechnology is expected to hit the market to the tune of 1 to 2.9 trillion dollars.

2 If all this comes to pass, the year 2015 will indeed go down in history as a golden year of the 21st century.

While beating cancer and hitting the market to that order are ambitious predictions, halving poverty is not. It is not ambitious not because it is difficult to achieve but because the goals themselves are modest.3 There is no pretence that poverty will be eradicated; the goal is only to eradicate extreme poverty and poverty of other degree is not on the radar. What is more is that it is not the total eradication of extreme poverty that is set as a goal but halving it. Former World Bank President, Paul Wolfowitz, could value the goals as no more than an 'incomplete guide' (Wolfowitz, 2007) concurring with analysts who take the MDGs as 'either an agreed agenda, a minimalistic agenda or an incomplete agenda for human development' (Vandemoortele, 2005, pp. 5–11).

Achieving all the goals will be a challenge that would be there well beyond the 2015 rendezvous. Assessing the pace of progress, former UK Prime Minister Gordon Brown warned that unless 'extraordinary effort' is exerted even some of these minimal targets will not be achieved before the end of the century (Channel 4 News, 2008). UN Secretary General, Ban Ki-moon vented his disappointment at the progress being made which he said is 'unacceptably slow' and facing setbacks due to climate, food and economic crises (UN Department of Economic and Social Affairs, 2010, p.3).

2 The US NSF estimate is $ 1 trillion creating 7 million jobs worldwide, IRGC, 2006, p.21. According to Lux Research, nanotechnology will be a $ 2.9 trillion market as early as 2014 and will make up 15% of the global economy. Miller, 2007, p.162; Edwards, 2006, p.38

3 The Millennium Development Goals (MDGs) are adopted by the UN Millennium Summit in 2000 and are enshrined in its Millennium Declaration. The eight goals are eradicating extreme poverty and hunger, achieving universal primary education, promoting gender equality and empowering women, reducing child mortality, improving maternal health, combating HIV/Aids, malaria and other diseases, ensuring environmental sustainability, and developing a global partnership for development. The progress towards the goals is measured through 21 targets and 60 indicators. UN Department of Economic and Social Affairs, 2010, p.74

Despite the modest ambitions and with only less than a third of the time left, achieving the MDGs as planned is proving less and less certain. And for this the niggardliness of the rich is to blame.4 On the contrary, technology is taking the credit for the achievements so far attained as is evident from mobile phone penetration which is well ahead of schedule.5 One of the targets through which progress towards the eight goals is measured is the provision of the benefits of new technologies especially ICTs. Nanotechnology is held to be a potential catalyst in achieving the goals and is directly correlated to seven of the eight goals (Salamanca-Buentello, 2005; UN Millennium Project, 2005).

Beyond the 2015 rendezvous, there are other less publicized yet crucial dates. Eventually, by the end of the century, the wealth that our technologies will enable us to create will be worth$1000 trillion according to the predictions of Bill Joy — a man rather known for his dystopian views on technological progress. US House Speaker Newt Gingrich (2001, p.25) helps us translate this figure in a more intelligible way using a familiar benchmark — the US economy: it is like adding 100 US economies. This in turn means that by the end of the century, everyone on the planet will be enjoying the same or higher living standard as contemporary US citizens.

These generous predictions do not go unchallenged though. A back-of-the-envelope calculation has it that in order for everyone to attain the living standards of US citizens, the resources of two other planets are required (Spohrer and Engelbart, 2004, p. 67). A rather sober calculation by the World Wildlife Fund (WWF) sets the figure at four planets (in Barry, 2007, p.233).

Whereas 'terraforming' and the colonisation of space is humanity's dream being pursued with ongoing endeavours in space exploration — and it is made an even more serious proposition with nanotechnology and the space elevator project (O'Mathuna, 2009, pp.25, 164) — the full subscription of the resources of two other planets at least in this century is utterly unrealistic.

The role of technology in economic development has always been a defining issue of the whole debate on development and underdevelopment. The debate dwells on the issue of technological

4 Some figures shade light on this state of affairs: According to the 2006 WTO World Trade Report 21 developed countries spent almost $250 billion on subsidies close to 1.4 % of their GDP. This is double the percentage GDP that Blair's Commission for Africa recommended as necessary for Africa to achieve the MDGs. South Bulletin, 2006, p.430. It has also come to light that Americans favour the elimination of all kinds of foreign aid. See Avi-Yonah, 2004, p.372.

5 Uniquely among the MDGs, the target to enable 50% of the world population to have access to phone services by 2015 has already been met as 80% are now within the reach of a mobile telephone service. Bishop, bbb2005.

See also UN Department of Economic and Social Affairs, 2010, p.4resources more than it dwells on natural resources. Ever since its inauguration on the world stage as a component of the Truman Doctrine, development has been but a highly contested concept. The deplorable situation the vast majority of humanity finds itself after half a century of development thinking and practice has failed to provide a solution to the enigma of development. As Wolfowitz (2007) said 'there is still a lot that we don't understand about what makes development work'. As a result, there has been fatigue and resignation on the part of a significant section of the development scholarship which has evolved into the 'post development' school that, on the whole, rejects the various incarnations of the concept of development (Pieterse, 2010, p 110). Other schools of thought added to the various nuances of development ranging from those that enhance the dominant neoliberal paradigm to those that claim to present viable alternatives to it.

While the development discourse has been dominated by growth, reflections on 'limits to growth' and on sustainability made it abundantly clear that unfettered growth is not after all a good idea. Economic growth in pursuit of development needed to be recast as 'sustainable development' in view of the ecological crisis humanity is facing. After much reflection on what will happen if current practices of economic growth are maintained for the coming decades and centuries, 'development' was re-authored as 'sustainable development' in the landmark UN document, Our Common Future, alias, the Brundtland Report (Escobar, 1995,pp.196- 7).

The launching of 'sustainable development' as a way of addressing the ecological crisis has not introduced much that is seen to be working. Growth remains the bottom line while environmental quality and social equity — that are supposed to be amending the unwanted consequences of unfettered growth — are continually deferred. The problem with sustainable development is that it is enceinte with incompatible ideas in a bid to reconcile profit-mongering as driver of growth with ascetic thrift for the sake of environmental preservation.

Now, with nanotechnology and the requisite global regulatory regime in place, it may be possible for the first time to contemporaneously pursue the dissonant sustainability trio, viz, economic prosperity, environmental quality and social equity. As changes in science and technology unravel new opportunities and challenges, the debate on development continues with new and broader dimensions. Nanotechnology is opening up a whole new avenue by which the renaissance of the 'development' concept in ways agreeable to both its critics and proponents alike may become far from tenuous.

Beyond efforts aimed at reconciling, or rather balancing, opposing pursuits, a new approach to address the quest of development is required. The development discourse needs to aggressively engage with the governance of the phenomenon of nanotechnology. Such engagement is not an absent activity as can be seen from efforts to correlate what is largely thought to be 'blue skies 'nanotechnology research to the here and now of the poverty alleviation effort as expressed in the MDGs.

The consensus on the MDGs, and through them on sustainable development, seems to offer away out of the enigma of development. The MDGs are 'the standard reference' to 'development' that momentarily put off the erudite wrangling on the essence of the entire project of 'development' (Pieterse, 2010). From the unusually overwhelming assent to the MDGs, it seems the world has reached a working consensus over the notion of development.

By the time the MDGs were adopted, poverty has become too real to be a subject of further global deliberation. In terms of having a set of common goals and priorities, the MDGs present a definitive agenda that would certainly serve well beyond their use-by date.

## No half measures: Nano-applications vs. Nano-proper

The momentous speech by Richard Feynman, the speech that is widely held to have started all the bustle about nanotechnology today, reserved a clause on developing countries.6 However, the more influential work that championed the particular issue on the nanotechnology and development interface has been that of Salamanca-Buentello et al (2005). Credit is attributed to this work for having ignited 'a lively debate' on the issue of Nano and development (Schummer, 2007, p.291). The debate was indeed lively with the authors being subjected to some of the most unforgiving critiques on many levels as their work continues to influence legislatures in some developing countries into committing significant funds for nanotechnologyinitiatives.7 Among the charges they still have to answer is their sweeping deduction in assuming the voices of a handful of experts to be coterminous with the voices of the people in developing countries (Invernizzi and Foladori, 2005, p.300).

6 Feynman's (1960) illustration on the advantages of miniaturisation cites the Brazil of the 1950's and 1960's just as a representative developing country to whose needs nanotechnology can be deployed.

7 Lawmakers in the Argentinean Congress seem to have been influenced by these scholars when they decided to fund nanotechnology research. Malsch, 2008

The other major yet less direct critique is their piecemeal approach. Salamanca-Buentello et al(2005) came up with a list of top ten applications of nanotechnology in areas of energy storage, production and conversion; enhancement of agricultural productivity; water treatment and remediation of disease, drug delivery systems, food processing and storage, air pollution and remediation, construction, health monitoring and vector and pest detection and control. Their failure to emphasise the generic feature of nanotechnology coupled with their preoccupation with various applications relevant for achieving the MDGs was at best an incomplete account of the potential of nanotechnology. The debate having degenerated into a debate on particular artefacts and applications, their detractors did not lose time to add nanotechnology in the list of technologies that were supposed to benefit the poor but were never seen to have done so (Hunt and Mehta, 2006, p.279). The lack of emphasis on the generic nature of nanotechnology and on its core ideas was such a grave oversight that even their subscribers were misled to limit their recommendations to the applications rather than nanotechnology proper — the generic enabling technology: Schummer (2007, p. 294) who credits the work for stirring up debate advises that developing countries should invest in selected applications rather than on 'such vague projects as nanotechnology overall'.

The matching of nanotech applications and development goals is a nice dive yet misses the pool — the potential of nanotechnology and its comprehensiveness. The correlation of certain Nano-applications to certain problems offers only a partial glimpse of the potential of nanotechnology in the way towards 'the final mastery over nature'.8 Countries engaging in Nano-initiatives allured by specific applications should not lose sight of the big picture. This overarching promise of nanotechnology should be the major interest in any Nano-policy. It is not the passive role of consumers that these countries need to play in the innovation process. The days of the transfer of black-boxed technology widely practiced under the aegis of liberal free trade regimes did little by way of facilitating technological learning in developing countries (UNCTAD, 2007, p.2). Avoiding a slide towards 'dependency on ongoing technological charity' by foreign nations and firms is what developing countries should be aiming at (Miller and Scrinis, 2010, p.120). African scholars are warning that Africa should not fall prey to the marketing ploys of technology developers pointing out that 'there is no developer of a new technology that will readily transfer his technology but every technology8 On the 'notion of finality' see Berne, 2006, p.215; also Fogelberg and Glimmel, 2003, p.3.developer provides literature on his technology for market growth' (Durojaiye, 2011). At any rate, even to use the technology as consumers, which may itself require a great deal of innovation, developing countries need to develop their research infrastructure just like the developed nations are doing with theirs.

Valuable discussion on the general concept of nanotechnology, and its impact on 'the broader socio-economic structures', was missed due to the excessive attention on specific applications (Miller and Scrinis, 2010, p.120). Nanotechnology is more than (the sum of) its applications.

There is more to it as an idea, a blueprint, and information than particular applications. History of technology teaches that this is the point that needs to be emphasised: Blueprints and ideas and not the stuff or applications are the essence of technology. Jeffrey Sachs (2006, p.41) argues that 'the essence of the Industrial Revolution was not coal; it was (the idea) how to use the coal.' Similarly, the assembly line that is associated with Fordism is mistakenly perceived as the key to mass production. It was the idea that 'the complete and consistent interchangeability of parts and the simplicity of attaching them to each other' that made the assembly line possible (Womack, 1990, p.26). Likewise it is the idea that individual atoms instead of crude chunks of matter can be used as interchangeable building blocks and not the black-boxed finished products that come as Nano-applications that make nanotechnology truly revolutionary. It is a meta-blueprint calling for a full distinct treatment in its own right irrespective of the various applications.

Such was the treatment nanotechnology is accorded in the report of the UN Millennium Project Task Force, Innovation: Applying Knowledge in Development, that carefully avoided generalisation from various applications. The report's conclusion on the relevance of nanotechnology was based not on a casuistic correlation of problems to their Nano-solutions but on the general features of nanotechnology and its promises.

Nanotechnology is likely to be particularly important in the developing world, because it involves little labour [sic], land, or maintenance; it is highly productive and inexpensive; and it requires only modest amounts of materials and energy. Nanotechnology products will be extremely productive, as energy producers, as materials collectors, and as manufacturing equipment (UN Millennium Project, 2005, p.70).

Considering the more common discussion of specific applications, the relative generality of the above statement is unquestionable. However, an even more radical claim can be and should be made as Fogelberg and Glimmel reason out:

It [the vision captured in the concept of nanotechnology] also promises the long-awaited answers to a burdensome set of urgent questions or needs in the world, thus addressing global problems persistently plaguing mankind such as pollution, physical disease, and material poverty. From engineering meta-perspective it has been argued that the roots of those mega-problems are in fact one and the same: the poor control of the structure of matter. (Fogelberg and Glimmel, 2003, p.11)

While the discussion of various applications may serve as a necessary antidote to the generality of the claims that are being made on behalf of nanotechnology, it only gives an incomplete and misleading picture of the potential of the technology. Nanotechnology offers the opportunity for a total control of nature — deciphering her code, and emulating and outpacing her performance. Nanotechnology converging with other technologies marks the transition from the 'Age of Discovery' to the 'Age of Mastery' whereby humans will be graduating into the 'active choreographers of Nature' (Kaku, 1998, p.5). Those who lay hands on this technology will certainly be in control of nature not only in terms of harnessing it to their advantage but also in terms of totally reconstructing it.

## 'Ready or Not!'— What Africa Needs to do?

Preparedness is a priority theme in the nanotechnology policy of the advanced countries. US law has provided for the establishment of the American Nanotechnology Preparedness Centre right at the outset while non-governmental organisations like Drexler's Foresight Institute have as their mission educating society on preparing for the nanotech era. (Schummer, 2006, p.425)

A report for the UK government has called for preparatory measures to be put in place before society is overwhelmed by the breath-taking advance of nanotechnology that is catalysing advances in S&T in general (BBC News, 2006).

Whereas the preparedness in the developed countries is essentially aimed at managing the changes brought about by the technology in the rich world environment, 'preparedness' for Africa could mean a totally different issue of survival. A more compelling reason for emphasising the need for Africa's readiness for the Nano-revolution is the devastating effect it may have on

Africa. Africa would not have the luxury of a choice of interested investors that will be willing to exploit its people and resources. Africa may find itself yearning for the good old days of arm-twisting by multi-nationals like the diamond giant DeBeers whose interest in Africa's diamond and cheap labour is set to diminish. African diamond that DeBeers has been marketing for more than a century is certainly to lose the market as purer yet cheaper diamonds 'cultured' in backyard garage workshops take over the market. (Maney, 2005). Africa cannot rely on mining its diamonds or on creating the market for 'organic diamonds'. The impact of the Nano-revolution on Africa is in no way limited to the loss of markets for certain commodities. The renowned neuroscientist Baroness Greenfield explains the dire consequences of falling behind the technological wave in terms familiar in African history. [The Vast Majority outside of the first world — of which Africa is the bottom] are endanger of not only of being disenfranchised from a vastly more comfortable way of life but also of being exploited and abused in ways more sinister, pervasive and cruel than even that witnessed by the worst excesses of the colonial past.(Greenfield, 2003, p.268)

The comforting proviso is that the Baroness has not dismissed the possibility of an alternative scenario whereby the capabilities developed by the new technologies can be deployed to bring an end to the binary world of the haves and have-nots (Id).

Measures to avoid such horrid eventualities need to be radical given the prevailing development practices. Summarising the development thinking and practice of the post-War decades, Juma highlights how technological innovation was kept strictly on the side-lines:

One of the most damaging legacies of [the divergence in developed countries own innovation policy and development cooperation programmes] was the consistent downplaying of technological innovation as a force in economic development. In fact many development agencies exhibited outright hostility to proposals that sought to integrate innovation in development cooperation strategies (Juma, 2010, p. xiv).The World Bank, until the recent swing, consistently failed to promote S&T in its developmentpolicies.9

Critics both without and within the Bank exposed the Bank's insincerity in helping developing countries build S&T infrastructures. A recent UNCTAD report revealed that this

9 Over a 25-year period (1980-2004), only 3.9 % of total World Bank lending has on average gone to strictly defined S&T projects. Crawford et al, 2006, p.33sustained neglect is still unceasing even in the latest projects of the Bank (UNCTAD, 2007, p.13). Current practices at the Bank like strategies embodied in Poverty Reduction Strategy Papers (PRSPs) hardly contain references to S&T (id).

What is more the bank's policies were even inimical to S&T endeavours by developing countries. Far from being encouraged to invest in research activities, their obligations to service their debts were the priority that they were constantly reminded of. Accordingly, in order to perform their debtor obligation, they were compelled to squeeze and slate the meagre and the less politically sensitive budget they allocate to S&T research (Dickson, 2003). The emphasis of development policy at the World Bank has been on the exploitation of natural resources and cheap labour. This was in contrast to the prescription for developed countries whose development is entirely tied to science, technology and innovation (UNCTAD, 2007, p.13). A more pertinent illustration of the importance of S&T for development is the experience of the newly developed countries that owe their lift-off from poverty to S&T (UN Millennium Project, 2005, p.xiv). The example of South Korea and Ireland is cited as particularly instructive and a question is posed as to why governments in least developed countries should not follow the same route (UNCTAD, 2007, p.14).

Despite the self-evident relevance of S&T and the official reiterations that S&T is the defining element of development, the belated awakening that developing countries should raise their research capabilities took place only recently (Juma, 2010, p.xiv) One of the changes in policy direction brought forward by the Wolfowitz presidency of the World Bank is the affirmation that developing countries will have to develop their own S&T base. As can be seen from some activities being undertaken by the bank, this change in policy direction is probably there to stay and unlikely to be as short-lived as the Wolfowitz presidency.

10Developing countries are now being told that investing in S&T is a make-or-break decision if they are to remain relevant to the global economy (Wolfowitz, 2007). More than 50 years of development practice has yielded little for the majority of the globe's population. Neither the invisible hand of the market nor the visible hand of state planning as practiced in communist

10 At the World Bank level, the launch of the 'Science, Technology and Innovation Global Forum' is indicative of a sustained policy change on the subject. See the 'STI Global Forum' website athttp://go.worldbank.org/DWODQ7E3EO. There is also a major rethink among donor countries emphasising science in development assistance. The Economist, 2004states brought acceptable results. The hope now rests on the 'invisible leg' of technology (The Economist, 2006).

While the emphasis on building research infrastructures in developing countries relates to S&T in general, the focus has been mainly on agriculture and only in certain selected countries (UNCTAD, 2007, p.13). References to other emerging technologies let alone nanotechnology are rare and there are even suggestions that developing countries need not engage new technologies like nanotechnology as there are widely available technologies not yet used by developing countries. The current enthusiasm on policy changes in respect of technology and development is of limited relevance to emerging technologies in general and nanotechnology in particular as it insists on making an exception for high-tech in favour of mature technologies apparently because the patents on them have expired.

11The usefulness of these mature technologies for developing countries is undeniable. It is however, limited for at least two major reasons: For one thing if developing countries are ultimately expected to integrate into the competitive global market, their failure is a preordained reality. They cannot rely on outdated technologies that patent-holders would have long made obsolete in expectation of the expiry of their patents. Such obsolescence naturally follows disruptive technologies like nanotechnology. By definition, a disruptive technology is one that 'renders the basic skills associated with the old technology useless and makes its infrastructure obsolete' (Romig et al 2007, p.1637). Moreover, the idea of a 'second rate technology' is implicit in the preference for mature technologies just as was the case with 'intermediate technologies' tried earlier (Edgerton, 2006, p.191).For the poor countries (excluding the newly rich developing countries) acquiring nanotechnology capability is still expensive, but so is maintaining the pre-Nano status quo. A Nigerian government minister spelled out what choice developing countries can afford to have: developing countries will not catch up with developed countries by investing in existing technologies alone. [In order] to compete successfully in global science today, a portion11 Juma (2007) explicitly excludes nanotechnology. See also the reference to cutting-edge technologies in UN Millennium Project, 2005, p.33. The UN Task Force recommends a S&T and innovation policy favouring mature technologies rather than 'cutting-edge technologies' but later the danger of 'path dependence' is raised and it all becomes an issue of directing policy at 'the right level'. (pp.32, 40) of the science and technology budget of every country must focus on cutting-edge science and technologies (Wood et al, 2007, p.15).

Secondly, the need for new technologies arises because there is some drawback of a prior technology. The ecological cost of the majority of the crude pre-Nano technologies is one of the most compelling reasons for an accelerated drive towards adopting nanotechnology. The advice encouraging the use of mature technologies side-lines the issue of sustainability. The core idea of nanotechnology and in particular the biomimicry that it heavily employs as its mode of operation aligns with the prevailing sustainability ethos. The scepticism that nanotechnology is not for developing countries is grounded on the appreciation that the present level of research capacity and funding in these countries is not conducive for nanotechnology initiatives to be effectively implemented. The message is that these countries should utilise their resources in areas where their relative strength lies with the implicit intention to maintain the global division of labour in which developing countries would continue as 'banana-exporting republics'.

It should be kept in mind that countries, rich and poor, are pursuing nanotechnology not for the purpose of tackling some pressing exigencies but mainly for the purpose of avoiding a deficit in the 'competency legacy that might well form the basis for the next Kondratieff or Schumpeterian economic long wave' (Romig, et al, 2007, p.1639). The comparative advantage that developing countries were to cash in as a result of the global division of labour and their natural endowments no longer exists and it is unlikely to revive in the technology-led future of the global economy (Pieterse, 2010, p.49). In the rapidly changing global economy, the idea of focussing on a static comparative advantage is not a saleable advice.

Regarding their research capacity, it is noted that it is not a divine rule that these countries cannot develop this capacity. Opportunities for developing countries to build their research infrastructures for nanotechnology and S&T in general exist in a manner never experienced before. Thus far, the flight of their educated population to the core countries, the so-called 'brain drain', has been a major cause of the stagnation and deterioration of their research infrastructures. There is now a way to look at the 'brain drain' in a positive way. While some countries like India are able to halt and reverse the process, others are finding ways to make the most out of it (Bound, 2007, p.9).The exodus of scientists to countries where their potential can be realised to the full is not an unjustifiable move unless one looks forward to their potential being wasted on sub-optimal performances. Rather than those brains being drained, they were being 'banked' overseas where they have the chance for further enrichment (id, p.8). The 'brain drain' is indeed a positive phenomenon and as Juma likes to recast it, it should be understood as 'global knowledge flow' (Juma, 2007). With the ICT revolution reaching the four corners of the globe the repatriation of knowledge has never been easier. It is characteristic of our time that it offers 'an unprecedented opportunity to develop, disseminate and share the benefits of technical innovation to more users more rapidly.' (UNEP, 2007, p.69) That is one reason for optimism about the possibility of a research infrastructure capable of handling nanotechnology in developing countries.

Africa needs to be ready for the long haul in pursuing nanotechnology. In this sense, readiness for the Nano-revolution involves a double effort on Africa's part: designing and implementing Nano-initiatives to harness the technology for development and, on the other hand, negotiating and securing safety-net mechanisms until such time that it scores a level of technological progress that places it on a par with the advanced regions. This double-pronged action fissures into several lines of action that need to be taken both by African governments individually and at the regional level collectively. It involves seemingly discrete yet closely related measures ranging from readjusting citizenship and immigration laws to reverse or otherwise exploit the 'brain drain' to budgetary and tax laws to source Nano-initiatives. The other less recognised yet vital measure is filling the post of a 'concept champion' in Africa. The post should be held ex officio by the highest echelons of political power. That was the role taken in the US by Bill Clinton referred to as the 'father of nanotechnology' (Jones, 2004) while in office. In Russia, it is held by Vladimir Putin throughout his presidency and premiership. Considering the shadow it will cast over the entire Nano-engagement process, filling the post of the 'concept champion' should be an immediate preoccupation. Ugandan President Yoweri Musevini has inadvertently acknowledged that taking up this responsibility is long overdue when he called for a more proactive science communication in respect of nanotechnology (Wamboga-Mugirya 2008). Likewise, what Ethiopian Prime Minister Meles Zenawi has to say in relation to ICT is true of nanotechnology and alludes to the call for leadership in the S&T effort as a whole: 'Because we are poor, we can't afford not to use ICT.'(quoted in Firth, 2005). While the political will to give due attention to the role of S&T in African development in general is in greater supply, the emphasis on nanotechnology in particular is not at the level required by the exigency the technology is triggering.

The time has come for Africa to aggressively engage with the governance of the phenomenon of nanotechnology. Such engagement is not an absent activity. Yet, the need to spur the momentum cannot be overstated. Africa is waking up to a world of opportunities that could enable it play catch-up which, given Africa's economic and technological predicament, could only mean 'leapfrogging'. A long overdue re-examination of the role of S&T in development is underway and is galvanising major policy changes within the multilateral development agencies and among donor countries. The UN declaration on the MDGs and the efforts to correlate nanotechnology research to poverty alleviation endeavours by the UN are proof of the changing emphasis on the technology-development nexus. Africa, on its part, is laying the ground for action. The Addis Ababa Declaration on Science, Technology and Scientific Research for Development (Assembly/AU/Decl.5 (VIII), January, 2007) is among the most recent expressions of renewed commitment in this direction. Beyond such broad and undifferentiated moves, Africa needs to further demonstrate its readiness to have a slice of the value chain that the nanotech revolution will create by specifically targeting nanotechnology on its own.

## Conclusion

Referring to the biotechnology debate, Yoweri Mussevini said the debate has 'gone on for a very long time while other parts of the world have moved to other technologies like nanotechnology' (Wamboga-Mugirya 2008). The debate is still stuck with what are soon to become technologies of yesteryears. It is not nanotechnology but ICT and biotechnology that are high on the agenda of the political leadership in Africa. Meles Zenawi was not speaking of nanotechnology when he said that '[t]he first mental block that we had to cross was the view that ICT is for the rich' (quoted in Firth, 2005). However, his statement is not without relevance for nanotechnology engagement. While, half a decade later, the 'mental block' about ICT may have been crossed, another mental block that nanotechnology is not for Africa has yet to be tackled. There should be no second thoughts on the need to have Africa on board the nanotech bandwagon. Africa needs to avoid the eventuality where it finds itself on the side of the Nano-divide that is so reminiscent of the digital divide but with far more sinister consequences — potentially more sinister than was witnessed under slavery and colonialism according to Baroness Greenfield (2003).

## References


Avi-Yonah, Reuven, 2004, 'Bridging the North/South Divide: International Redistribution and Tax Competition', 26:1 Michigan Journal of International Law 371

Barry, J., 2007, Environment and Social Theory, 2nd Ed., Routledge

BBC News, 2006, 'Robots could demand legal rights.' at http://news.bbc.co.uk/go/pr/fr/-/1/hi/technology/6200005.stm. (last accessed 13/10/10)

Berne, Rosalyn, 2006, Nano talk: Conversations with Scientists and Engineers about Ethics and Belief in the Development of Nanotechnology, Lawrence Erlbaum Associates, Publishers

Bishop, M., 2005, 'Loose Talk Saves Lives', 31 Development, 01 November 2005, also available at http://www.developments.org.uk/data/issue31/loose-talk.htm (last visited04/02/08)

Bound, Kristen, 2007, India: The Uneven Innovator, Demos, available athttp://www.demos.co.uk/files/India_Final.pdf (last visited 16/11/10)

Cameron, Nigel, 2007, 'Toward Nanoethics?', in Cameron, Nigel and Mitchell, Ellen, 2007, Nanoscale: Issues and Perspectives for the Nano Century, Wiley& Sons Inc.

Channel 4 News, 2008, 'Brown back peddles on global poverty', 06 May 2008, available athttp://www.channel4.com/news/articles/politics/domestic_politics//brown+backpeddles+on+global+poverty/2144947 (last visited 08/05/08)/

Crawford, M., et al, 2006, Review of World Bank Lending for Science and Technology, 1980-2004, Education Department Human Development Network, World Bank, available at http://www.usp.ac.fj/worldbank2009/frame/Documents/Publications_global/Review_WB_lending_ST_80-04.pdf (last accessed 21/0910)

Davies, Clarence, 2008, Nanotechnology Oversight: An Agenda for the New Administration, available at http://207.58.186.238/process/assets/files/6709/pen13.pdf (accessed 20/04/2011)

Dickson, D., 2003, 'Does the World Bank really care about science?', available athttp://www.scidev.net/editorials/index.cfm (last visited 09/02/08)

Durojaiye, Bode, 2011, 'Experts brainstorm on potential application of Nanotechnology in developing countries', The Nation, 24/07/2011, available athttp://www.thenationonlineng.net/2011/index.php?news=13401 (last accessed 16/08/11)

The Economist, 2006, 'Power to the People', 11 March 2006

The Economist, 2004, 'Southern Comfort, Eastern Promises: Third World Biotechnology', 11December 2004, available at http://www.economist.com/node/3471269 (last accessed21/10/10)

Edgerton, D., 2006, The Shock of the Old: Technology and Global History since 1900, Profile Books

Edwards, 2006, p.38 Edwards, Steven, 2006, The Nanotech Pioneers: Where Are They Taking Us, Wiley-VCH

Escobar, A., 1995, Encountering development: The Making and Unmaking of the Third World, Princeton University Press

Feynman, Richard, 1960, 'There is Plenty of Room at the Bottom: An Invitation to Enter a New Field of Physics' available at http://www.zyvex.com/nanotech/feynman.html (last visited21/8/2010)

Firth, Holden, 2005, 'Ethiopia Leaps into the Information Age' Times Online, 12 August,2005, available athttp://technology.timesonline.co.uk/tol/news/tech_and_web/article554550.ece (last accessed12/08/11)

Fogelberg, Hans, and Glimmel, Hans, 2003, Bringing Visibility to the Invisible: Toward a Social Understanding of Nanotechnology, Göteborgs Universitet; also available athttp://www.sts.gu.se/publications/STS_report_6.pdf (last visited 20/01/07)

Gimzewski, Jim and Vesna, Victoria, 2003, 'The Nanoneme Syndrome: Blurring of Fact and Fiction in the Construction of a New Science', 1 (1) Technoetic Arts: A Journal of Speculative Research 1

Gingrich, Newt, 2001, 'The Age of Transitions' in Roco, Mihail and Bainbridge, William(eds), 2001, Societal Implications of Nanoscience and Nanotechnology, National Science Foundation

Greenfield, Susan, 2003, Tomorrow's People: How the 21st-Century Technology is Changingthe Way We Think and Feel, Penguin Books

Hunt, Geoffrey and Mehta, Michael, 'What Makes Nano-technologies Special?' in Hunt, Geoffrey and Mehta, Michael, 2006, Nanotechnology: Risk, Ethics and Law, Earthscan

Invernizzi, Noela, and Guillermo Foladori, 2005 'Nanotechnology and the Developing World: Will Nanotechnology Overcome Poverty or Widen Disparities?' 2:3 Nanotechnology Law &Business 294

Jones, Richard, 2004, Soft Machines: Nanotechnology and Life, Oxford University Press

Juma, Calestous, 2010, 'Foreword' in Conway, Gordon and Waage, Jeff, 2010, Science and Innovation for Development, UK Collaborative on Development Sciences (UKCDS) available at http://www.ukcds.org.uk/_assets/file/book/science_innovation_book_lowres.pdf (last accessed 11/08/11)

Juma, Calestous (ed.), 2005, Going for Growth: Science, Technology and Innovation in Africa, The Smith Institute

Kaku, Michio, 1998, Visions: How Science Will Revolutionise the 21st Century and Beyond, Oxford University Press

Khan, Jennifer, 2006, 'Nano's Big Future', National Geographic, June 2006

Kurzweil, Ray, 2003, 'Testimony of Ray Kurzweil on the Societal Implications of Nanotechnology', available at http://www.kurzweilai.net/testimony-of-ray-kurzweil-on-thesocietal-implications-of-nanotechnology (accessed 18/03 2011)

Lee, Maria, 2010, 'Risk and Beyond: EU Regulation of Nanotechnology', 35:6 European Law Review 799

Loder, Natasha, 2005, 'Small Wonders: A Survey of Nanotechnology', The Economist, 1January 2005

Malsch, Ineke, 2008, 'Nanotechnology in Argentina' at http://www.malsch.demon.nl/ http://www.malsch.demon.nl/Nanotechnology%20in%20Argentina.htm (last accessed26/07/09)

Maney, Kevin, 2005, 'Nanotechnology is everywhere', USA Today at http://www.usatoday.com/money/industries/technology/maney/2005-05-31-nanotech_x.htm(last accessed 12/10/10)

Mashelkar, RA and Borde, Sushil, 2010, 'Value for Money and for Many', Technology Review India, February 2010 available at http://www.technologyreview.in/business/24448/page1/ (last accessed 24/11/10)

McHale, Jean, 2009, 'Nano medicine in the EU: Some legal, ethical regulatory challenges' 16:1Maastricht Journal of European and Comparative Law 65

Miller, Georgia and Scrinis, Gyorgy, 2010, 'Nanotechnology and the Extension and Transformation of Inequity' in Cozzens, Susan and Wetmore, Jameson (eds), Nanotechnology and the Challenges of Equity, Equality, and Development: Yearbook of Nanotechnology in Society, Vol. 2. Springer

Miller, Sonia, 2007, 'Regulating Nanotechnology: A Vicious Circle' in Cameron, Nigel and Ellen Mitchell, 2007, Nanoscale: Issues and Perspectives for the Nano Century, Wiley, HobokenNiosi, Jorge and Reid, Susan, 2007, 'Biotechnology and nanotechnology: Science-based enabling technologies as windows of opportunity for LDCs?' 35:3 World Development 426NNI, 2010, 'What is Nanotechnology?' at http://www.nano.gov/html/facts/whatIsNano.html(last visited 16/06/10)

O'Mathúna, Dónal, 2009, Nanoethics: Big Ethical Issues with Small Technology, Continuum Pieterse, Jan, 2010, Development Theory (2nd ed.), Sage Publications

Romig, A., et al, 2007, 'An Introduction to Nanotechnology Policy: Opportunities and Constraints for Emerging and Established Economies', 74 Technological Forecasting & Social Change 1634

Sachs, Jeffrey, 2006, The End of Poverty: Economic Possibilities for Our Time, The Penguin Press, N.Y.

Salamanca-Buentello, Fabio, et al, 2005, 'Nanotechnology and the Developing World', 2:5PLoS Medicine, available at http://medicine.plosjournals.org/perlserv/?request=getdocument&doi=10.1371/journal.pmed.0020097 (last accessed 26/04/11)

Schummer, Joachim, 2007 'The Impact of Nanotechnologies on Developing Countries' in Allhoff, Fritz, et al (eds), 2007, Nanoethics: Examining the Societal Impact of Nanotechnology, Hoobken, NJ: Wiley Schummer, Joachim, 2006a, 'Societal and Ethical Implications of Nanotechnology: Meanings, Interest Groups, and Social Dynamics,' in Schummer, Joachim and Baird, Davis, Nanotechnology Challenges: Implications for Philosophy, Ethics and Society, World Scientific Publishing, New Jersey 413

South Bulletin, 2006, 'Will the Doha Round Deliver for the MDGs, Editorial,' 130, September,2006

Spohrer, James and Engelbart, Douglas, 2004, 'Converging Technologies for Enhancing Human Performance: Science and Business Perspectives' in Roco, Mihail and Carlo Montemagno (eds), 2004, The Coevolution of Human Potential and Converging Technologies,1013 Annals of the New York Academy of Sciences 50

UNCTAD, 2007, The Least Developed Countries Report 2007, UNCTAD

UN Department of Economic and Social Affairs, 2010, The Millennium Development Goals Report 2010, New York

UNEP, 2007, 'Emerging Challenges: Nanotechnology and the Environment', Geo Year Book,2007

UN Millennium Project, 2005, Innovation: Applying Knowledge in Development, UN Millennium Project Task Force on Science, Technology, and Innovation, Earthscan

Vandemoortele, J., 2005, 'Ambition is golden: meeting the MDGs', 48 (1) Development: Journal of the Society for International Development 5

Wamboga-Mugirya, Peter, 2008, 'Museveni urges science journalists to plug knowledge gap', available at http://www.scidev.net/en/news/museveni-urges-science-journalists-to-plugknowled.html (accessed on 12/08/11)

Wolfowitz, Paul, 2007, Speech delivered at the Science Technology and Innovation Global Forum in Washington D.C. on Feb. 14, 2007 available athttp://go.worldbank.org/IMH19FJG10 (last visited 18/02/10)

Womack, J. et al, 1990, The Machine that Changed the World, Simon & Schuster Wood, Stephen et al (eds), 2007, Nanotechnology: From the Science to the Social, ESRC

Wood, Stephen et al (eds), 2003, The Social and Economic Challenges of Nanotechnology, ESRC

# CHAPTER 11

## The Implications of Problems of Good Governance in FDI Flow and Transfer of Knowledge to Ethiopia

*Semahagn Gashu*

### Introduction

The unrestricted flow of capital and knowledge to developing countries has been the most critical factor for realizing economic development in these countries. South Asian countries such as Malaysia, China and Japan economies have shown remarkable success in creating a knowledge-based economy through attracting large amount of FDI (Foreign Direct Investment) and transfer of knowledge (Chandra and Yokoyama, n.d) .The major factor that contributed for such tremendous flow of capital and knowledge to the region is through the creation of enabling environment that attracts FDI. Though there are different opinions as to what constitutes 'enabling environment' that may be conducive for attracting FDI, the main determinants of FDI include: size of markets, infrastructure, macroeconomic stability, product and labor distortions, incentives, and integration schemes, methods of foreign enterprise participation, attitudes, and overall business environment(Ögütçü, 2002), the institutional framework should ensure a good flow of knowledge between scientific research and technological applications, as well as a good flow of information among researchers and users(Chandra and Yokoyama, n.d).

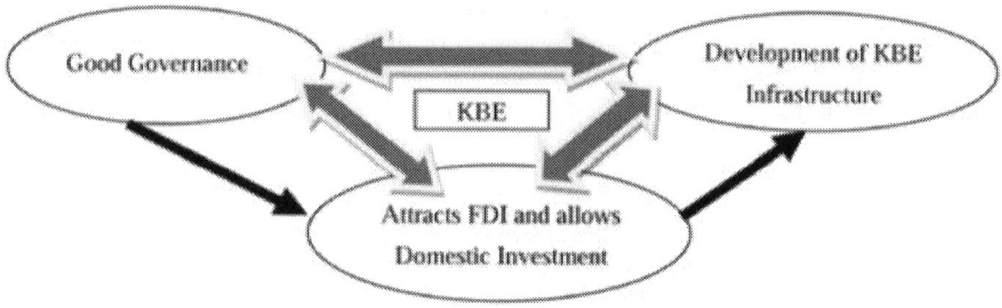


*Frame work on role of good governance in attraction of FDI and development of Knowledge Based Economy (KBE): developed by Chandra and Yokoyama*

A number of recent studies identify the importance of governance as a determinant of inward FDI flows. Strong argument have been put forward that asserts without 'good governance' structures, the poor and the developing nations cannot reduce poverty and contrarily bad governance is being viewed as the main cause behind the ills confronting these societies (Uddin and Joya, 2003).For instance, the United States Agency for International Development (USAID) that widely engaged development activities in developing countries stress that democratic governance is the independent causal factor that improves an economy's performance(Abdellatif, 2003).Speaking on the role of economic development of Africa, Nobel prize winner Wale Soyinka quoted saying good governance is an essential element for development while dictatorships are a crime (ttp://www.scidev.net,2010). But there are also studies that claim there is no clear correlation between good governance and economic development. Some researches reach very mixed conclusions suggesting that the links between the good governance reform agenda and development are weak (UNCTAD, 2009 Report).

Despite such skepticisms, there is now a wider consensus that asserts promoting good public governance by increasing transparency and predictability of laws and regulations and consistency in their enforcement, improving efficiency of procedures and encouraging higher standards of public service contributes not only to a better regulatory environment for business but also to the attractiveness of an investment location (Abdellatif, 2003).

Specially there is evidence that non-democratic countries with a low quality of governance cannot sustain their economic growth in the long run without good governance(Abdellatif, 2003). The correlation between good governance and economic growth has been especially vital in East Asia where good governance played an important role in attracting FDI, which helped them develop knowledge-based economic infrastructures. There are several reasons why the quality of institutions may matter for attracting FDI :by raising productivity prospects, good governance infrastructures may attract foreign while poor institutions can bring additional costs to FDI due to illegal acts such as corruption and FDI is especially vulnerable to any form of uncertainty, including uncertainty stemming from poor government efficiency, policy reversals, graft or weak enforcement of property rights and of the legal system in general(Abdellatif, 2003).

The fundamental challenge of emerging economies is, therefore, ensuring effective governance that provides an enabling environment of competitive education system and highly qualified human resources, excellent information and communication technology infrastructure (ICT) and capable scientific infrastructure for innovations (Chandra and Yokoyama, n.d). This challenge has been especially evident in sub-Saharan countries that are characterized by poor governance and corruption. Despite endowed with resources that may bring about economic miracles in the region, due to poor governance, the region is one of the lowest in terms of attracting FDI and transfer of capital and knowledge. In sub-regional context, Ethiopia's volume of attraction of FDI is even below the average in Sub-Sahara Africa. According to the 2011 World Investment Report released by the UN Conference on Trade and Development UNCTAD, Ethiopia's share of FDI inflows to Africa is just two percent[43].The reports also shows that Ethiopia's share of FDI in its gross fixed capital formation has declined by an annual average of 14.8 percent over the 1995-2004 period to a mere 3.2 percent in 2010 which is below average for Africa that stood at 15.9 percent ( UNCTAD, 2011).

43 The amount of FDI flow to Ethiopia has continued to decline, from an annual average of US$240 million between 1995 and 2004, to US$184 million in 2010," reads part of the report.

The challenge to Ethiopia's attraction of FDI and knowledge transfer is not only failure to share significant flow of international FDI but also its challenge to control brain drain and attract the capital and transfer of knowledge from Ethiopian Diaspora living around the world. Ethiopia is experiencing one of the worst brain drains of any country in the world with the country losing about 75% of its highly skilled professionals over the past ten years (www.onlineuniversities.com, 2011). Furthermore, though hundreds of thousands of Ethiopians are residing in different parts of the world, their contribution in terms of transfer of knowledge and capital into the country is very limited. There might be different factors for such limited FDI flow to the country and failure to tap on Diaspora capital and knowledge. Despite existence several interrelated factors, it is submitted that one of the major factor that seem to have prevented flow of international FDI and the majority of educated and skilled Ethiopians from contributing their part in the economic development of the country is lack of good governance. In this paper, I argue that one of the most important factors that limited significant flow capital and technology into the country is historical institutional and administrative backwardness as well as wrong policies that have been adopted in the last four decades. Such persistent wrong policies have seriously weakened the country's administrative and institutional capacity. The paper consists of the following parts. The first part deals with principles of good governance, followed by sections on development of administrative agencies and polices as well as problems of good governance in the country. Finally some concluding remarks are made in light of improving good governance in Ethiopia.

## Principles of Good Governance

The idea of "good governance" was initially introduced into international policy debates in the late 1980s following the realization that "getting the institutions right" was as important as "getting the prices right" to the success of policy reforms (UNCTAD Report, 2009). In broad terms, governance encompasses laws, regulations and public institutions that determine the extent of economic freedom in a country, the security of private property rights, the costs to the private sector of complying with government regulations and legislation, the competence and efficiency of the civil service in carrying out state activities that, in turn, affect the efficiency of private sector enterprises, the transparency of the legal system and the honesty of government officials (Globerman, et al.2004).But meaning and scope of good governance is contestable and diverse.

Different authors and multinational organizations have defined the concept of good governance in different ways. The European Commission defined good governance as 'the transparent and accountable management of all a country's resources for its equitable and sustainable economic and social development' (UNCTAD Report, 2009).Organization for Economic Cooperation and Development (OECD) on the other hand limits good governance in relation to creating enabling environment for economic development: 'Good governance is established when public institutions act efficiently, providing an enabling environment for economic growth and development. Good governance requires the improvement of accountability and transparency of public sector agencies, concomitant with the effective fight against corruption. The effective performance of democratic institutions, including legislatures, and the fight against corruption, are central elements of good governance' (Uddin and Joya, 2003, p.10)

Some other international agencies such as British Overseas Development Administration (ODA) interpret good governance in a broader perspective:

(1)"The *legitimacy* of government: government, which depends on the existence of participatory processes and the consent of those who are governed ;(2.) The *accountability* of both the political and official elements of government for their actions, depending on the availability of information, freedom of the media, transparency of decision making and the existence of mechanisms to call individuals and institutions to account;(3. ) The *competence* of government to formulate appropriate policies, make timely decisions, implement them effectively and deliver services; 4. Respect for *human rights and rule* of law, to guarantee individual and group rights and security, to provide a framework for economic and social activity and to allow and encourage all individuals to participate" (Uddin and Joya, 2003, p.15).

Good governance essentially has the following basic elements: participation[44]rule of law[45], transparency[46], responsiveness[47], consensus orientation, [48]Equity,[49]effectiveness[50],accountability[51], strategic vision[52] and predictability.[53]

## Overview of Development of Administrative Institutions and Problems of Policy Frame work in Ethiopia

Compared to many of the African states, Ethiopia had embarked on the establishment of modern administrative institutions earlier at the down of $20^{th}$ century. Despite the establishment of the first modern ministries in 1907 was conducted under the shadow of traditional administrative system that made the significance of the new institutions shabby, the measures that were un-

---

44 All men and women should have a voice in decision-making, either directly or through legitimate intermediate institutions that represent their interests. Such broad participation is built on freedom of association and speech, as well as capacities to participate constructively.

45 Legal frameworks should be fair and enforced impartially, particularly the laws on human rights.

46 Transparency is built on the free flow of information. Processes, institutions and information are directly accessible to those concerned with them, and enough information is provided to understand and monitor them.

47 Institutions and processes try to serve all stakeholders.

48 Good governance mediates differing interests to reach a broad consensus on what is in the best interests of the group and, where possible, on policies and procedures.

49 All men and women have opportunities to improve or maintain their well-being.

50 Processes and institutions produce results that meet needs while making the best use of resources.

51 Decision-makers in government, the private sector and civil society organizations are accountable to the public, as well as to institutional stakeholders. This accountability differs depending on the organization and whether the decision is internal or external to an organization.

52 Leaders and the public have a broad and long-term perspective on good governance and human development, along with a sense of what is needed for such development. There is also an understanding of the historical, cultural and social complexities in which that perspective is grounded.

53 Predictability refers to the existence of clear policies and a legal framework for investment, predictable rules and regulations, and their fair and consistent application Predictability also implies that regulations should be minimized and simplified to the greatest degree possible

dertaken by Emperor Menelik and subsequently by Emperor Haileseassie had been points of departure to the archaic style of feudal administration. The introduction of ministerial offices had also contributed a lot in terms of ensuring continuity of government administration (Bahru, 2002). Emperor Haileselassie took the modernization process of governance one step further when he declared the first constitution in 1931. Though the constitution largely used to legitimize the traditional power of the emperor, the constitution was instrumental in terms of introducing institutions such as parliament and civil rights of citizens. The role of Italian occupation between 1936-41 should not be also over looked in terms of its contribution in introducing modern construction, architecture, business and administration.

Deeper measures of modernization began to be undertaken immediately after the restoration of independence in 1941. In 1943 two imperial orders set up eleven ministries and the office of the Prime Minister (Bahru, 2002). In somewhat slow process of decentralizing power, the Emperor assigned the prerogative to select ministries to the prime minister in 1966. In related development, tax reform, commercial agriculture, industry and financial institutions were flourishing throughout the decades following liberation. The Emperor was in particular instrumental in term of expanding education from primary to territory education. The 1955 revised constitution was also enacted in an effort to catch up with international and national socio-political developments. Though such legal reforms could not change the power equation in the country, the trend was an indication that the country was transforming into capitalist system albeit slowly and with stiff resistance from the regional nobility.

The partial, slow and shallow modernization of the Ethiopian socio economic system satisfied neither the nobility nor the new intelligentsia that was created through the expansion modern education. Beginning from the end of 1960's university students began openly challenging the legitimacy of the monarchy and the repressive land tenure system. Unfortunately, the students' movement was dominated by the leftist Marxist ideology. Duet to such radical move, more than any African state, the implication of Marxist ideology in terms of shaping the country's political discourse had been devastating. Not only the impact of Marxist ideology had been instrumental in terms of overthrowing the historical imperial rule in the country but also it has dominated many of the social, economic and political policies of the country in the last four decades. The far reaching consequence of this ideology to the country's political discourse is more evident in the land tenure system and the 'self-determination right of nationalities' that gripped the country's political land escape leading the country into unknown and dangerous trenches.

Furthermore; the most destructive political practice that was introduced into the Ethiopian political culture since the revolution is the obsession with destroying everything of the past. Though traditional feudal Ethiopia was hierarchical society where by the societal gap between the nobility and the ordinary or the rich and poor had been sharp making life for the majority unbearable, it had also its own positive values. Probably the most important traditional value that could have contributed for the transformation of the country's economic development, if maintained and modernized, was the value attached to private property in traditional Ethiopia. Despite the laws were applicable largely to the nobility and rich individuals, there had been relatively strong protection for private property during the Imperial period compared to the devastation of the rights during the regimes that replaced the feudal system. As it has been evidenced by the economic transformation of societies in Western Europe, the most important factor that contributed for the development of capitalist economic structure is effective regime for protection of private property. There has never been any economic transformation made without effective protection of private property.

The various measures undertaken by the *Dreg* during its early stage had been instrumental in terms of culminating the very life line of the nobility and the emerging middle class. In March 1975, *Derg* nationalized all rural and urban land and extra houses, practically ending every aspect of the right to property. Though the political significance of these measures was appealing to the masses and revolutionary student groups, its impact in terms of transforming the country's socio-economic structure was little. Though the peasants theoretically became holders of the land following the land reform, they became tenants of the government effectively arresting any initiative and investment over the land. The *Derg* legacy over control of land has not been changed after the second revolution in 1991. This has made commercial agriculture or private farm investments almost nonexistent in the country. Though the country is endowed with fertile land and big rivers, the country could not be self-sufficient in food production let alone producing surplus products for export. As far as the land policy could not be amended in a way that gives security to peasants to invest on the land with their full potential and domestic commercial farms are encouraged, the country's real economic transformation and transfer of knowledge into the agricultural sector may not be realized.

Though end of the Cold War has brought about economic and political liberalization in many parts of the world, its impact to the Ethiopian political and economic discourse is little. Though *Derg* enacted Mixed Economic Policy at the end of its downfall, it was too little and too late to mobilize capital and transfer of knowledge into the country's economy. Soon after, ending 17 years of civil war, an insurgency group called Ethiopian Peoples' Revolutionary Democratic Front (EPRDF) controlled state power and the wheels of the second revolution started to roll down. Immediately after control of power, a Transitional Charter that recognized multi-party system, freedom of the press, self-determination right of nationalities and civil and political rights of citizens was enacted. Furthermore, free market economic policy was declared which was symbolic in terms of partially nullifying the nearly two decades of socialist economic system. Following the suit, many domestic investors and some sections of the Diaspora began to participate in the economic sector. The partial commitment undertaken by the regime towards free market economy coupled with political and security dilemma could not, however, mobilize capital and transfer of knowledge that renovates the impoverished economic base of the country.

The partiality of the measures related to ensuring economic freedom is apparent in different ways. The fact that EPRDF in general and its dominant wing, Tigrean People's Liberation Front ( TPPLF) in particular, is basically a Marxist party, the liberal economic measures seem to have been largely undertaken in order to adjust to the Post Cold war international scenario. This can easily be attested from the political and economic policy of the regime. The most important *Derg* economic policy that has been maintained by the new regime is the land policy. The control of land in the name of the people has rendered the most abundant resource in the country insignificant, though with properly designed policy, it could be source of capital in a country that lacks capital for investment and infrastructure. In addition to this, the policy of the regime in relation to the liberalization of important economic sectors in telecommunication, financial and electrical power sectors is indicative of the restrictive liberalization of the Ethiopian economic policy frame work.

The political system and the problems in relation to building democratic system have also their own implications in terms of transforming the Ethiopian economy. Since the enactment of ethnic federal system in 1995, the quest for economic prosperity has encountered more complications due to such structural and political developments. In terms of structural problem, the administrative power of the government has been divided between the Centre and the periphery. Due to the fact that country had not effective administration in the past, creating nine additional administrative regions is practically multiplying inefficiency. In addition to this, due to the political reality that has been created since 1991, local administrations controlled by local elites who abused their powers by claiming investment or provision of government positions are available only to the 'owners' of the region excluding 'new comers'. Though this approach has been significantly improved after the 'reformation' of EPRDF in 2001, the confusion and impact it has left has contributed a lot in terms of limiting flow of investment to the regions. Thus, though the political significance of ethnic federal arrangement is obvious, the system has created sense of division rather than unifying the people and mobilizing the country's resources to change the abject poverty and misery in the country. Specially, 'ethnicenterpreauners' of different ethnic groups have been using the opportunity to maximize their profits through instigating ethnic hatred and division.

In addition to this, there has never been any national ideology in the country that may serve as a foundation for any political, economic and social policies in the country. Though EPRDF, presumably due to the end of cold war, appear to have embraced values of liberal democracy and freedom of the market, there have always been mixed policies and actions. For instance, though the Ethiopian constitution contains one of the most generous political and civil rights of international standard, it has also incorporated the concept of 'rights of nations and nationalists' and fussy separation of power principles that have nothing to do with liberal constitutional values. In addition to this, despite the incorporation of liberal values in the constitution, the regime publicly claims that Ethiopia does not need liberal democracy but rather 'revolutionary democracy'. Though EPRDF has not defined the concept in any of its policy documents, the various measures taken by the party suggest that it is mobilization of the masses at the grassroots for economic, social and political transformation under the leadership of the vanguard party.

In the aftermath of general elections in 2005, revolutionary democracy has got a new Developmental State dimension. In the last few years rhetoric of the government is bringing economic miracles with the active involvement of the state in all economic activities. Contrary to the political philosophy of the regime, opposition parties and the majority of the country's middle class seem to prefer a liberal democratic political and economic system adapted to the country's specific situations. The absence of any consensus on the country's future destiny among the various political groupings and the elites will continue to pose challenge to come up with a road map that may mobilize resources for attracting capital and knowledge that may transform lives of ordinary Ethiopians.

The existence of problems in relation to political and economic structure of the country does not mean that there have never been efforts made to improve the socio-economic structure of the country. Primarily, the government has been implementing agriculture led economic policy that aimed at boosting agricultural output from the small peasant farms though providing primary agricultural extension assistance to the farmers. In addition to this, a number of policy measures have been taken to encourage investment through incentives and tax exemptions. In order to increase the efficiency of the administrative system, civil service reform and legal reform programs have been undertaken. An attempt has also been made to improve the tax system of the country and establishing Ethics and Anti-Corruption Commission. Such efforts have also brought limited and sporadic results. In the last few years, investors from countries such India, China and Turkey have been engaged in investment activities and few members from the Diaspora have also been using the opportunity to invest in the country. In addition to this, there have been increased efforts to expand the construction of roads, schools and health facilities. But compared to the potentials of the country and the size of capital and knowledge needed to transform the country's poor economic structure, such fragmented efforts are not satisfactory.

## Challenges to Ensure Good Governance in Ethiopia

In light of ensuring good governance, initial steps that were taken during the inception of the new regime 1991 have been significant and encouraging in terms of democratizing the Ethiopian polity. The participation of opposition parties in the political discourse, the introduction of independent media, decentralization and adoption of the federal and parliamentary system were initial important steps that are unprecedented in Ethiopia's long history. Other democratic institutions such as the judiciary, electoral commission, human rights commission, ombudsman and other democratic institution have also been established during the last two decades. Despite there are critics who argue concepts like 'good governance' and 'economic liberalism' were taken into the EPRDF vocabulary in 1991 to please the international diplomacy and the donor community (Aalen, 2002), the introduction of these institutions is indispensable in realizing transition to democracy in the country.

The constitution further provides for the protection of different democratic rights such as the right to hold opinion, thoughts and free expressions, freedom of assembly, public demonstration and the right to petition, right to association, freedom of movement, and rights of citizenship (Articles 29-33, FDRE constitution). The constitution further provides for the structure and separation of the three branches of state. Federal law making power is granted to the bicameral parliament of the House of Peoples Representatives and the House of Federations, executive power is granted to the Prime Minister and the Council of Ministers and establishment of an independent judiciary (Articles 53, 72, 78 FDRE constitutions).In addition to this periodic general elections have been taken place four times (1995, 2000, 2005 and 2010). Thus, from formal perspectives, it seems that Ethiopia is an emerging democracy and departing from its authoritarian past. Impressed by such formal rhetoric, many international organizations such as Freedom House and most western states ,until recently, refer the country as an 'emerging democracy'.

On the other hand, independent researches have confirmed the institutional development in the country is nominal. According to the 2010 Ibrahim Index of African Governance Ethiopia ranks at 35th out of the 53 African countries. Ethiopia's economic freedom score is 50.5, making its economy the 144th freest out of 183 countries in the 2011 Index (2011 Index of Economic freedom). The report further noted 'Ethiopia underperforms in many of the 10 economic freedoms. The business and investment regime is burdensome and opaque. The quality and efficiency of government services are poor and made worse by the weak rule of law and pervasive corruption. State distortions in prices and interest rates undermine monetary stability' (2011 Index of Economic freedom). In related report, Ethiopia ranks 120th out of 180 countries in Transparency International's 2009 corruption index (Transparency International's 2009 corruption index).As recent Human Rights Watch report reveals 'democracy's technical framework will remain a deceptive and hollow façade so long as Ethiopia's institutions lack independence from the ruling party and there is no accountability for abuses by state officials (Human Rights Watch, 2010). There have also been widespread violations of human rights and increasingly silencing of political dissent in the country. In addition to this, the vital political decisions are made in the informal sphere, behind the façade, in circles and networks of a neo-patrimonialist nature, impervious to what institutions like a parliament or a high court say (Abbink, 2006).

There are different factors for absence of good governance in the country. Some of the factors are structural while others are related to ideologically motivated government structures and policies that encourage patronage rather than merited principles. One of the major problems that seem to have prevented realization of good governance in the country is the absence of democratic culture in the country's long history. The country had undergone considerable part of its history under traditional feudal rule that was characterized by absolute loyalty that legitimizes the exploitation of the poor.[54] Under the feudal system, Ethiopian social fabric was characterized by gross inequality between the largely aristocratic elite—consisting of landowners, lords, nobles, the royal family, government officials, and elements of the clergy—and the impoverished peasantry.[55] There was pattern of social interaction that sustains a strictly hierarchical stratification of society, where one is constrained by a large, invisible, but rigid system of collective sanctions, to obey the 'orders from above'(Yilmaz and Venugopal, 2008;Vaughan and Tronvoll, 2003). The patriarchal order was and still is visible in every aspect of Ethiopian life, including food and dining (Solomon Addis, n.d).

In such feudal system, there was no any democratic culture and system of accountability. As Alemseged noted 'Ethiopia is still a paternalistic society where its institutions from the smallest, the family, to the largest unit, the state, are pervaded by authoritarian values of obedience. Orders, however offensive, are implemented. Loyalty is held in higher esteem. Such proclivity to complete subservience is the result of people's powerlessness'

54 'Traditional adages such as *kesew merto leshumet, keinchet merto letabot* (only a selected few are worthy of officialdom, so is the finest wood that is chosen for tabot [the replica of Mosses' Tablet that is central for Ethiopian Christianity]) This, and saying such as ketat tat yibelital (all fingers are not equal), which means all men are not born equal, reflects the entrenched belief in uncontested authority' (Solomon Addis ,www.forumonpublicpolicy.com/summer09/archivesummer09/getahun.pdf)

55 The social order was to be respected in every aspect of life .Herbert Vivian, who was in Ethiopia at the turn of the 20thcentury, remarked,"...what is specially to be observed among the Abyssinians is the way in which they cling to feudal traditions...The older men eat first, then the younger men, then the women, then the men-servant, and finally the female servants." (Solomon Addiswww.forumonpublicpolicy.com/summer09/archivesummer09/getahun.pdf)

( Alemseged,2009) .Furthermore, as an old saying in Ethiopia goes 'He who does not „eat" while in power, will regret it when he is out' indicating that appointment into the public service is not perceived as serving the public rather it to the advantage of the appointee. Such attitude is still evident at every level of authority. Even though the country has undertaken two revolutions in the last four decades, firmly entrenched feudal power relations and repressions are still well and alive rendering good governance efforts futile.

The other major challenge for realization of good governance is abject poverty and high rate of illiteracy. According to recent Multi-Dimensional Poverty Index, 90% of Ethiopians are poor (UNDP Human Development Reports, 2010). Due to high rate of poverty and illiteracy Ethiopian social structure is characterized by lack of a strong and economically rooted middle class that may burden the transition to democracy and good governance. As noted by Tekola (2011):

> When we consider the social and economic situation in Ethiopia, we find absolutely dismal social conditions, and a starvation-economy. There is no way a viable liberal democracy or Marxist-Leninist systems of governments would work under such social and economic conditions in Ethiopia. Ethiopia is far too barren to grow seedlings of any foreign democratic or Marxist-Leninist government system. I think the best solution is to work with the system that had been with us for centuries and creatively improve and adopt that system to meet modern demands.

In the absence of the basic needs among the public, the quest for good governance is luxury and poverty coupled with high rate of illiteracy will continue to taunt the democratization process of the country.[56]

The other major challenge to realize good governance in Ethiopia is the introduction of leftist Marxist ideology into the country's political discourse

56 'The idea of establishing American or European type democracy is as difficult or is as far removed as establishing Marxism-Leninism in our Ethiopian setting. To state the obvious, either system requires certain degrees of economic, educational, and technological advanced base. Either system requires a well established literate culture. A high degree of social cohesion pulling toward the same goals would also make such social programs viable and possible. The infrastructure of both human networking in associations and public activities and the physical material infrastructure of roads, railway systems, air transport systems, in inland navigation water ways et cetera are all vital for such advanced political development'( Tecola, 2011)

in 1960's. As a matter of fact, the ideology is still firmly maintained by the new rulers of the country. The leftist ideology is strictly attached to undemocratic practices such as extreme loyalty to party discipline, restricting rights of citizens and economic freedoms as well as party apparatuses are more powerful than formal governmental institutions. In such controlled system of government, the decision making process is secretive and there is no any effective mechanism to hold government officials accountable.

The other barriers to ensure good governance in the country are related to the political system and institutions of the country. Primarily, elections are one of the mechanisms used by the electorate to hold their leaders accountable for their actions. Without having a political system that ensures citizens hold government authorities accountable through free and fair elections, realizing good governance remains to be a myth. Since the fall of the *Derg* and the dissolution of the single party political system, Ethiopia has undertaken four general elections. But all the elections that have so far been conducted are marred by violence, vote rigging and failed to reflect the genuine will of the people (Tronvell, 2009). Such electoral frauds have increasingly undermined the ability of citizens to change their leaders through the ballot box and make them legally accountable.

The other basic foundation for ensuring a system of good governance is existence of effective separation of power and checks and balances between the organs of state. When there is effective system of checks and balances in the system, authorities shall be accountable and their activities shall be transparent. In light of such core principle, the absence of effective check and balances is the other challenge to improve good governance in Ethiopia. Though Ethiopian constitution vests the highest authority with the House of People's Representative, since the elections held so far have not been genuine, the parliament is largely dominated by a single political party. In addition to this, since there is strict democratic centralism with in EPRF, members of parliament are always expected to act in accordance with party lines rather than being accountable to the electorate or to their conscience. Such extreme dominance has left the executive organ without any accountability. Not only is the parliament unable to control the executive, it has also become rubberstamp of the actions of the executive organ.

The highest executive powers of the federal government are vested in the prime minister and in the council of ministers (Art.72, FDRE constitution). The constitution provides different powers to the executive, particularly to the prime minister. The prime minister has quite extensive powers, akin to those of presidents in presidential systems. This has made the power of the prime minister uncontrolled and unchecked by the executive, legislative judiciary and other federal or regional institutions. In addition to this, the most important aspect of the executive is that major decisions are not necessarily made by formal state institutions. Largely, important decisions are made in the inner party circle that mainly consists of few party leaders rather than formal state institutions. This has made the state-party separation vague in every channel of government especially at lower levels of government. The party network may even pass decision on the rights and duties of the government employees that make the government networks obsolete. [57] Due to high degree of patronage relations in the party network, there is no serious accountability system against the major political actors and they are not transparent in their dealings. What may express the degree of secrecy of the system is the prime minister's response to parliament to the question raised on how many Ethiopian troops were killed during the intervention in Somalia. He responded that 'since EPRDF has no tradition of publicly declaring such loss of life, it is not necessary to state the number.'

In addition to the absence of system of accountability and transparency in executive organs, they remain to be ineffective and the civil service is not free from political influence (Getachew and Common, 2006). All the regimes manipulated the civil service to advance their own political agenda rather than committing the civil service to professional integrity and quality of service. The trend has continued up to date. Nowadays, it has become an open secret that civil servants are recruited in light of their political loyalty

57 For instance, in January 2010 EPRDF members in Ethics and Antic-Corruption Commission have decided that the deputy Commissioner return back the extra government residential house and mobiles to the government.( Reporter, Wednesday 19, 2010)

rather than their competence (International Crisis Group, 2009).Filling civil service vacancies with party members and supporters is in contradiction to the civil service law that provides that the only criterion to fill vacancy is the qualification of the person (Proclamation No 515/ 2007). In the last few years, the trend of compromising quality for loyalty has been in the increase in the various government departments especially at lower levels, leading the degeneration of the civil service.[58].

The other important institution essential to ensure rule of law is existence of independent and strong judicial organ. In light of this, the FDRE constitution provides for the independence of the judiciary (Art.78, FDRE Constitution).Despite *de jure* independence of the judiciary, there are problems of political interference, ineffiency and corruption. Former judges and members of the legal community indicated that there have been instances of delicate political cases being assigned to perceived pro-government benches and political interference in civil and criminal trials. Since 2005, judges have fled the country alleging government interference, arrested, threatened, intimidated, pressured to resign, transferred to remote locations or somehow removed from active judicial duties (SIDA, 2008 ). In addition to this, the limited budget allocated to the judiciary and lower incentives to the judges has also resulted in high turnout of judges that further weakened the judiciary.

The mass media are often referred to as the fourth branch of government because of the power they held in terms of oversight function they exercise[59]. Access to information is essential at least for two reasons: First, it ensures that citizens make responsible, informed choices rather than acting out of ignorance or misinformation and secondly, information serves a "checking function" by ensuring that elected representatives uphold their oaths of of-

---

58 civil servants at the federal and regional level reveals that the most important criteria to join the civil service is loyalty to the ideology of the ruling party where as qualification is of secondary issue (Reporter Sunday, 26 September 2010)

59 Thomas Jefferson, for all his bitterness against journalistic criticism celebrated the press "Were it left to me to decide whether we should have a government without newspapers or newspapers without government, I should not hesitate to prefer the latter."

fice and carry out the wishes of those who elected them (USAID,1992). In the Ethiopian context, when EPRDF came to power in 1991, it adopted a national charter that recognized freedom of speech and a law was issued on freedom of the press addressing under special law designed to regulate the press. Soon after the issuance of the law, different newspapers, magazines and books hit the market. The trend was very welcoming since the action taken to open up the sector to independent media was a historical moment for this was unthinkable during the past regimes.

But the distribution of the private press was very limited to Addis Ababa and other major cities. Radio remains the principal news medium in predominantly rural Ethiopia. But private electronics media is seriously restricted. Only few FM private radios are permitted to operate in the capital. Many of those established are allegedly pro-government or shy away from issues they feel unpleasant to the government. Licenses have never been granted to short wave radio and television media that could reach the general public in rural areas. The government renders different justifications for the delays, though the main reason behind is the regime's lack of confidence in meeting the demands of well-informed citizens. All government printed and electronics media have continued to be tightly controlled by the regimes being used as instruments of propaganda. The lack of professionalism, finance and abuse of the publishers of the newly emerged media has been also a challenge to development of full-fledged media in the country. Since there was no experience in the past, the personnel that engaged in the private press lacked the appropriate professionalism related to balancing news items or cross checking the source of the news. In addition to this, frequent arrests and trials of editors, reporters and publishers, and the subsequent closure of many publications including journalists fledging the country has seriously weakened the independent media.

## Conclusion

To bring about significant economic transformation, unrestricted flow of capital and knowledge is crucial. To achieve this goal, ensuring good governance is an essential precondition. A transparent, accountable, responsive governance where rule of law reigns is an important ingredient of achieving economic development. Due to the absence of good governance in Ethiopia, transfer of capital and knowledge is limited. The major bottlenecks that have hindered good governance from taking roots in the country are manifold: repressive political culture, abject poverty and illiteracy, absence of democratic institutions, wrong political and economic policies shaped by leftist ideological tendencies and absence of strong independent media. In order to mobilize international FDI and facilitating transfer of knowledge from abroad, including the Diaspora, an electoral system that is acceptable to the public and the political forces is the primary step to words building confidence and ensuring legitimacy of the system. Furthermore, since many of the existing political and economic policies are shaped by failed leftist ideology, doing away with traces of such polices from the political and economic discourse is an important step forward to realize good governance in the country.

In addition to this, there has to be efficient checks and balances between the legislature, executive and judicial organs of government. As far as the relations between these organs is remains blurred, there exists an executive organ that has unlimited power that may lead to corruption. In relation to the executive organs, there has to be also clear separation between party and state functions which has made accountability increasingly complicated. An independent judicial organ that reviews the acts of administrative agencies is also essential for realizing rule of law in the country and an independent media instrumental in controlling corruption and facilitating transparency.

## References

Aalen, Lovise, 2002. Ethnic Federalism in Dominant Party State: The Ethiopian Experience 1991-2000. Bergen: Chr. Michlsen Institute

Abbink, Jon, 2006 .Discomfiture of Democracy? The 2005 Election Crisis in Ethiopia and its Aftermath. Journal of African Affairs, 105,173-179

Abdellatif , Adel, Good Governance and Its Relationship to Democracy & Economic Development, Global Forum III on Fighting Corruption and Safeguarding Integrity, Seoul 20-31 May 2003

Alemseged Abbay, 2009, Diversity and democracy in Ethiopia, Journal of Eastern African Studies, 3:2, 175 — 201

Brain gain and good governance key to development, posted at http://www.scidev.net/en/opinions/brain-gain-and-good-governance-key-to-development.html

Bahru Zewede, 2002. A History of Modern Ethiopia 1855-1991.Ohaio University Press

(CIDA)Canadian International Development Agency, 2008, Independence, Transparency and Accountability in the Judiciary of Ethiopia at http://www.abbaymedia.com/pdf/nij_ethiopian_judiciary_assessment.pdf

Chandra, S Debnath and Yokoyama, Kenji, The Role of Good Governance in Knowledge Based Economic Growth of East Asia- A study on Japan, Newely Industrialized Economies, Malaysia and China. Available at http://www.ritsumei.ac.jp/acd/re/ssrc/social/kiyou/kiyou22/22-02.pdf

10 Countries Facing the Biggest Brain Drain at http://www.onlineuniversities.com/blog/2011/07/10-countries-facing-the-biggest-brain-drain/

Ethiopian Reporter Sunday, 26 September 2010 at http://www.ethiopianreporter.com/

Getachew Hailemariam and Common, Richard, 2006. Civil Service Reform in Ethiopia: Success in two ministries at http://www2.hull.ac.uk/hubs/pdf/memorandum59.pdf

(HRW)Human Rights Watch, 2010.'One Hundred Ways of Putting Pressure ', Violations of Freedom of Expression and Association in Ethiopia at http://www.hrw.org/node/89128http://nazret.com/blog/index.php/2011/07/27/ethiopia-s-foreign-direct-investment-declines

(ICC) International Crisis Group, 2009. Ethiopia: Ethnic Federalism and Its Discontents, Africa Report N°153 – 4

Ögütçü1, Mehmet, 2005.Good Governance and Best Practices for Investment Policy and Promotion, UNCTAD workshop on Efficient and Transparent Investment Promotion Practices: The Case of LDCs, Geneva, 6-7 June 2002

Quéré, A Bénassy .Coupet, M and M, Thierry. 2007. Institutional Determinants of Foreign Direct Investment, The World Economy, doi: 10.1111/j.1467-9701.2007.01022.x

Solomon Addis Getahun, Ethiopia in the New Millennium: Issues of Democratic Governance at www.forumonpublicpolicy.com/summer09/archivesummer09/getahun.pdf

The constitution of the Federal Democratic Republic of Ethiopia, Federal Negarit Gazeta,1st year No. 1, 21st August 1995.

Tecola W. Hagos, This way Ethiopia: Constitutional Monarchy or Liberal Democracy? http://www.tecolahagos.com/Ethiopia_constitution.htm

Transparency International's 2009 corruption index http://www.telegraph.co.uk/news/newstopics/mps-expenses/6589735/Transparency-Internationals-2009-corruption-index-the-full-ranking-of-180-countries.html

Tronvoll, Kjetil, 2011. The Ethiopian 2010 Federal and Regional Elections: Re-establishing the One-Party State. African Affairs, 110(438), s 121-136

Uddin, Mohammed and Joya .2007.Development through Good Governance: Lessons for Developing Countries, Asian Affairs, Vol. 29, No. 3:1-28

UNCTAD (United Nations Conference on Trade and Development), The Least Developed Countries Report 2009, The State and Development Governance, UNCTAD/LDC/2009 at http://www.unctad.org/en/docs/ldc2009_en.pdf

UNCTAD (United Nations Conference on Trade and Development), Trade and Development Board, Commission on Investment, Technology and Related Financial Issues Expert Meeting on Good Governance in Investment Promotion, Geneva, 1–3 November 2004

UNDP (United Nations Development Program), Governance for Sustainable Human Development, Policy Document, at http://mirror.undp.org/magnet/policy/chapter1.htm

UNDP (United Nations Development Program), Human Development Reports http://hdr.undp.org/en/reports/global/hdr2010/news/title,20523,en.html

(USAID) United States Agency for International Development, 2009. The Role of Media in Democracy: A strategy Approach at http://www.au.af.mil/au/awc/awcgate/usaid/media_in_democracy.pdf

Vaughan, Sarah and Tronvoll, Kjetil, 2003. The Culture of Power in Contemporary Ethiopian Political Life, Stockholm

Yilmaz, Serdar and Venugopal, Varsha, Local Government Discretion and Accountability in Ethiopia, The World Bank "Obstacles to Decentralization: Lessons from Selected Countries" Conference Andrew Young School of Policy Studies, International Studies Program, Georgia State University September 21-23, 2008

# CHAPTER 12

## When Lightweights Beat Heavyweights

*Tadesse Biru Kersmo*

### Introduction

The newly industrialized countries of Asia have been growing at rates never encountered in history. We have observed, repeatedly, when new competitors beat well-known, deeply rooted and time-honored giant firms or corporations. Many of the most successful entrepreneurs were poor or financially weak at the beginning. How could such things occur? How could they win the battle against their powerful, rich and well-experienced rivals? Generally, under what conditions could small enterprises beat giant corporations?

Many people conceive market competitions in terms of a two-person zero-sum game like boxing in which the winner takes all and the heavyweights always have the highest probability of knocking out their competitors. On the contrary, market competitions are games played by more than two persons or groups (clubs); and the number of players makes a lot of difference to the outcome of the game. As the number of players increases, the probability that the heavyweights will win the contest diminishes and that of the lightweights increases. Moreover, market competitions are not necessarily zero-sum games in which there is no possibility of cooperation between the players themselves. Rather, market competitions are positive-sum games in which both competition and cooperation go together and even the loser gets a stake. The following model game substantiates the above argument that not always the giants win.

## The Model

Assume that three antagonists, Mr. X, Y, and Z are engaged in a three-way duel. Assume further that Mr. X is a poor shot with only 30 per cent chance of hitting his target while Mr. Y is a much better shot, achieving 80 per cent accuracy; Mr. Z is a perfect shot - which never misses. Each person knows his own and the rival's strength and reputation. Who has the highest chance of survival? Is there any possible strategy for Mr. X, the weakest of all, to win this death-game?

Professors A. Dixit and B. Nalebuff in their popular book "Thinking Strategically: The competitive Edge in Business, Politics, and Everyday life" (P. 329-331) state the rule of the game and the likely outcomes as follows:

## The Rule

*There are two rounds. In the first round, each player is given one shot: first X, then Y, and then Z. After the first round any survivors are given a second shot, again beginning with X, then Y, and then Z.*

## The Likely Outcomes

To start the game, Mr. X is given a pistol with a bullet. At whom should he aim? If he shoots at Y and hits, the next turn is Z's and Mr. Z certainly will kill X. This is the worst strategy for Mr. X since his chance of survival is 0%. If Mr. X shoots at Mr. Z and hits, then it's Y's turn. Mr. Y will shoot at Mr. X and the probability that Y will win the game in the two rounds is 0.912 (91.2%). Obviously this is not an attractive option either. If X misses his first shot, no matter at whom he aimed, then it's Y's turn. Mr. Y will certainly aim at Mr. Z because eliminating Z increases his chance of survival. If Y kills Z, then in the second round Mr. X will have at least a 30 per cent chance of survival.

The third outcome is better than the other two. Why then does not Mr. X fire up in the air in the first round? By doing so *he will force the giants to kill each other!* If Mr. X is wise enough to choose this strategy his chance of survival will be 41.2 per cent, that of Y - 56 per cent, and that of Z, the "super giant", only 14 per cent.

This game has many important theoretical and practical applications. Thus, this paper attempts to further the analysis of the game.

This game model can be improved by raising the following six questions:

For three entities there are six ways of ordered arrangements; (X,Y,Z), (Y,X,Z), (Y,Z,X), (Z,X,Y), (Z,Y,X) and (X,Z,Y). Professors A. Dixit and B. Nalebuff discussed only the first, (X,Y,Z), order. What are the possible outcomes in the other five orders?

If outcomes differ in different orders, then the choice among orders cannot be arbitrary. Who, how, and on what bases should an order be chosen?

Is there any possibility for side-payments, exchange, or cooperation between any two players to eliminate the third?

Can the rule of the game be amended?

In what ways can the outcomes be affected if the game is played for more than two rounds?

How could this model be applied if more than three players play the game?

This paper tries to solve five of the six above-stated problems; moreover, it also tries to give some guidance for further investigations for solving the sixth problem in addition.

## Orders

The possible outcomes are dependent on the order that determines who should shoot first, second, and last. Table 1 summarizes the likely outcomes in terms of probability of survival in each order when all players move strategically.[60]

60 In order to make the discussion clear six charts (one for each order) are attached in the Annex..

| Order | Probability of Survival (%) | | | Remarks |
|---|---|---|---|---|
| | X | Y | Z | |
| X,Y,Z | 41.2 | 56.0 | 14.0 | (X) ∩ (Y)=11.2 |
| Y,X,Z | 41.2 | 52.64 | 14.0 | (X) ∩ (Y)=7.84 |
| Y,Z,X | 41.2 | 52.64 | 14.0 | (X) ∩ (Y)=7.84 |
| Z,X,Y | 30.0 | 0.0 | 70.0 | |
| Z,Y,X | 30.0 | 0.0 | 70.0 | |
| X,Z,Y | 30.0 | 0.0 | 70.0 | |

**Table 1: All Possible Outcomes for Rational Players**

Close observation of table 1 can help us substantiate the following statements:

According to the likely outcomes, we can classify the six orders into three groups: group A- (X,Y,Z), group B-(Y,X,Z) and (Y,Z,X), and group C- (Z,X-,Y), (Z,Y,X), and (X,Z,Y).

The position of X (that is, should his turn be before or after Y, or before or after Z) *does not* determine his own probability of survival. Consequently, he is indifferent to his own position. But, and this is an important point, he is not indifferent to the positions held by the other two; *he should always prefer those orders in which Y moves before Z, no matter what his own position is.*

The position of Y in relation to Z (or that of Z in relation to Y) is what determines the likely outcomes of the game for *all* participants. Specifically, Z will be better off if he moves before Y but *both X and Y* will be worse-off. On the other hand, if Y moves before Z, *both X and Y* will be better off.

From the above discussion, it should be clear that it is in *X and Y's common interest* that one of the orders in group A or B is chosen for the three-way duel. However, the degree of preference is not equal; Mr. Y's degree of preference is by far higher than that of Mr. X's. *This implies that Mr. X is in a better bargaining position.*

## IV. Choice Among Orders

Since orders affect the likely outcomes, for any rational players, the choice among orders should be part of the game itself.

Thus, any rational player would be expected to exert substantial effort to affect the choice among orders. The problem will not vanish even if we assume that some external subjects such as a jury will make the choice. In this case, we may expect that each member will try to lobby or influence the decisions of the jury.

Let the choice among orders be an endogenous variable. The following points can be easily derived under the consideration that the players themselves choose the order:

There is not one order that can be chosen by all the three players unanimously.

If Mr. X is a dictator, either of the two orders in group B will be chosen. If Mr. Y is a dictator, the order in group A will be chosen. If Mr. Z is a dictator any of the three orders in group C will be chosen.

At the mid-point between the two extremes of collective decision-making rules (i.e. between unanimity and dictatorship) stands the simple-majority rule. Under this rule, either of the orders in group B will be chosen by both Mr. X and Mr. Y and thus will pass by 2:1 vote[61].

Let's now concentrate on the outcomes of the simple majority rule. This means that either (Y,X,Z) or (Y,Z,X) is the order to be chosen.

## V. Game with Side-Payments

If side-payment or exchange is introduced in our model, the outcomes will be changed substantially. As is shown in Table 1, if order (Y,X,Z) or (Y,Z,X) is to be chosen, the most unfortunate person will happen to be Mr. Z, the sharp shooter. Is there any possibility for him to change the order without violating the simple majority rule? We know that Mr. X is indifferent to his own position. It is the position of Mr. Y in relation to Mr. Z that concerns him. This and all other factors already discussed make Mr. X's vote decisive, pivotal. Any order preferred by Mr. X will pass by 2:1 vote. Why then does not Mr. Z try to buy Mr. X's pivotal vote? This can be done, for

61 Simple majority rule seems to be the most acceptable condition under the hypothetical case; however, under real conditions like economic, political, etc. "duel" dictatorship may be the most likely rule.

instance, in the following way. Mr. Z will go to Mr. X and say: "I offer you the following deal. If you vote for (Z,X,Y) or (Z,Y,X) orders, I will eliminate our common enemy, Mr. Y, in my first shot; then it will be your turn. You will shoot at me. If you kill me you will be the absolute winner. If you miss me, then it's my second turn. You know that I am a sharp shooter, but since you helped me by voting for my favorite orders, I promise you that I will fire up in the air (this is the price for your vote) thus both of us will survive." Mr. X should accept this offer because it improves his probability of survival, in this particular case, it even guarantees him life! If the game is played in accordance with the deal, the probability that X will survive is 100 percent; that of Y is 0 percent, and that of Z is 49 percent. That is both Mr. X and Mr. Z will benefit a lot from this deal at the expense of Mr. Y.

However, once side-payment is allowed and knowing that Mr. Z will certainly try to alter the choice of orders by buying Mr. X's pivotal vote, it is unlikely that Mr. Y will passively await his fate. He will also come to Mr. X with a counter offer for his vote for (X,Y,Z) order. The procedure is similar to the one described above.

The following table summarizes the likely outcomes when side-payment (exchange) is allowed:

| **Orders** | **Probability of Survival (%)** | | | **Remarks** |
|---|---|---|---|---|
| | X | Y | Z | |
| X,Y,Z | 86.0 | 56.0 | 14.0 | When Y trades with X |
| Y,X,Z | 86.0 | 39.2 | 14.0 | " " " |
| **Y,Z,X** | **86.0** | **39.2** | **14.0** | " " " |
| Z,X,Y | 100.0 | 0.0 | 49.0 | When Z trades with X |
| Z,Y,X | 100.0 | 0.0 | 49.0 | " " " |
| X,Z,Y | 100.0 | 0.0 | 70.0 | " " " |

**Table 2 Outcomes with Side Payments Allowed**

One may note from table 2 that orders (X,Y,Z) and (X,Z,Y) have a unique characteristic: they improve the well-being of Y or Z, respectively, without making X worse-off (*pareto-* improvement)[62]. This means that there is always one player who wants X move first. This makes Mr. X even stronger in determining the choice among orders and, ultimately, the likely outcomes as well.

The analysis proves that in one-time three-player game with side payment allowed *Mr. X, the weakest of all, is the most likely survivor*. He may even guarantee himself life. Good news for the miniatures!

We have already proved that miniatures may win battles against giants. However, since rivals do not necessarily trust each other, there is a problem of enforcing the contracts (deals). This problem calls for an amendment of the rule of the game.

## VI. The Rule of the Game

Let's assume that Mr. Z has traded with Mr. X and therefore (Z, Y, X) order has been chosen. This means that after Z killed Y, X will fire at Z. If X hits Z, the game is over and X will be the only survivor. However, there is a 70 per cent chance that X may miss Z when, if it occurs, then according to the deal, will lead to Z fire up in the air because this is the "price" Z has agreed to pay for X's pivotal vote in choosing the order.

Does this sound realistic? Being antagonists, how can Mr. X trust Mr. Z? What if Mr. Z broke his promise and killed Mr. X? Of course, Mr. X cannot take revenge after being killed. Without an effective mechanism that may guarantee enforcement of promises, the sort of exchange we have discussed makes little sense, if any. This problem calls for an amendment to the rule of the game. The amended rule should read something like the following:

62 A distribution is said to be a *Pareto-efficient* if and only if it cannot be redistributed in such a way as to make at least one member of the group better-off with- out making any member worse-off, compared to the status quo ante. A distribution that makes at least one member better-off without making any other members worse-off is a *pareto-improvement*.

*There are two rounds. In the first round, each player is given one shot, according to the order chosen by the players themselves. After the first round, any survivors are given a second shot, again according to the same order as in the first round. The game will continue until only one player is left alive or a point of exchange is reached if such a thing exists.*

A short description of the amended rule may be necessary.

It is possible that the players will approve the amended rule unanimously. Since this rule will be amended before the choice among orders is made, neither Y nor Z will be certain who will benefit more. Rather each of them will remain hopeful that under the new rule he will be in a better position to buy X's pivotal vote. Thus, both Y and Z will vote "yes" for the amended rule[63]. Mr. X is, of course, the person most interested in adopting the amended rule.

Even if we assume that the players possess perfect knowledge and thus there is no uncertainty considering the likely effect of the amended rule, the rule nevertheless will be passed by 2: 1 vote.

Under the amended rule promises are fully enforced. For example, in (Z,Y,X) order, Z kills Y, then X shoots at Z. If X kills Z, the game is over and Mr. X is the only survivor. If he misses, again the game is over. A pistol with a bullet will not be given to Z since he has already sold off his chance of playing further.

There is only one point in each order where the game may come to an end in the way described at 3 above. The point is in the area marked for exchange in each chart.

The amended rule brings about new patterns of likely outcomes. The following table summarizes the likely outcomes with a side-payment allowed under the amended rule.

63 This reminds us John Rawls' "The veil of uncertainty and ignorance". According Rawls, adopting just and fair rules through collective decisions is easier if members of the collective cannot anticipate in what ways each member will be affected by the rules.

| Order | Probability of survival (%) | | |
|---|---|---|---|
| | X | Y | Z |
| (X,Y,Z) | 86.0 | 56.0 | 14.0 |
| (Y,X,Z) | 86.0 | 56.0 | 14.0 |
| (Y,Z,X) | 86.0 | 56.0 | 14.0 |
| (Z,X,Y) | 100.0 | 0.0 | 70.0 |
| (Z,Y,X) | 100.0 | 0.0 | 70.0 |
| (X,Z,Y) | 100.0 | 0.0 | 70.0 |

**Table 3 Outcomes under the Amended Rules**

Now, under the amended rule with side-payment allowed, all the first three orders in table 3 (hence after "group I") yield the same set of likely outcomes. Similarly, the last three orders (hence after "group II") yield another set of likely outcomes. Thus, if the players are perfectly rational, the choice is *not* among six orders but only between the *two groups* only.

Which of the two groups will be chosen? Since Mr. has the decisive vote any option preferred by him will pass by 2:1 vote. If the game is considered to be a two-rounds only game, which will not be repeated sometime in the future between the survivors (in those cases where two of the three players survive), then Mr. will rationally vote for the orders in group II.[64]But, what if the game should continue between the survivors?

## VII. Continuity

Mr. X will rationally choose any order in-group II, if the game is a one-time game. However, if there is a possibility that the game may be repeated sometime in the future between two survivors, the choice becomes more difficult.

In the I group orders, the probability that both X and Y will survive is 56 per cent, and this is the probability that the game may be repeated sometime in the future, if any one of the I group orders has been chosen for the first round game. In the second round, (future) game there will be two orders: (X,

64 Note that repeating the game and playing it for more than two rounds are basically the same.

Y) and (Y, X). Under the same rule of the game, (X, Y) yield 36.16% chance of survival for Mr. and 65.8% for Mr.; and order (Y, X) yields only 8.8% chance of survival for Mr. X and 93.16% for Mr.

In the II group orders the probability that both X and Z will survive is 70 percent, and this is the probability that the game may be repeated sometime in the future, if any one of the II group orders has been chosen for the first round game. In the second round (future) game there will be two orders: (X,Z) and (Z,X). Under the same rule of the game, (X, Z) yields 30% chance of survival for Mr. X and 70% for Mr. Z; and order (Z, X) yields 0% chance of survival for Mr. X and 100% chance of survival for Mr. Z.

In second round games Mr. X will be better off if he moves first, no matter against whom he is going to play the game. However, since there is no privilege for X, the order should be chosen by tossing a coin, that is to say that each order has a 50% chance of being chosen.

Table 4 summarizes the likely outcomes in second-round (future) games among survivors of the first round game.

| **Orders** | **Probability of Survival (%)** | | |
|---|---|---|---|
| | X | Y | Z |
| (X, Y) | 36.16 | 65.80 | - |
| (Y, X) | 8.80 | 93.16 | - |
| (X, Z) | 30.0 | - | 70.0 |
| (Z, X) | 0.0 | - | 100.0 |

**Table 4 Outcomes of the Second-round Game**

Note that if any one of the orders in group I has been chosen for the first round game, the probability that the game may be repeated is 56%. Then in the second round game, the chance that any of the two orders may be selected is 50% each. Thus, the probability of survival of the players under the amended rule with a side-payment allowed in two-round game (or in a one-round game with four-round shots) is 0.3 + [0.56 x 0.5 (0.3616+0.088)] = 0.425888 i.e. 42.55888% for Mr. X; 0+[0..56 x 0.5 (0.658+0.9316)] =0.445088 i.e.

44.5088% for Mr. Y; and 14 +0=14% for Mr. Z.

However, if any one order in group II has been chosen, then the chance of survival of each player after playing two round games will be 0.3+[0.7 x 0.5(0.3+0)] = 0.405 i.e. 40.5% for Mr. X; 0% for Mr. Y; and 0 +[0.7 x 0.5(0.7+1)] =0 .595 i.e., 59.5% for Mr. Z.

Therefore, should the game be repeated one more time, Mr. X will rationally choose one of the orders in group I. By doing so, Mr. X will secure 42.55888% chance of survival however he still runs a 1.96% risk that the game may be repeated for the third time.

The following table summarizes the whole analysis of the game.

| **Condition** | **Orders that are likely to be chosen by the players under majority rule** | **Ultimate chance of survival (%)** | | | **Remarks** |
|---|---|---|---|---|---|
| | | X | Y | Z | |
| A one-time game without side-payments | (Y,X,Z) or (Y,ZX) | 41.2 | 52.64 | 14.0 | From table 1 |
| A one-time game with side-payments allowed under the amended rule. | (Z,XY), (Z,YX) or (X,Z,Y) | 100.0 | 0.0 | 70.0 | From table 2 |

| | | | | | |
|---|---|---|---|---|---|
| A two-time game without side-payment | (Y,X,Z), or (Y,Z,X) | 35.5814 | 50.994 | 14.0 | Table 3 and 4 plus more calculations.<br><br>P(X)^P(Y)=0.58084 |
| A two -time game with side payments allowed under the amended rule | (X,Y,Z), (Y,X-,Z) or (Y,Z,X) | 42.53888 | 44.5088 | 14.0 | Table 3 and 4 plus more calculations.<br>P(X)^P(Y)=1.0976% |

**Table 5 Summarizing Table**

Note that even when the game is played for four rounds (i.e. a two-time game), still Mr. X, the miniature has a greater chance of survival than Mr. Z, the super-giant!!!!!

Here we come to our final conclusions:

Being the weakest does not necessarily mean becoming the looser in war-games. And also being the strongest does not guarantee winning competitions.

In one-time games (for e.g. in spontaneous street fights where side-payments are almost impossible), it is better not to be the strongest and not to behave as such.

It is better to be the weakest in one-time street fights where there are verbal insults and knuckle clenching before the real fight begins. Verbal insults and knuckle clenching are often used to provoke the other(s) to start the fight, so determining the order.

Cooperation exists even in such war-games as the one we have just analyzed. However, as the game is played several times, the incentive for further cooperation seems to diminish.

## VIII. Applications

Students of economics, politics, and sociology may interpret the game just analyzed in different ways. The following are mentioned just to show how it might be done:

Students of economics may use the model to describe the nature of competition in oligopoly markets where some firms are big while others are small. Small-scale firms take advantage of rivalry between the giants and may end up as absolute winners.

Students of political science may also use the model to describe the nature of international politics. For example, many argue that poor countries were better-off during the Cold War, when the conflict between the giants was at its climax.

Students of institutional sociology can benefit a lot from this model. We have proven that a minor change in the rules of the game (i.e. the institutions) changes the likely outcomes substantially.

The game effectively demonstrates that there are two distinct types of choices: choices between orders and choices between strategies in a given order. Choices between orders are what James Buchanan, a Nobel Prize Laureate in economics, call them *creative choices*, or *constitutional choices*, or *choices of higher order* while choices between strategies in a given order are what he called *reactive choices*, or *post-constitutional choices*, or *choices of lower order*. The analysis proves that the likely fate of the players depends not only on their skill and the strategy chosen by them but also by the order; thus the choice among orders is part of the game itself. A choice between alternative political and economic systems (for example, a choice between a market and command economy) can be conceived as a choice between orders.

Of course, one can analyze many aspects of real life with the help of the model. This task, however, is left to the reader, himself/herself.

## IX Conclusion

Fortunately, the world does not belong to the giants only. Giants are not fully secured; they run a high degree of risk of being engaged in conflict with rival giants the outcome of which may benefit the miniatures. No one, for instance, can be certain that Microsoft will secure its current market position for a long period of time to come. Similarly, no one can be certain that Europe and North America will remain economic and political centers of the world for many years to come.

Miniatures may beat the giants. Sometimes weakness is an asset if utilized wisely. Miniatures (i.e., the poor, the weak, the small,…), therefore, should not give-up; rather, they have to choose their best strategies carefully.

## References

Buchanan, James and Tullock, Gordon: *The Calculus of Consent: Logical Foundation of Constitutional Democracy.* The University of Michigan Press. 1967.

Buchanan James: *The Economics and the Ethics of Constitutional Order.* The University of Michigan Press. 1994.

Dixit A. and B. Nalebuff : Thinking strategically: *The Competitive Edge in Business, Politics, and Everyday Life.* Norton. 1994.

Rawls, John: *A Theory of Justice.* Harvard University Press 1971.

Tadesse Biru "Redirecting Theory in Political Science." *Spoleczenstwo I Kultura. Oficyna Akademeka,* Warszawa, 1996.

## Chart I: Extended Game Tree - Order XYZ

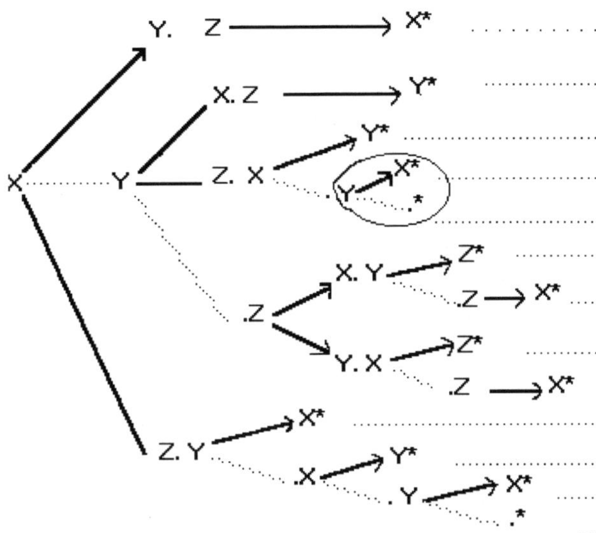


| Survivors | Prob. of survival % 1 | 2 | 3 | 4 |
|---|---|---|---|---|
| | 30.0 | - | - | - |
| Z | 56.0 | - | - | - |
| X | 16.8 | 24.0 | 24.0 | 24.0 |
| Y | 31.36 | 44.8 | - | - |
| Y, X | 7.84 | 11.2 | 56.0 | 56.0 |
| Y | 11.2 | - | - | - |
| Z | 2.8 | - | - | - |
| X | 4.2 | 6.0 | 6.0 | 6.0 |
| Z | 9.8 | 14.0 | 14.0 | 14.0 |
| Y | 24.0 | - | - | - |
| X | 1.8 | - | - | - |
| X | 3.36 | - | - | - |
| Y, X | 0.84 | - | - | - |

## Note

The 1st column shows the survivor(s)

The 2nd column shows the outcomes of the game when it is played without side payments.

The 3rd column shows the outcome when side payment is allowed.

The 4th column shows the outcomes when the game is played according to the amended rule.

The positions where exchange of probability of survival is possible are marked in each chart.

## Chart II: Extended Tree - Order YXZ

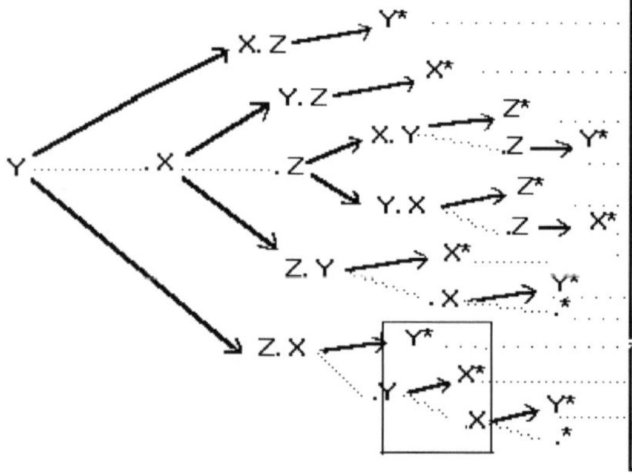


| Surviv-ors | Prob. survival (%) 1 | 2 | 3 | 4 |
|---|---|---|---|---|
| Z | 80.0 | - | - | - |
| Z | 6.0 | - | - | - |
| Y | 11.2 | - | - | - |
| Z | 2.8 | - | - | - |
| X | 4.2 | 6.0 | 6.0 | 6.0 |
| Z | 9.8 | 14.0 | 14.0 | 14.0 |
| Y | 4.8 | - | - | - |
| X | 3.6 | - | - | - |
| X, Y | 8.4 | - | - | - |
| X | 24.0 | 24.0 | 24.0 | 24.0 |
| Y | 44.8 | 44.8 | - | - |
| X | 3.36 | 3.36 | 16.8 | - |
| X, Y | 7.84 | 7.84 | 39.2 | 56.0 |

## Chart III: Extended Game Tree - Order YZX

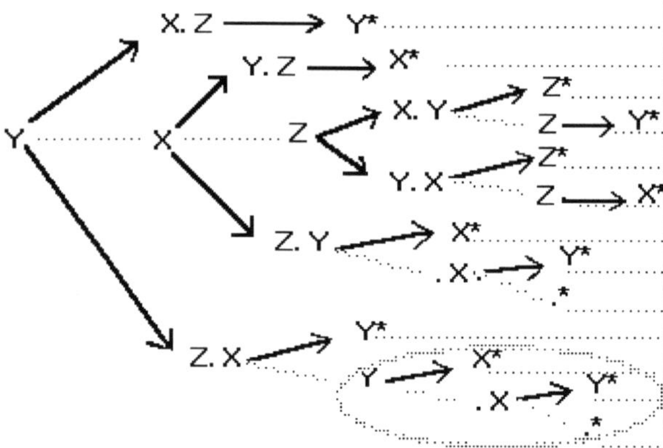


| Survivors | Prob. of survival (%) 1 | 2 | 3 | 4 |
|---|---|---|---|---|
| Z | 80.0 | - | - | - |
| Z | 6.0 | - | - | - |
| Y | 11.2 | - | - | - |
| Z | 2.8 | - | - | - |
| X | 4.8 | 6.0 | 6.0 | 6.0 |
| Z | 9.8 | 14.0 | 14.0 | 14.0 |
| Y | 4.8 | - | - | - |
| X | 3.6 | - | - | - |
| X,Y | 8.4 | - | - | - |
| X | 24.0 | 24.0 | 24.0 | 24.0 |
| Y | 44.8 | 44.8 | - | - |
| X | 3.36 | 3.36 | 16.8 | - |
| X,Y | 7.84 | 7.84 | 39.2 | 56.0 |

## Chart IV: Extended Game Tree - Order ZXY

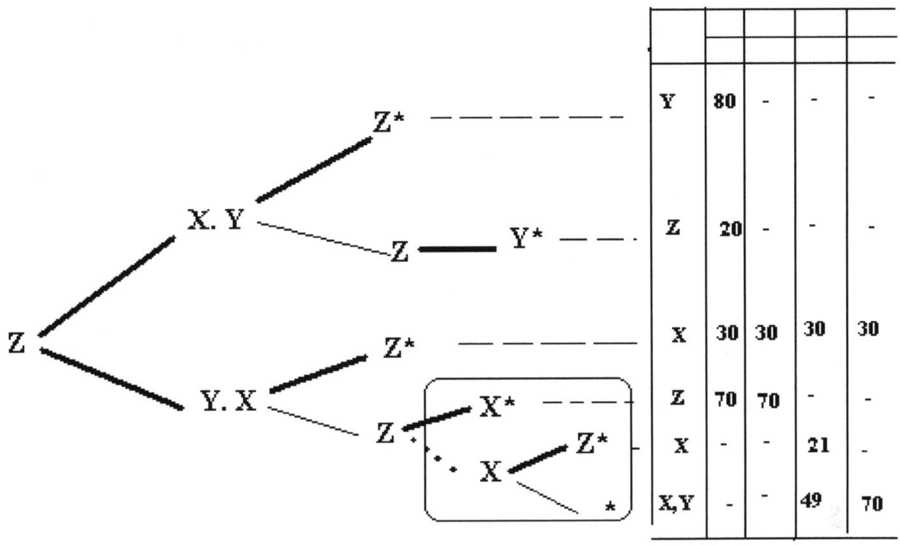


## Chart V: Extended Game Tree - Order ZYX

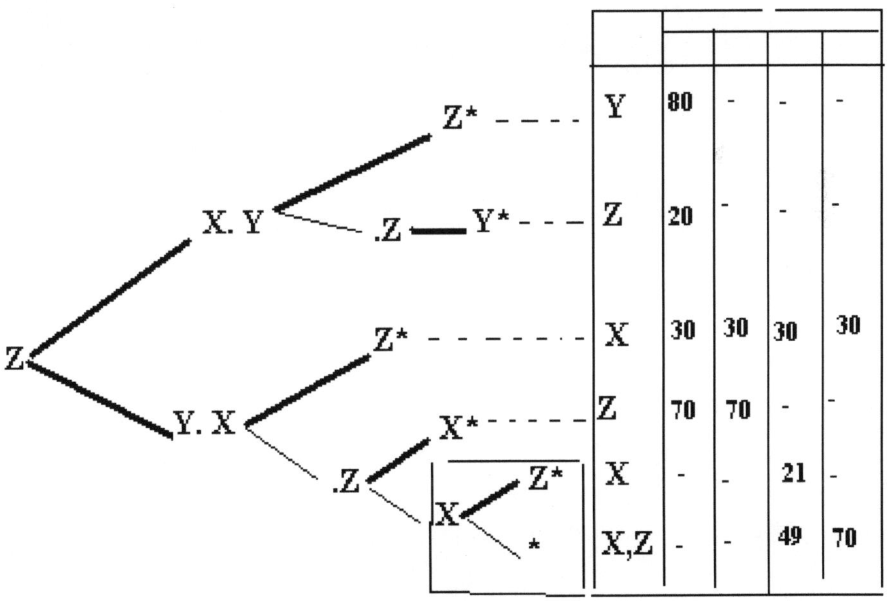

## Chart VI: Extended Game Tree - Order XZY

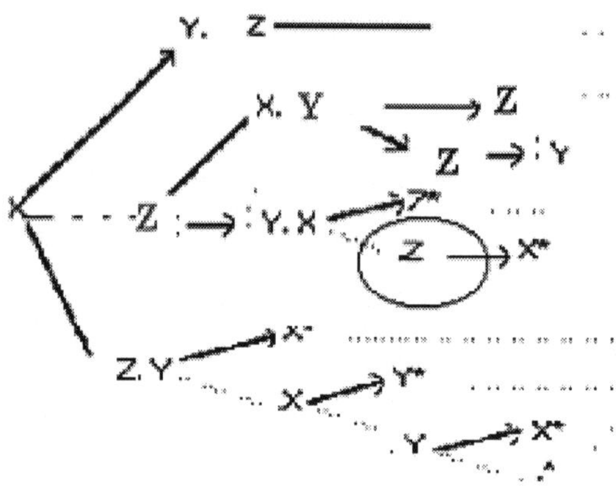


| Survivors | Prob. of survival % 1 | 2 | 3 | 4 |
|---|---|---|---|---|
| Z | 56 | - | - | - |
| Y | 30 | | - | - |
| Z | 14 | | | |
| X | 21 | 30 | 30 | 30 |
| Z | 49 | 70 | | |
| XZ | - | | 70 | 70 |
| Y | 24 | - | - | - |
| X | 1.8 | | | |
| Y | 3.36 | | | |
| Y, X | 0.84 | - | | |

# CHAPTER 13

## What Could Have Predicted the North African and the Middle East Uprising?

*Yared Haile-Meskel*

### Introduction

In a book entitled "The True Believer:-Thought on the Nature of Mass Movement", Eric Hoffer, an American self-made philosopher, wrote, "when people toil from sunrise to sunset for bare living, they nurse no grievance and dream no dream" (Hoffer, 1951, p27). He used this to describe how abject poverty becomes an impediment for desire to change. Instead he states, "[f]or men to plunge headlong into an undertaking of vast change, they must be intensely discontented yet not destitute" (Hoffer, p.11). He emphasised "discontent is likely to be higher when misery is bearable; when conditions have so improved that an ideal state seems almost within reach" (P.27). This view is not in line with prevailing view of revolution from days of Aristotle to Marx, where the emphasis is placed on discontent caused by hardship and exploitation.

The interpretation of Hoffer's view is that uprising and discontent is more likely to happen when the economic reality of a nation has improved. This improvement can be measured using economics parameters such as GDP/capita. This means Hoffer's statement could have a profound implication on forecasting, risk management, insurance and investment decisions. It means, as the economy goes through transition, there will be a region where people are discontented and yet not destitute, which means that they are more likely to start a revolution than when they were under bondage of abject poverty. Accepting this hypothesis puts macroeconomic parameters at the heart of social and political changes to bring about a paradigm shift how risk is perceived.

The focus of macroeconomic theory is analysis of the way in which deliberate government policies can influence the overall "economic health" of the country. (Economic Dictionary, 2011). That implies Government's economic policies affect the well-beings of citizens and their satisfaction. This in return influences how people view and react to decision makers. In democratic countries people use ballot boxes to bring about changes, but in countries where people have no similar rights, they might resort to violent or non-violent means to change decision makers.

In this article these assumptions were tested using economic indicators of the North Africa and the Middle East (MENA) countries to see if uprisings were caused by economic factors. The questions asked are:

What causes uprising and which variable is most closely associated with the propensity of a population to rise against the existing order, such as GDP/Capita, income distribution (GINI), corruption, or democratic scores, and

What economic variables do investors need to give close attention to when scanning for FDI opportunities?

## Reliability of Forecasting

Investors make decisions by forecasting future returns after factoring risks associated with a particular investment in a particular country. As a result, the equity market, resource exploration companies, national security agencies, insurance companies and private investors spend trillions of dollars on analysing Political, Economical, Social, Technological, Environmental and Legal risks of nations (PESTEL). This is often done by analysing the past data to find patterns and signals to predict the future, hoping that the past repeats itself in predictable ways. But reliable prediction is not an easy job. Stieglitz (2008) mockingly states "economists are better at predicting the past than the future"

In particularly, the accurate prediction of big and rare events, where there is little or no historical data, is very difficult. Taleb (2007) argues "[t]he inability to predict outliers implies the inability to predict the course of history, given the share of these events in the dynamic of events". That is why analysts, economists and security experts often miss big impacts, such as, September 11, the credit crunch, and uprising in the MENA.

The problem is, these events are so profound that they catch investors unprepared and cause businesses to collapse. For example, companies investing in Libya, Tunisia, Syria, Yemen or Egypt may have been advised that this was a safe area to invest considering the last 40 years data of stability. Now these companies may dismiss their risk consultants but that does not compensate them for the losses incurred as result of force majeure.

Social phenomena are even more difficult to predict since various factors could turn the direction of events in many unpredictable ways. This is captured by metaphor of the butterfly effect, in chaos theory. It is exactly what happened in Tunisia when a street vendor, whose merchandise were confiscated and he felt humiliated enough to set himself on fire in protest. This action appears to be a call for justice and fairness, and it reflects the underlying economic and political reality of the country that drove the street vendor in to desperate action.

Hence, this article tries to explore what underlying, economic, social, political patterns are distinctly shared between MENA countries that are going through an uprising. This is done by examining economic, political and social data in Gapminder.

Post uprising, various analysts have come on the media to attribute the uprising to demography change, emergence of social media, dictatorship, endemic corruption, global warming driven commodity price increases (Johnstone and Maze, 2011), rising unemployment, increasing inequality and other factors. These hindsight rationalisations were also put under scrutiny.

## Methods

From Gapminder some of the economic and social indicators of Tunisia, Egypt, Bahrain, Yemen, Syria, Iran, Jordan and Libya were plotted and compared with Sub-Saharan African countries and democratic countries like India, Jamaica, China, Brazil, Italy, South Africa and Canada to provide a comparative perspective of democracy and corruption.

Based on this analysis an argument have been put forward for managers and investors as to what indicators they need to give special attention to in order to avoid costly FDI.

## Analysis and Discussion

### GDP per Capital and Life Expectancy

GDP/capital shows the comparative living standard of nations. It shows whether a people of a certain country are poor, middle income or rich. This measure is often associated with perceived satisfaction of citizens. Monshipouri (2004) quoting Collier (2003) states that "[e]conomic development is an essential vehicle to prevent rebellion and to build the conditions in which groups engage in conflicts or conflicts resolution through normal political means". Based on these assumptions, politicians often associate revolution, terrorism and other major social upheaval to poverty. That is probably why, Tony Blair (Oct 2, 2001) made a passionate speech to tackle terrorism and global poverty in the aftermath of September 11. This "widespread view that poverty creates terrorism" is criticised by Abadie (2004).

To see whether poverty is a contributory factor for dissatisfaction and uprising in MENA countries, GDP/capita and life expectancy of MENA countries were compared with Sub-Saharan African and Asian countries.

Figure 1 shows MENA countries that have been going through uprisings have the highest GDP/capita and life expectancy for more than 70 years. The exception is Yemen, with less than $1000 per capita and life expectancy similar to India. Egypt has the second biggest economy with GDP of 188 billion dollars, next to Saudi Arabia, among MENA countries (World Bank, 2009). Egypt's GDP is comparable to Israel and Malaysia, however it's GDP per/ capita is around $2000 as a result of its larger population.

Bahrain, Libya, and Saudi Arabia have GDP/capita higher than $10,000 which ranks them far above Brazil and Russia, (World Bank, 2009). In general, Tunisia, Syria, and Egypt are middle income countries with GDP/ capita higher than China, Ukraine, India, Pakistan and most sub-Saharan countries. Most Sub-Saharan countries live in much worst conditions with GDP per capita less than 500 USD and life expectancy less than 55 years. If poverty is the driving force it should have been these Sub-Saharan countries that should have experienced uprisings. The irony is it the most affluent countries like Bahrain, with GDP per capita around 26,000 USD, felt "discontent but not destitute" to go on uprising and risk their lives. Therefore GDP/capital data of MENA countries suggests that poverty is the least likely cause of the uprisings.

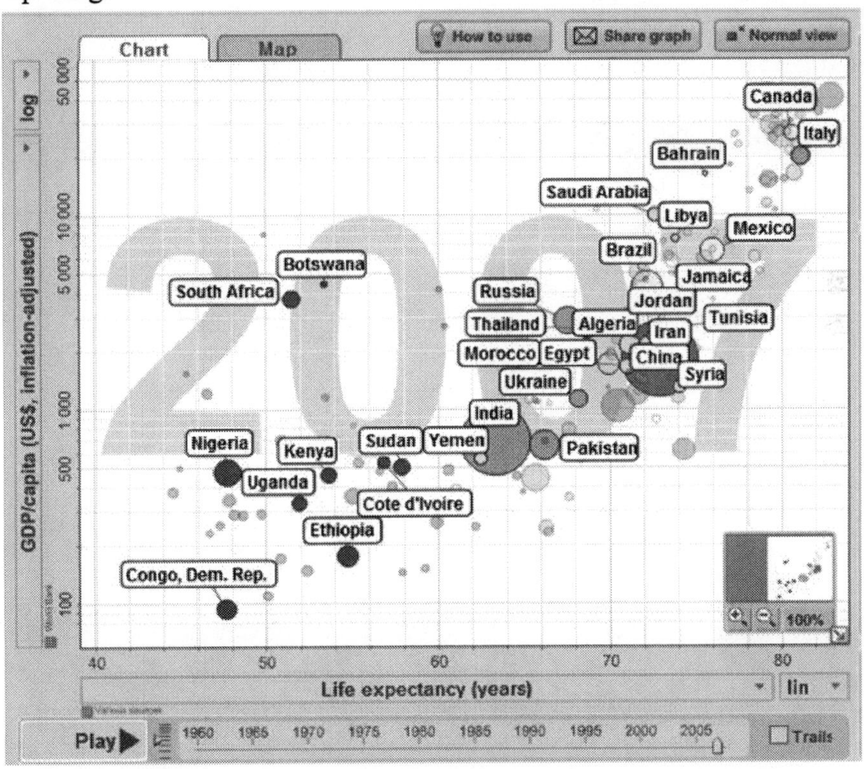


**Figure 1. GDP per capita vs. life expectancy www.bit.ly/lb05us**

## GINI, Inequality Index

Exclusion of poverty, as measured by GDP per capita, does not necessarily mean that there was not economic hardship. GDP per capita does not give the full picture if there is high inequality between the rich and the poor. GINI is a measure of income inequality. If there are macroeconomic policies that favour "the few", inequality could grow and the disadvantaged may hold grievances to demand for change or support revolution. Figure 2 shows the GINI coefficients of a selection of countries between 1980s and 2005.

Contrary to the popular stereotype of the "rich Arab oil sheiks" and "the Arab masses", the inequality in MENA countries is not as bad as some Sub-Saharan or other emerging economies. Egypt, for example, is one of the countries with the lowest GINI index compared to South African, Kenya, Mexico, China, Russia and Zimbabwe. Tunisia ranks in the middle with comparable inequality to China and Russia. The inequality in Yemen is low and similar to Italy and India. Unfortunately there is no GINI data available for Libya, Bahrain, Algeria and Saudi Arabia in the Gapminder or other UNDP HDI publications. However, the fact that Egypt and Yemen are among the few countries with very low GINI indexes, suggests that inequality could not have been the major factor for the uprising. Furthermore this uprising did not lead to demands for redistribution of wealth or nationalisation as often happens in socialist inspired revolutions.

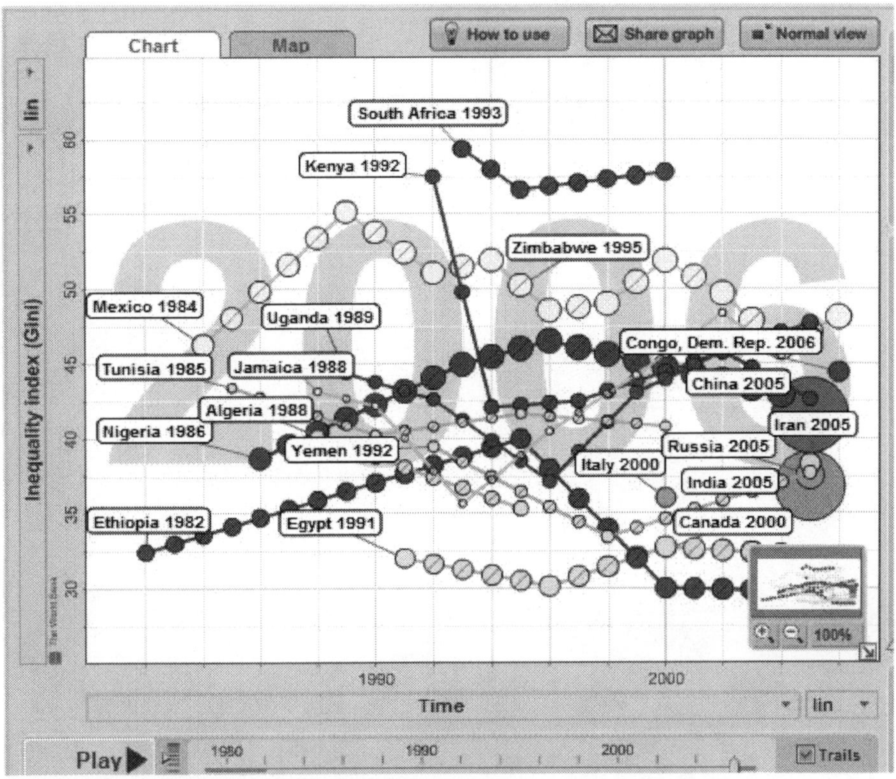


**Figure 2. Inequality index across time www.bit.ly/hvt6Hn**

## Corruption and democracy

Wilhelm, (2002) states that "[c]orruption is increasingly argued to be a barrier to development". Figure 3 provides a glimpse of corruption indexes and Democratic scores.

In Figure 3, the higher the corruption index, the less the level of corruption in a country. In contrast higher democratic scores suggest a more democratic country. Canada for example is the least corrupted and most democratic country with both indexes having values of 10. Most countries that have been going through uprisings have negative democracy scores. Only Algeria of the MENA countries has a marginal positive democratic score of +2. However, when it comes to the corruption index, these countries are spread across the y axis. For example, Saudi Arabia, Bahrain and Jordan and

Tunisia are less corrupted than all BRICS countries. In comparison India, South Africa, Brazil, Greece and Italy, which have higher democratic scores, fall below Saudi, Bahrain and Tunisia on the corruption index.

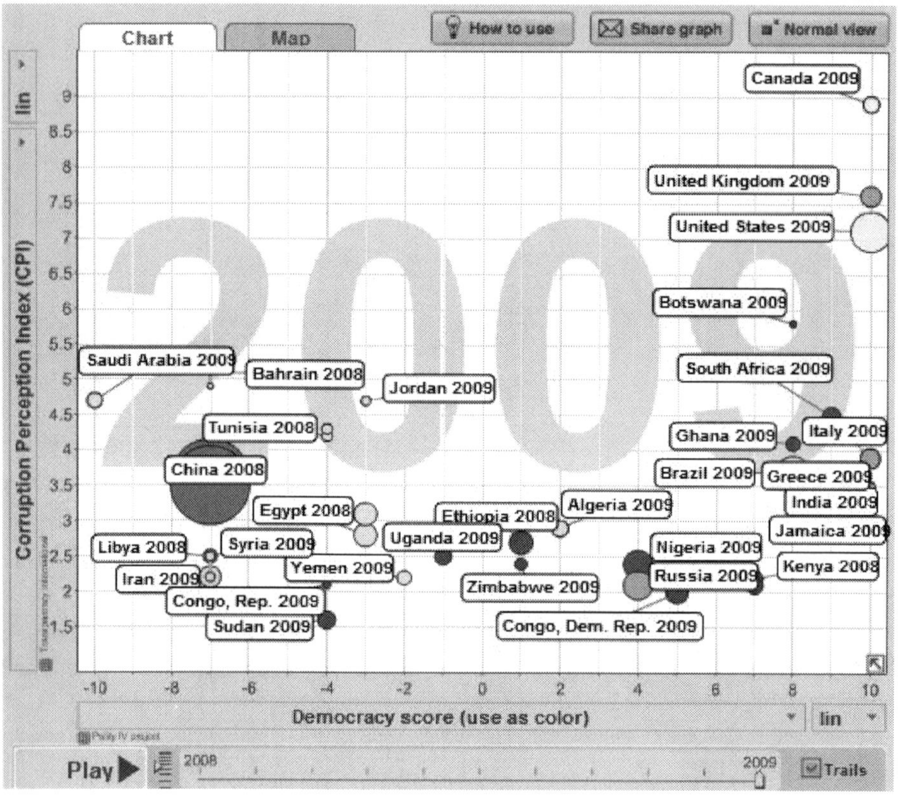


**Figure 3. Corruption Perception Index (CPI) vs. Democracy Score www.bit.ly/jdlmgA**

China is as undemocratic as Syria and Libya and much corrupt than Saudi, Tunisia, and Bahrain.

From Figure 3, more than the other variables, the democracy score appears to suggest a clear pattern in predicting the uprising in MENA countries. All the countries that have been going through an uprising score less than minus 2 on the democracy score. One could say countries like Sudan and Iran face the same threat of uprising. In Sudan protests were started but maybe due to preoccupation with Southern Sudan independence the protest did not gain momentum. Iran has put down the protest quickly since it was well prepared from the last uprising. The fact that the uprising did not succeed this time does not reduce the risk as long as the democratic index remains the same.

The other point that can be drawn from Figure 3 is democracy might not eliminate corruption. India, the largest democracy, Italy, Greece, Jamaica and Brazil have high scores on the democratic scale but the level of corruption is much worse than Saudi Arabia, Tunisia, Bahrain and Jordan.

Despite high degree of corruption in BRIC counties, BRIC has been a favourite destination of FDI. This means investors do not usually get deterred by corruption as long as the return on investment outweighs the cost of corruption. A good example is China. China is undemocratic and corrupt and yet it has remained a favourite destination for FDI for US, Japan, UK and European companies. The supply side of corruption might have also contributed to the problem.

Figure 3 also shows countries may not have to be democratic to be less corrupt. High income in undemocratic countries like Qatar, Bahrain, and Saudi Arabia has managed to keep corruption at a minimum in the absence of a free media or checks and balances.

From the developed world Italy and Greece are the exception with high democratic scores but very low corruption indexes.

## GDP per capita versus Democracy

Plotting GDP per capita against the democratic score clearly isolates the countries that are going through uprising as can be seen in the left hand corner of Figure 4. Those countries have been going through economic transformation to improve their living standards but this growth was not matched by political liberalisation and democratic rights. Maslow's hierarchy of needs shows once human's physiological, safety and belonging needs are fulfilled, what follows is the need for self-stream, achievement, confidence and respect. With high access to education, medical care and long life expectancy, it would not be a surprise for the new generation of MENA countries to demand democratic rights, respect and a say in national affairs. These rights are better served through a democratic system and it is not a surprise that the demand was for democracy rather than the long feared authoritarian system lead by theocrats. So no wonder that the new generation of Bahrain, Libya, Syria and Tunisia chose to die fighting to have a say in their one own affairs, compete for political power and office, and have equal rights than to put up with authoritarian rules where their material and security needs are met. This has a profound implication for other developmental states like China and strategic analysis designed to combat fundamentalism. For China as the prosperity spreads to the inland and when the new generation start taking higher standard of living for granted and wants to move up on the hierarchy of needs, there is a high probability of a challenge to the prevailing order and an uprising. For strategic analysts the fear that this change might lead to theocratic lead authoritarian rule becomes less relevant.

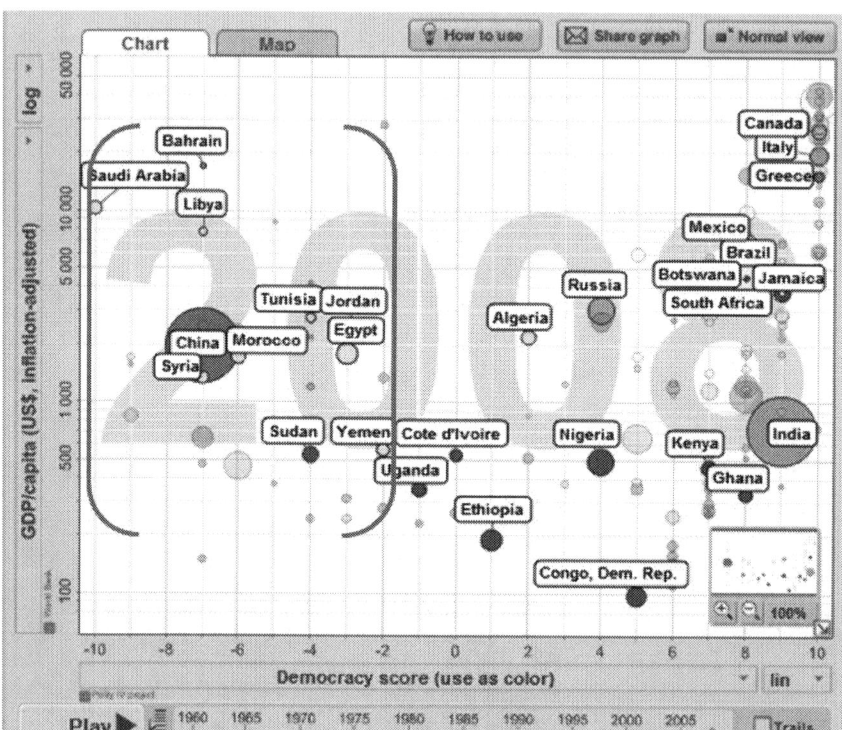


**Figure 4. GDP per capita and Democracy score** <u>**www.bit.ly/kS2BBX**</u>

## Internet penetration

Many post uprising analysts have attributed the uprising to social networks. Some main stream media have gone as far as referring to the events as "the Facebook revolution" (Bloomberg, BBC, CNN, Feb, 26, 2011). Social media reaches places where there is infrastructure for internet connections. Those countries that can afford to build the infrastructure have higher penetration of the internet. Small size countries with high population density also tend to have higher internet penetration. There is no doubt that the Internet has brought the wider world into the living room of people and this may have helped to communicate the start of protest in Egypt. The first Facebook protest was called in June 2010, which caught no mainstream attention and did not lead to widespread support. The second one was called on 25th of January, 2011 (CNN). This is after the Tunisia uprising. No doubt the Internet had created a platform for sharing tacit knowledge about democratic rights, freedom of expression and the ability to communicate and network with similar minded people. This may have empowered citizens. Tunisia's uprising was the trigger in December 2011, and resulted from an incident in a provincial town and not from an internet campaign.

The assumption that the Internet contributed to the uprising was tested using data on Internet and Facebook penetration statistics. Some of MENA countries have high internet penetration but the others have very low. Bahrain (88%), Iran (43%) Saudi Arabia (38%) Tunisia (34%), Egypt (21%), Morocco (33%) and Syria (17%) have high internet penetration but Libya with 5.5% and Yemen with 1.8% are the lowest (Internetworldstatats, 2011)

The Facebook statistics also show a similar pattern. Bahrain (41%), Tunisia (22 %) and Egypt (8.8%) have high penetration but Yemen and Libya once again have the lowest, with less than 2% of the population having Facebook membership. Looking at these exceptions and irregularities it would be very difficult to say that the Internet and Facebook penetration were common factors which had major impacts on the Libyan and Yemeni uprisings.

## Implications of macroeconomic and social data for FDI

For investment decisions the future value of money has to be higher than the net present value of money invested today. This means forecasting, though unreliable, is unavoidable. To avoid loss, a huge amount of money has to be spent in forecasting the future and the risk involved. The question asked is what variables are better predictors of political risks?

Companies investing overseas often look at political, economic, social, technological, environmental, and legal factors (PESTEL). Under the economic factor, there is a general perception that the higher the GDP and the better the living standard, the safer the place to invest. As a result most North African countries were attracting FDI in tourism and resource exploration. Oil exploration companies like BP rushed back in to Libya to make a deal after the economic embargo lifted. This shows the level of confidence that they had in stability in Libya. But these assumptions fail when these middle income countries begin to be engulfed by social uprisings making the basis of forecasting models flawed.

Most economic variables of MENA countries show higher living standards, higher life expectancy, and higher GDP per capita, with the exception of Yemen. These countries had healthy economic development on the back of oil and tourism. Though there is corruption, it was not as bad as most of Sub-Saharan African and BRIC countries. In fact, on the corruption index even Italy, a member of G8, appears to stand well below the countries facing uprising. Nevertheless what MENA countries share in common are authoritarian regimes. Authoritarian systems never deterred FDI. In fact, Hayashi (2010) argues in favour of developmental states to bring about better returns then democracies. Quoting Haggard (2004), Hayashi state that "one may claim that authoritarianism could solve problems that democracy cannot... Industrialisation requires massive resource mobilization in the direction of capital accumulation, in other words, from consumption to investment. Developing countries are generally not well equipped with the necessary capacity to carry out this process in a democratic way, but authoritarian regimes could, in part, succeed in achieving their aims by silencing the opposition"

This assumption did not deter companies from investing in Asian developmental states in the 1970s and 1980s and currently in China. However, this perceived stability needs to be questioned in the light of uprisings in MENA countries. The rushes to get in to Libya, by all means, have left many multinational companies accepting losses as a result of force majeure. This shows, at some point of development, that citizens of developmental states demand for their democratic rights and this appears to be more compelling than the satisfaction earned from better living standard.

Leamer (2009, pp.27) states GDP only measures the market value of goods and services and not people's happiness or satisfaction. This shows that fulfilling the material needs of the people is not enough. Libyan's and Bahrain's have high income per capita, longer life expectancy, universal education and rank high on most human development indices. But these people were not satisfied with their GDP/capita or comparative privileges based on oil revenue. When the chance arises people did not hesitate to rise up at the risk of losing their lives.

This suggests that the democratic index is probably the most significant factor that investors and managers need to look at before making investment decisions. Though authoritarian countries appear to give a semblance of control and stability, an uprising could be around the corner. As observed in Tunisia the trigger could be a minor incident making it very difficult to predict or control.

The other issue that stood out is, GDP growth is not an insurance against uprising. It has been a fashionable thing for authoritarian rulers to call their country a developmental state adopting the Chinese model. The assumption behind this is, as long as they keep the economy growing they do not have to bother about political liberalisations or giving up power.

What had happened in the MENA countries shows that these demands are universal and shared by all people. In fact once the citizen's materials needs are fulfilled they will move up on the hierarchy of needs to demand their democratic rights and they hate to be told what to do or to mind their own business by the authorities. This contrast could be seen by comparing

MENA countries with Sub-Saharan countries. African countries have similar authoritarian rulers but the vast majority of citizens live below poverty lines to free themselves from day to day struggle to demand for democratic rights and freedom of speeches. That is why the conflict in poor countries is localised in urban centres, where the income per-capita is much higher than the rural population. But the urban class is much smaller to have a critical mass to drive a change. That is why despite similarities with the MENA countries, the African countries have not yet come out on to the street.

Hoffer wrote in 1951 (pp.27) stating "It is not actual suffering but the test of better things which excites people to revolt…Our frustration is greater when we have much and want more than when we have nothing and want some. We are less dissatisfied when we lack many things than when we seem to lack but one thing" but this theory did attract much attention due to a lack of empirical evidence His statement was based on the works of Alex de Tocquellive (1888) whose made observations of the economic improvement in France before the French Revolution and one-third of agricultural land ownership by peasants in Russia prior to the revolution there (Hoffer, 1951, pp.27-28)

This leads to a conclusion that GDP growth, improved living standards, longer life expectancy, universal education and health care are not insurance for authoritarian leaders to stay in power. In fact, meeting the basic economic needs is going to move people up the hierarchy of needs and make them demand their freedom and democratic rights.

## Implication for Developmental States

The rise of China into an economic powerhouse while remaining under authoritarian rule has made the model attractive for those who want to remain in power. By bringing economic progress and infrastructure expansion, many Sub-Saharan African leaders hope that people will put up with undemocratic governance. This assumption is put to the test by the uprising of the MENA countries. On the back of their oil economies, most of the MENA countries have achieved significant progress in improving the living standards of their people. MENA countries have a high rate of universal education, access to healthcare, improved housing provisions and 70 years life expectancy. This rapid improvement of living standards did not stop the citizens of those countries from aspiring to have a democratic system. In fact, it appears that one thing that they did not have and were prepared to die for is the right to have a say in their national affairs, which means democracy.

The experience of MENA shows countries in Sub-Saharan countries that are going through rapid economic growth without corresponding political liberalisation are bringing the demand for democracy much closer, rather than delaying it. It is also common for opposition forces in those countries to become a victim of the same assumption. They often tend to discourage development activities or in some cases sabotage development projects for fear that the government might take the credit to extend its rule. The data in this analysis shows that opposition by discouraging economic development is not going to shorten authoritarian rules but extend it.

## Conclusion

Revolutions have complex causes and one must always have reservations about predicting on the basis of economic variables but it worth considering factors attributing to revolutions. In addition, correlation does not prove causation [Goodwin, 2011]. From the above discussions and analysis leads to the following conclusions.

GDP growth and increasing living standards in authoritarian countries is not a sign of stability. Instead it could be an indicator of pending revolution so investors must not be seduced by growing GDP.

The democratic score is the most significant variable that could have predicted the uprising in the MENA countries. Though it appears under control, those countries like China may present a bigger threat to investors since the demand for democratic rights appears to follow GDP growth. Once the material needs are fulfilled people will move up the hierarchy of needs and be discontent with their lack of freedom and democratic rights.

Despite the overselling of Facebook by the media as the cause of uprising, countries that have being going through uprising , like Libya and Yemen have the lowest access to Facebook and the internet.

Democratisation is probably the only way of creating long-term stability, security for investment and sustainable development. As the GDP figures march ahead of the democratic scores a crisis is more likely to happen than when there is no economic progress.

Countries like China need to match economic changes with political liberalisation to meet the material and psychological needs of their citizens. The discrepancy between the two will only lead to dissatisfaction and an uprising similar to that experienced in the MENA countries.

## Reference:

Abadie, A, 2004, "Poverty, Political Freedom, and the Roots of Terrorism"; [online], Harvard, http://www.hks.harvard.edu/fs/aabadie/povterr.pdf, Accessed on April 12, 2011

Romesh Ratnesar, 2nd June, 2011, Bloomberg; http://www.businessweek.com/magazine/content/11_24/b4232062179152.htm, Accessed on June 29, 2011

BBC, 11th of February, 2011, "Did social media create Egypt's revolution, http://www.bbc.co.uk/news/world-middle-east-12435550, Accessed on June 29, 2011

Blair, T, Speech, October 2, 2011 [online] American Rhetoric online speeches bank, http://www.americanrhetoric.com/speeches/tblair10-02-01.htm, Accessed 12 April, 2011

Collier, P.; et al., 2003, "Breaking the Conflict Trap: Civil War and Development Policy" 91 (2003)

De Tocqueville, A, 1888, On the State of Society in France Before the Revolution of 1789, London: John Murray, p.149

Economic Dictionary, 2011, [Online] http://www.babylon.com/define/35/Economics-Dictionary.html, Access on April, 14, 2011

Goodwin, P.; 2011, Private communication, University of Bath School of Management

Hoffer, E, 1951, "The True Believer: Thought on the Nature of Mass Movement", Harper & Row Publishers, PP. 11.

Hoffer, E, 1951, The True Believer: Thought on the Nature of Mass Movement", Harper & Row Publishers, PP. 27-28

Johnstone, S; Mazo, J, 2011, Global Warming and the Arab Spring, Survival, 53: 2, pp. 11-17

Kahn, J. and Weiner, T., 2002, "World Leaders Rethinking Strategy on Aid to Poor," *The New York Times*, 18 March 2002, sec. A(1), p. 3.

Leamer, E.E.; 2009, Macroeconomic patterns and stories, Springer; pp. 27

Monshipouri, M.; 2004; Book review: Fareed Zakaria, The Future of Freedom: Illiberal Democracy at Home and Abroad, Human Rights Quarterly, Vol. 26. No. 1, pp. 211-215

Oxford Dictionary of Economics, 2011 [Online], http://www.enotes.com/topic/Microeconomics, Accessed on 13 April, 2011.

Stiglitz, E. J., 2008, "It Doesn't Take Nostradamus", *Economists' Voice*, www.bepress.com/ev November, 2008

Taleb, N, N, 2007, The Black Swan, Penguin Book Ltd, pp. xx

UNDP, 2009, Human Development Index, [online], http://hdrstats.undp.org/en/indicators/161.html; Accessed on 10 April, 2011

Wilhelm, G. P.; 2002; "International Validation of the Corruption Perceptions Index: Implications for Business Ethics and Entrepreneurship Education", *Journal of Business Ethics*, vol. 35, pp. 170-189

# CHAPTER 14

## Exploring Governance of Lake Tana Fishery: Interactive Perspective on Governance

*Dessalegn Molla Ketema, Nickolas Chisholm and Patrick Enright*

### Introduction

Following the secession of Eritrea in 1993, Ethiopia lost access to an estimated 1,011 km of Red Sea coastline. Since then Ethiopia is a landlocked country and its main fish resources are from Lakes, rivers and reservoirs. Despite being considered as the "water tower" of Horn of Africa and endowed with huge water resource potential, the current status of the Ethiopian fishery sector contribution to GDP is considered as marginal. And yet, aquaculture is recognized as an alternative means of achieving food security and poverty reduction in rural areas, and is now considered an integral part of rural and agricultural development policies and strategies in Ethiopia.

Even though there is no, specific fishery policy framework, a range of national development documents identify fisheries as being a sector in need of support to enable it to increase production and contribute to the food security of the growing population. As reported by ACP (African, Caribbean and Pacific Group of states) Fish II project, currently the bulk of the fish catch originates from 4 lakes, namely: Lake Tana (25%), Ziway and Langano (19%), Chamo (18%) and Abaya (12%) of the national total production. Despite catches on these lakes declining in recent years there is still widespread belief in the country that potential exists to raise the annual production to over

49,000 tons per year possibly through the development of new constructed reservoirs, under-exploited river fisheries and aquaculture (ACP, 2012).

Lake Tana is the largest freshwater body in Ethiopia, which covers 50% of the total inland water bodies in the country. It is an important source of fish both for the people immediately around the lake and elsewhere in the country (Awulachew *et al.*, 2009). Lake Tana contributes 20-31% of the total fish potential of the country and is used for commercial fishing (Wondie, 2010). Though, fishing in Lake Tana is believed to be started around $18^{th}$ century by the *Negede woyto* community, modernization of the fishery sector started recently. Commercial fishery in the Rift valley started in the 1950s, where as in Lake Tana it started very late in the 1980s.

Research efforts with regard to the biological, abundance and marketing aspects of Lake Tana fishery have been done for the last two or three decades. But they were inadequate to illustrate the human-nature interaction and the trends in resource degradation and anticipate possible solutions especially in reference to institutional arrangements and fishery resource governance. The governability of this sector is however not recognized by the potential stakeholders in the sub-basin. Governance in fisheries is usually understood as the sum of the legal, social, economic and political arrangements used to manage fisheries. It has international, national and local dimensions. It includes legally binding rules, such as national legislation or international treaties as well as customary social arrangements (FAO, 2005-2012). Governance however is not merely something governors do, but comprises the totality of the interactions between those governing and those governed-it is itself an interaction (Bavinck *et al.*, 2005, Kooiman *et al.*,2005). It is an ongoing social process that is constantly negotiated and achieved among relevant stakeholders (Wilkes, 2005).

In recent time, interactive governance has become an important approach for understanding multi-level governance issues in natural resource management and governance in general fisheries and coastal governance in particular. Interactive governance is a perspective that focuses on understanding the characteristics of the natural and social systems that are being governed, the governing systems, and their interactions (Jentoft, 2007).The integrated and comprehensive lens that the interactive governance perspective brings to the concerns, challenges and problems related to fishery resource systems and human interactions helps broaden the possibility and opportunities for better solutions. It is also a prescriptive tool in that it helps determine which governing responses and interventions, including institutions and policies, need to be put in place in order to improve the overall fishery governance (Chuenpagdee, 2011). Hence, a fisheries governor aiming to put governance into action should first examine the governability of the fishery" (Mahon *et al.* 2005)

Therefore, this research was designed to critically analyze and understand the system to be governed (fishery sector, local community and other stakeholders), the power and interests of multi-stakeholders in fish resource governance, the governing systems (the governors and the rules of the game) and the governance interaction (realities on the action arena). The study focuses on status of fish resource, stakeholder involved, institutional framework that support the fishery sector and the knowledge and perception of stakeholders towards the rules and regulations.

## Research Methodology

### The study area: Lake Tana

At 12°N, 37°15′E, and 1,830 m altitude, Lake Tana is situated on the basaltic plateau of the north-western highlands of Ethiopia. It is the largest freshwater lake in the country and the third largest in the Nile Basin which contributes about 85% of the Blue Nile water. It covers an area of 3,050 $km^2$. The catchment area of the lake at its outlet is 15,321 $km^2$ of which about 20% is covered by the lake itself. More than 40 rivers and streams flow into the lake, but 93% of the water comes from four major rivers: Gilgel Abbay, Ribb, Gumara and Megech. The mean annual inflow to the lake is estimated to be 158 $m^3s^{-1}$ (i.e. 4,986 $Mm^3y^{-1}$). The mean annual outflow is estimated to be 119 m3s-1 (i.e. 3,753 $Mm^3y^{-1}$) (SMEC, 2008). The total population in the lake catchment was estimated to be in excess of 3 million in 2007 (CSA, 2003). The largest city on the lake shore, Bahir Dar, has a population of over 200,000 and at least 15,000 people are believed to live on the 37 islands in the lake. Most islands in Lake Tana are small, but two of them are larger (*Daga estifanos*, and *Dek*, which used to be the seat of Ethiopian emperors). The majority of the population depends for their livelihoods on rain fed agriculture. Fishery is an integral part of their livelihood both for household consumption and income generation.

## Research Methods

In order to acquire the data in regarding research questions, primary, secondary, qualitative as well as quantitative techniques of data collection techniques were applied. Data for this study were drawn from five main sources: (a) a set of case studies conducted in selected fishery cooperatives; (b) Households survey (c) in-depth key informant interview and series of focus group discussion (d) a documentary analysis and (e) stakeholder identification and analysis workshop. The study adopted mainly a qualitative research methodology, and employed semi-structured interviews and Participatory Rural Appraisal (PRA) methods (key informants, semi-structured interviews, informal group discussions, focus group discussions). Following household survey data collection, data were coded and entered in to SPSS software version 17. Simple descriptive statistics such as mean, standard deviation, frequency were used to analyze the socio economic variables of the respondents. Historical trend analyses of major changes in the fishery sector were used.

## Results and Discussion

### The system-to-be governed: fishery sector and its stakeholders

#### Socio-economic features of the community

The Amhara National Regional State (ANRS) is one of the 9 ethnically based administrative regions of Ethiopia. The mainstay of the regional economy is agriculture, characterized by subsistence crop production. Next to farming, animal husbandry is the most important activity in most agro-ecological zones of the region. The socio economic conditions are characterized by low GDP per capita and high levels of illiteracy. The illiteracy rate is very high with the urban population having better literacy rates than rural dwellers and males having about twice the literacy rates of females (WME, 2009).

The household survey result shows that 50% of the sampled households are illiterate and almost 67 % of the sampled households are dependent on crop and livestock production as a primary livelihood source. More than 20 % of the respondents are engaged in an integrated crop, livestock and off farm activities as a form of livelihood. The other livelihood options include crop integrated with fishing, wood collection, and charcoal production. Fishing around Lake Tana and other water bodies in the region is increasing in importance and consumption with an estimated annual off-take of 20 thousand tons.

The lake fishery has employed more than 4,500 persons who are directly dependent on the major activities of fishing, marketing, and processing for their livelihood. It is also contributing in giving employment opportunity to women and other landless people like ex-soldiers other than the fishers. There are also about 5,800 occasional fishers who fish for house consumption (Eshete, Unpublished). With the advent of dramatic population growth, all land based resource scarcity becomes a critical problem for the rural land tillers. Every new family demands land for cultivation. However, unless they do inherit from their parents, it is not easy to access land. Therefore, they seek to adapt by taking on other supplementary livelihood options. Fishing, charcoal production, and serving as casual laborer both in rural and urban areas are the main supplementary forms of income generation.

A typical household in the study area owned an average of 5 livestock (excluding poultry). Cultivable land is a critical problem in the study area. More than 70% and 58% of the respondents owned and cultivated less than 1 hectare of land respectively. Households in the study area are predominantly male headed (77%) with a fewer than (23%) of headed by female. The majority of female head were either divorced (30.4%) or widowed (45.7%). The remaining 23.9% of the female head were married or re-married women who have their own land and lead their family. The minimum and maximum number of family members in the household was 1 and 12 respectively. The Average household size is 5 persons per household which is greater than the national and regional average, 4.7 and 4.3 persons per household respectively. The respondent's age ranges from 20 to 87 years. The predominant ethnic group within each of the five districts chosen for the household survey is Amhara. Even though, they are small in numbers the *Negedewoyto* community are also found in Bahir Dar and alongside of Lake Tana shore.

| Characteristics | Number | Percentage |
|---|---|---|
| **Sex** | | |
| Male | 154 | 77.0 |
| Female | 46 | 23.0 |
| **Level of education** | | |
| Illiterate, no schooling | 99 | 49.5 |
| Adult Education | 53 | 26.5 |
| Grade 1-4 | 27 | 13.5 |
| Grade 5-8 | 16 | 8.0 |
| Grade 9-10 | 1 | 0.5 |
| Other(Religious-priests) | 4 | 2.0 |
| **Primary livelihood source** | | |
| Crop+ Livestock production | 134 | 67.0 |
| Crop+ livestock +off farm | 41 | 20.5 |
| Crop+ Fishing | 7 | 3.5 |
| Crop production | 13 | 6.5 |
| Livestock production | 5 | 2.5 |

**Table 1: Socio-demographic characteristics of the respondents (N=200)**
Source: Own survey result, 2011

## The *Negede Woyto* community and history of fishing in Lake Tana

The *Negede Woyto* community is considered to be an indigenous people around Bahir Dar, and the southern shore of Lake Tana. Members of this community are also residing in the North and South Gondar zones adjacent to Lake Tana. The community Elders claimed that their descendants were originally from Egypt. According to the 1994 national census of Ethiopia, the population of the *Negede* was estimated 1,677 but they are not mentioned in the preliminary report of the 2007 census. Currently, they predominantly live in kebeles 03, 11(Abay Mado), 13 and 16 of Bahir Dar city and are not provided basic social services. However, along with the establishment and continued expansions of the town, they have been evacuated from their original settlements and pushed to various areas of the town.

Regarding their social status, they are inferior group in the society and the other community groups look down and exclude them. They claim themselves as Muslim but the Muslim community does not recognize and accept them as a member of Muslim proper. But they have their own mosque in one of the kebeles. In most cases, their children are not attending formal schools but are engaged in simple informal activities like collecting garbage, firewood, street vending and shoe shining. Economically they are among the poorest section of the society. Traditionally, they were fisher and hunter men in and around Lake Tana (Care Ethiopia, 2011). They are endowed with traditional knowledge of hunting, fishing and construction of hand driven boats. They rely on papyrus reeds for much of their livelihood; using it to construct canoes for lake traders and other people who move around the lake by boat. They also use the reeds to make household utensils (such as baskets and mats) which they sell in local markets, both for utilitarian purposes and as tourist souvenirs (McCartney *et al.*, 2010).

They are known for their hand driven boats made of a dried stalk of papyrus, and attractive grass type native to the shores of the lake. At present they have engaged in unproductive informal activities to generate income for their subsistence. This includes fishery, shaping stone grinding mills, thatching *(sifet)*, and collecting fire wood, collecting and disposing home produced garbage. All the home based handicrafts are made using traditional/indigenous skills and tools passed from generation but they do not enable them to be competent and secure enough income for better life (Care Ethiopia, 2011).

Fishing in Lake Tana believed to have been started around 18$^{th}$ century by the *Negede woyto* community and then other poor members of the communities gradually adapted the activity. Though it is the largest freshwater body in Ethiopia, modernization of Lake Tana's fishery did not occur until the end of the 1980s. Before 1986 Lake Tana fisheries was made up by a predominantly subsistence reed boat fishery, operated by the *Negede* people (De Graaf *et al.*, 2006). The fishermen, limited in their mobility, only had access to the shore areas, using locally made fish traps, hooks and small gillnets (15–20 m).

In 1986 motorized boats and modern, nylon gillnets were introduced as part of the Lake Tana Fisheries Resources Development Program(LTFDP) which was initiated by the Ethiopian Ministry of Agriculture, the Ethiopian Orthodox Church and two Dutch NGOs (ISE-Urk and ICCO-Zeist) (Reyntjes *et al.*, 1998, Anteneh, 2005). This created new opportunities for the fishermen, extending their fishing area from the shore to deeper, offshore waters and, more importantly, to distant river mouths. Since then, the commercial gillnet fishery of Lake Tana developed rapidly and total annual catches increased from 39 tons in 1987 to 360 tons in 1997 (Wudneh, 1998). Despite the fact that they pioneered to Lake Tana fishery, for unknown reason the *Negedowoyto* community have not participated in the process of modernizing the fishery sector in the study area. They were excluded from participating in fishery cooperatives.

| Time period | Predominant fishers | Types of technologies used | Fishing ground | Rules and regulations of fishing |
|---|---|---|---|---|
| Before 1986 | The *Negede woyto* community | -Reed boat<br>-Local fish traps and hooks<br>-Poisonous plants | Limited to the Lake shore areas | Informal/ customary social arrangements/ self-governed institutions |
| Between 1986-2003 | -*Negede woyto*<br>-Modern fishers | Modernized boats and nylon gill nets | -Lake shore areas<br>-Deeper offshore waters<br>-Distant river mouths | There was no formal rules and regulations |
| 2003 to present | Organized fishermen | -Modernized boats and nylon gill nets<br>-Illegal gill nets imported from Sudan | -Deeper offshore waters<br>-Distant river mouths | - Proclamation and Regulation enacted<br>-But not yet implemented<br>-No directive since then |

**Table2: Major changes in Lake Tana fishery**

Source: Own survey result, 2011

### The status of fish resource: Declining trends in fish stock and production

Over recent years fish demands has grown significantly and the food habit of the population has been also changed. As a result, fishermen are motivated to harvest more fish to fetch a good price. The introduction of new fishing technology was expected to result in a rapid increase in utilization of the fish resource (Wudneh, 1998). As indicated by Nagelkerke (1997), being not well exploited at the time when commercial fishing started in 1980s, there has been a temporary rise in the yield. But gradually as the fishing pressure increases, fish production is decreasing. Moreover, if fishing continues to be intense at the time of spawning, the fish populations are expected to decrease rapidly. Members of the Tana haik No.1 fish cooperatives also reported that in the past 7 years, fish production and marketing has been declining (figure 1). As a result, they are struggling to secure and sustain their livelihood. Those who are not able to cope up with the situation, they are planning to leave the sector and searching for other livelihood options such as sand mining and casual labor in the nearby towns (Bahir Dar, Gondar, Woreta etc.). Too much fishing pressure is believed to be causing overexploitation of fish stocks and threatening the livelihood of the local fishermen.

In general, after the introduction of more efficient gillnets compared with traditional fishing gear, unregulated fisheries have had severe impact on the stocks of the Lake and rivers in spawning, lake-dwelling cyprinids (De Graff *et al.*, 2004). As reported by Vijverberg *et al.*, (2009), the abundance of endemic *Labeobarbus* species in Lake Tana decreased dramatically by ca 75% in 10 years' time (1991–2001). With the advancement of the fish sector in the last two decades, it can be imagined what is happening to the fishery resource in the past 10 years (from 2001-2011). The destructive fishing operations during the spawning season (August–September) in river mouths and upstream on the spawning grounds, alteration and destruction of spawning habitats is still going on which ultimately resulted in the loss of some fish species and extinction of 7 of the 15 endemic *labeobarbs*.

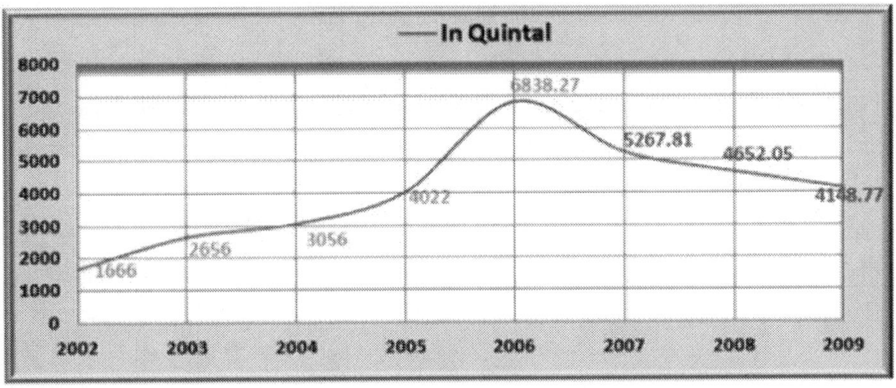


**Figure 1: Trends of fish production and marketing(2002-2009)**
Source: Calculated from Tana Haik No.1 fishing cooperative data sheet, 2011

## Who has a 'stake' and 'power' in Lake Tana fishery sector governance?

The word "stakeholder" was first recorded in 1708 as "a person who holds the stake or stakes in a bet" (Ramirez, 1999). Freeman (1984) defines a stakeholder as "any group or individual who can affect, or is affected by, the achievement of corporation's purpose. The understanding of stakeholders' involvement in the Lake Tana fish sector is an important step in bringing them into governance, using their competencies and capacities as necessary, and ensuring they are heard and have influence (Bavink *et al.*, 2005). This section depicts the stakeholders' landscape that evolved from stakeholder identification and analysis workshop which was carried out by a multidisciplinary team of experts from different stakeholder group (governmental, research, academic and NGOs).

Based on the mandate and responsibilities given by the Law of the Land (The Ethiopian Constitution), international treaties and customary laws of the specific community in the jurisdiction of Lake Tana Basin, stakeholders that have an interest in and influence over fish resource use, management and governance were identified, their stake values and power levels were assessed and evaluated. However, this identification cannot guarantee the exhaustive lists of stakeholders in the Lake Tana's jurisdiction. Because, stakeholders can change over time, new stakeholders can enter a resource management and governance system, while others may lose their role or interest (for instance the *Negede woyto* community). The assessment is based on the assumption that under complex and very dynamics systems, stakeholder's stake value and power level may be changed at any point in time, so that the current assessment is only representing the existing scenario. Accordingly, the nature of their mandate or interest, the bases of their stake, the stake value and the power level to take action can be also more dynamic in the wake of resource scarcity undetermined demand for those resources, a dynamic institutional and structural adjustments.

Policy analysts have long been aware of the importance of interest groups in the policy process; and the need to characterize and categorize levels of interest and power which influence, and therefore impact on, particular policies (Brugha and Varvasvsky, 2000). *Power* to influence policies or institutions stems from, the control of decisions with positive or negative effects. Stakeholder power can be understood as the extent to which stakeholders are able to persuade or coerce others into making decisions, and following certain courses of action. Power may derive from the nature of a stakeholder's organization, or their position in relation to other stakeholders (for example, line ministries which control budgets and other departments). Other forms of power may be more informal (for example, personal connections to ruling politicians) (IIED, 2005).

Stakeholders have very different degrees of power to control decisions that have effects on policies and institutions, and they have different degrees of 'potential' to contribute, or 'importance', to achieving a particular objective. As indicated in table 3, a high score on both variables (stake value and power level) will make them a clear candidate for becoming involved in the fishery governance decision-making process. Such stakeholders are likely to be the first to be consulted or represented. Variance in score among stakeholder groups may determine their relative influence and their formal status within the governing system. A low score on one attribute may be compensated by a high score on others. Thus, stakeholders may have less stake value and/or legitimate concerns, yet still enjoy a powerful governing position. Such a situation might easily challenge the participatory process and question the design of the governing system (Jentoft, 2007).

The stake-value-power analysis reveals that stakeholders that have higher level of stake value and high power to secure their stake (e.g.: Bureau of Agriculture) are likely have an extreme impact on the management and governance of the fishery sector. However, even when a stakeholder places a high value on their stake (whether positive or negative) but has a low power level to secure the stake in the resource system may not serve their interests. (e.g.. Youths). Conversely, a more influential stakeholder with perhaps only a moderate interest in a particular outcome may still have greater power to secure this outcome (e.g.. Institute of Biodiversity Conservation and Research)

| No | Stakeholder | Nature of the stake | Stake value | Power level | Total | Impact on CPRs |
|---|---|---|---|---|---|---|
| 1 | Bureau of Agriculture (BoA) | Regulation/Administrative | 5(+) | 6 | 30 | Higher |
| 2 | Bureau of Water Resource (BoWR) | Regulation/Administrative | 5(+,-) | 5 | 25 | Higher |
| 3 | Administration (Local to Federal) | Regulation/Administrative | 4(+,-) | 6 | 24 | Higher |
| 4 | Bureau of Env. Prote. Land Use and Administration (Bo-EPLUA) | Regulation/Administrative | 5(+,-) | 4 | 20 | Higher |
| 5 | Ethiopian Wildlife Dev. and Conservation Authority ( EWDCA ) | Regulation/Administrative | 3(+) | 5 | 15 | Medium |
| 6 | Local Community | Livelihood/Cultural value | 5(+,-) | 3 | 15 | Medium |
| 7 | Fishers | Livelihood/Cultural value | 5(+,-) | 3 | 15 | Medium |
| 8 | Academic and Research Institute | Knowledge generation and transfer | 3(+,-) | 5 | 15 | Medium |

| | | | | | | |
|---|---|---|---|---|---|---|
| 9 | Institute of Biodiversity and Conservation | Conservation and protection | 3(+) | 4 | 12 | Medium |
| 10 | Abbay Basin Authority | Regulation/Administrative | 3(+) | 4 | 12 | Medium |
| 11 | International Organization (FAO, UNESCO, RAMSAR etc.) | Cultural Values and Conservation | 3(+) | 4 | 12 | Medium |
| 12 | Investment Bureau | Regulation, Administrative | 3(-) | 4 | 12 | Medium |
| 13 | Ethiopian Electric Power Corporation (EEPCo) | Public Service/profit making | 4 (-,+) | 3 | 12 | Medium |
| 14 | Youths groups | Livelihood/Cultural value | 5(+,-) | 2 | 10 | Medium |
| 15 | Ethiopian Orthodox Tewahido Church (EOTC) | Livelihood/Religious value | 3(+) | 3 | 9 | Low |
| 16 | Fish Production and Marketing Enterprise (FPME) | Economic/Profit making | 4(-,+) | 2 | 8 | Low |
| 17 | Private Investors | Economic/Profit making | 4(-) | 2 | 8 | Low |
| 18 | Boat Owner | Livelihood/Profit making | 4(+,-) | 2 | 8 | Low |
| 19 | Lake Tana Transport Enterprise | Economic/Profit making | 2(+,-) | 3 | 6 | Low |
| 20 | Municipality | Administrative | 3(-) | 2 | 6 | Low |
| 21 | Civil Societies and Local NGOs | Non-profit making | 3(+) | 2 | 6 | Low |
| 22 | Women's group | Livelihood/Cultural value | 5(+,-) | 1 | 5 | Low |
| 23 | Bureau of Tourism and Culture | Administrative/Cultural value | 2(+,-) | 2 | 4 | Low |

**Table 3: Stakeholder analysis of fish resource management and governance in Lake *Tana* sub-Basin.**

Source: The outcome of the stakeholder identification and analysis workshop, 2011

Note: Stake value ranges from 1 up to 5 and power level from 1 up to 6 was assigned for each stakeholder and the value of the stake to the stakeholder was multiplied by the power

to take action, the result is an indication of the stakeholder's likely impact on the common pool resource use, management and governance.

**Stake Value**

| | |
|---|---|
| Critical | 5 |
| Essential | 4 |
| Necessary | 3 |
| Desirable | 2 |
| Non-essentia | 1 |

**Power level**

| | |
|---|---|
| Control - complete | 6 |
| Very significant | 5 |
| Influence - significant | 4 |
| Moderate | 3 |
| Low | 2 |
| Appreciation - lowest | 1 |

Accordingly, Bureau of Agriculture (BoA), Bureau of Environmental Protection Land Use and Administration (Bo-EPLUA), Bureau of Water Resource (BoWR), and Administration (from national to local level) had a particularly strong and legitimate claim to govern the fishery sector in Lake Tana sub-basin, because they ranked high on stake value-power level assessment and totaled the highest score. At the same time, stakeholders like Ethiopian Wildlife Development and Conservation Authority (EWDCA), Local community, fishers etc., had a medium impact on the governance system. This implies that, this information proved particularly useful in the policy and planning process, especially as a way to convince other stakeholders that the rights, responsibilities and accountability of these stakeholders had to be respected, monitored and integrated into the management and governance system.

Even though they possess high interest in and low power on fishery governance, local communities, fishers, and youths residing near to Lake *Tana*, who have the *de jure* [65]and *de facto* rights or claims over using and managing the CPRs are the most important stakeholders to the resource system. Their interests in fishery are mostly driven by economic incentives or livelihood security. Under the power vested on the 1995 Constitution of Federal Democratic Republic of Ethiopia (FDRE), Article 40(3) all 'Ethiopian people and the State have exclusive right to ownership of rural and urban land, as well as of all natural resources, presumably including fish resource. Therefore, their customary and ownership rights also helped them to own and/or access these resources.

Some stakeholders are interested because of administrative and social obligations to manage and govern the natural resource system. Power to take action and secure the outcomes could be vested politically or as a result of their social/financial capital the stakeholder has. For example, Bureau of Agriculture (BoA) has been vested a power by the supreme law under Proclamation No. 110/2007 and by Council of Amhara National Regional State to conserve fish biodiversity and its environment, cultivate fisheries resource with appropriate fishing equipment as well as prevent and control over exploitation of the fisheries resource. The Amhara National Regional State, Fisheries Resource Development, Protection and Utilization Proclamation enforcement, Regulation No, 50/2007 Article 3 sub article 4 states that;

*"The Bureau may particularly or fully cause the activity of fish production be terminated, as deemed necessary, at any water body whenever it believes that the potential of the fishery's resource may be disturbed and thereby be extinct, the area is ascertained to be that of delivery and reproduction, especially during breeding seasons, or where there happens to be a dwindling of the species of fish thereof".*

65 *De jure:*means according to law, by right, legally
*De facto:* means actual, exercising power or serving a function without being legally or officially established.

Generally at national level, The Ministry of Agriculture (MoA) has both development and natural resource management responsibilities (vested under Proclamation No. 380/2004). Relating to natural resources management, MoA has the responsibility to prepare policy on land use and draft legislation on forestry and wildlife. Bureau of Agriculture is the key natural resource management institution at regional level. It is responsible for the management of land, forest, wildlife and biodiversity resources (Proclamation No. 110/2007). The supremacy of the constitution as fundamental principle states that all international agreements ratified by Ethiopia are an integral part of the law of the land (Article 9 (4). Therefore, implementing agencies of the international treaties are also parts of the governance system. Therefore, international stakeholders such as FAO and UNESCO can affect fish resource management and governance through their interests in issues like ecosystem and biodiversity conservation. Basically their power over the resources is driven by their financial capital, international agreements and conventions that individual countries ratified.

## The governing System: The rules of the game and the governors

The policy, legal and institutional framework of the overall Ethiopian Natural Resource management and governance system comprises: The systems of formal national and international laws, regulations, treaties and procedures, and informal conventions, customs, and norms, which broaden, mould, and restrain socio-economic activity and behavior of multi-stakeholders. The Federal Democratic Republic of Ethiopia (FDRE) Constitution is the supreme law of the land that governs all laws in Ethiopia. As a result, a number of proclamations and supporting regulations were made that contain provisions for the protection, management and governance of natural resources and the environment which reflect the principles of the Constitution. A substantial body of policies and policy instruments are already in place with a direct or potential bearing on natural resource governance, watershed management and environment. However, despite the existence of proclamations and regulations at national and regional level, there is no specific policy framework and directives which addresses the fishery resource and to enforce enacted proclamations and regulations. The broad natural resource policy of the country is serving as an umbrella of the other specific resources.

## Institutions governing fishery resource

The Lake Tana Sub-Basin fishery administration and governance is characterized by lack of stakeholder participation and insufficient government commitment which leads to overexploitation and resource degradation. Before 2003, there were no specific formal rules that governed the behaviors of fish resource users. However, the introduction of more efficient gillnets in 1986 compared with traditional fishing gear, unregulated fisheries and an increasing number of fisherman have had severe impact on the fish resource in the Lake Tana and its tribute river mouths. As the number of fishermen and motorized/reed boat increases, competition and conflict over the fish resource also increased which ultimately led to decline of the fish stock (both in species and abundance). The existing broad institutions governing the natural resource were not in a better position to tackle fish resource degradation, depletion, minimize and/or prevent conflict among stakeholders, and secure sustainable livelihoods for the local fishermen and community. Hence, crafting a new and/or modifying the existing institutions have become a paramount need for the fish sector in general.

As a result, in 2003 the Fishery Development and Utilization Proclamation No. 315/2003 was enacted at Federal level to effectively use, manage and govern the fishery resources throughout the country. This is the only proclamation in Ethiopia that legally describes and elaborates capture fishery and aquaculture at the federal and regional level. Following the national proclamation, the council of Amhara National Regional State enacted its own Fisheries Development, Prevention and Utilization Proclamation No. 92/2003. The objectives of these proclamation are almost similar: to conserve fish biodiversity and its environment as well as to prevent and control over-exploitation of the fisheries resource; to increase the supply of safe and good quality fish and to ensure a sustainable contribution of the fisheries towards food security; and to expand aquaculture development (Article 3). The regional proclamation covers the same area as the national policy, but has an additional objective relating to the creation of employment opportunities in fishing communities. It also states that information, including research findings, should be made available to the fishing communities. After three years, Regulation No. 50/2007 was enacted to enforce the Fisheries Resource Development, Protection and Utilization Proclamation.

Concerning environmental protection, both proclamation states that the concerned organs of the Federal or Regional Governments shall ensure that development programmes and projects are drawn up in such a way that they will not have direct or indirect negative impact on the fisheries resource constituted in the basin where the programmes or projects are intended to be implemented. According to the Amhara region Proclamation Article 8 (2), the concerned body shall firstly ensure that different factories and other similar institution established around in any basins found in the region will not bring damage on the basin, the fisheries resource and all things in the water.

On the contrary, during the time of this study there was no limitation on the number, type of gill nets and fishing technology used. Because, directive to enforce the proclamation and regulations were not yet prepared and implemented at a regional level. There is no formal rule that govern the local fishermen.

The fishery resource development, protection and utilization proclamation enforcement regulation No 50/2007 Article 4/4 states that the sizes of meshes of gill nets that may be used in any water body shall, being varied and determined depending on the species of fish, and its developmental condition as well as the type of water body, be over 8cm in width. However, an illegal fishing nets with less than 8cm are being introduced from Sudan and causing a rapid depletion of the fish resource. Until the time of this survey, there was no initiatives from the concerning bodies to control the situation.

## The governing bodies

The House of Peoples' Representatives (HPRs) is the highest legislative authority in all matters (including natural resource governance) assigned by the Ethiopian constitution. With regard to natural resource governance, the HPRs, are given power to enact laws on the utilization of land and other natural resources (presumably fish resource), of rivers and lakes crossing the boundaries of the national territorial jurisdiction or linking two or more States; Article 55 (2) (a). The Ministry of Agriculture (MoA) is the principal ministry for fisheries and other aquatic life issues including the Institute of Ethiopian Agricultural Research (IEAR), Environmental Protection Authority. The Ministry is responsible for the development and management of the fisheries and IEAR undertake research in response to fisheries management and development needs. The Federal system of government in Ethiopia use regional Fisheries Experts in the Regional Governments with significant fisheries. Though not effective, at the user level members of fishers co-operatives are also charged with local decision making in fisheries and these co-operatives form the community-level basis of collaborative fisheries management in their respective jurisdictions.

## Modes of fishery governance in Lake Tana sub-basin

The interactive governance model recognizes three modes of governance; hierarchical, co-governance and self-governance, all of which are possible, depending on the properties of the systems that are being governed, the capacity of the governing systems, and the quality of their interactions. *Self-governance* refers to situations in which actors take care of themselves, outside the purview of government, where management authority and decision-making power rest within resource users organizations. It is also embedded within the societal realm of societal interferences, where individuals, families, groups, organizations, and even societal sectors govern themselves. (Bavinck *et al.*, 2005, Kooiman *et al.*, 2008, Chuenpagdee, 2011). Hierarchical governance is the conventional governance mode, characteristic of the interactions between a state and its citizens. It is a top-down style of intervention, with steering, planning and control as key concepts, which are expressed in instruments such as laws and policies (Bavinck *et al.*, 2005). In the co-governance perspective, parties co-operate, co-ordinate and communicate 'sideways', without a central or dominating governing actor. It is a situation where governments, resource users and/or community groups work collaboratively or engage in partnership arrangements to care for the ecosystems (Chuenpagdee, 2011).

Prior to the introduction of modern/commercial fishery in Lake Tana, traditional self-governance were purely the dominant governance mode over the hierarchical (centralized) and co-management. The *Negede woyto* community was the dominant fisher where they used to govern the fishery sector traditionally. Management authority and decision-making power rest within resource users. Still self-governance may occur at user level particularly among members of fishery cooperatives, where they can use their bylaws and local agreements to settle disputes and conflict of interests among fishers. After 1986, co-governance arrangements have been started, where organized fishery cooperatives and Bureau of Agriculture (BoA) share co-governance responsibilities. Various civil societies such as Ethiopian Orthodox Tewahido Church and two Dutch NGOs (ISE-Urk and ICCO-Zeist) were also involved in the governance system. However, since 2003 the government is trying to change the governance mode and structure into hierarchical. Even though, it is not implemented yet, they are trying to impose the Fishery Development and Utilization Proclamation and its regulation in to the fishery sector, so that BoA vested the power to control and monitor the overall fishery activities in the Lake Tana jurisdiction and other water bodies of the region. According to the national Fishery Development and Utilization Proclamation No. 315/2003, the fishery inspector, who is authorized by the Ministry of Agriculture or by the concerned Regional Authority has been vested the power to implement the proclamation, regulations and directives. Surprisingly, even after 10 years, fishery inspector was not assigned and directives were not prepared. And yet, the transition from the co-governance to hierarchical mode of governance became a long process which made the fishery sector unprotected and uncontrolled.

## Realities on the action arena: The governance interaction

Governance interaction between the systems to be governed (in this case fishery sector and its stakeholders) and governing system (the rules of the game and the governors) takes place at different level of the governance hierarchy spanning from the constitutional to the operational level at grassroots level. The process inquires to what extent the governing system is interacting with the system to be governed to insure the sustainable fish resource use, management and governance.

An examination of the relationship between the condition under which policy and legal instruments are devised at constitutional level and institutional contexts within which they are applied at operational level reveals that a top down governing structures and non-participatory policy and rulemaking process hindered a smooth relationship and plausible communication between the governing system and the system to be governed. Despite, the enactment of proclamation and legislation in 2003 at national and regional level, it took 3 years to endorse the regional regulation in 2006. And yet, there is no implementation directive to execute the rules and regulations at user level. It delayed for 3 years and in 2007, the Regional Parliament approved the Regional Fisheries Guideline. At a national level, guidelines have now been developed and submitted to the council of Ministers. It is surprising that it is still not endorsed at a Federal level where it affected the timely development of the Amhara Region Guideline.

## Fish resource access and withdrawal right

Institutions, such as property rights (the structure of rights to resources and the rules under which those rights are exercised) are mechanisms people use to control their use of the environment and their behavior toward each other (Bromley, 1991). Property rights consist of two components: the rule and its enforcement mechanism. The rules may derive from state law, customary law, user group rules, and other frameworks. Enforcement of statutory law is usually the responsibility of the state, which means that the rights ground on formal laws (Crewett *et al.*, 2008). The Ethiopian government is following a public trust doctrine, a legal principle dating back to Roman law in which state holds certain resources particularly land and other land based resources in trust for its citizens, prohibiting any transfer of those resources to private interests. Based on this premises the FDRE Constitution, which is considered as the supreme law of the Land (Article 40) vested an exclusive power of right to ownership of rural and urban land as well as all natural resources presumably including water, fish and wetlands to the state and the Ethiopian people. Constitution 1995 Article 40 (3) states that;

*"The right to ownership of rural and urban land, as well as of all natural resources, is exclusively vested in the State and in the peoples of Ethiopia. Land is a common property of the Nations, Nationalities and Peoples of Ethiopia and shall not be subject to sale or to other means of exchange".*

In addition, according to the regional Fishery Regulation for commercial fishing every fisherman has to have a fishing license to access and harvest the fish resource. Any fish harvester shall, having acquired a lawful license, have the right to harvest fish either permanently or temporarily, in an individual, group or associational capacity (Regulation No 50/2007, Article 6/2). The license will incorporate the type and amount of fishing gear, the boat number and size and the number of days someone should fish on the lake. However, any person dwelling nearby a water body have the right to capture up to 3kg fish using a single hook for his personal or household consumption without license provided that he/she may not avail same to market and sell (Article 6/3). The concept of "bundles of rights" as it is developed by Schlager and Ostrom (1992) is useful for our purpose to analyze the distribution of rights among different user groups and between individuals under the current tenure regime in Lake Tana Sub-Basin. Access and withdrawal are operational-level property rights, whereby access is the right to enter a given CPR system, and withdrawal involves the permission to extract products from the resource.

| Level of right | Property right | Stakeholder Group | | | | |
|---|---|---|---|---|---|---|
| | | The State | Local community | Organized commercial fishers | Private | Others (NGOs) |
| Operational | Access | √ | √ | √ | √* | √* |
| | Withdrawal | √ | √** | √* | √* | √* |
| Collective choice | Management | √ | √ | √ | ×* | ×* |
| | Exclusion | √ | ×? | × | × | × |
| | Alienation | √ | × | × | × | × |

**Table 4: Property right issues for fishery resource in the Lake Tana Basin**

Source: The outcome of Stakeholder analysis workshop and Focus group discussion, 2011

## Key

√ - Permitted

√*- Access and/or withdrawal of fish resource is allowed if and only if the stakeholders have a license

√**- Withdrawal is permitted without license only for household consumption up to 3kg per day

×? - In some cases communities may have the right to exclude others (except the government) from their commons

×*- In some cases stakeholders may be involved in management process up on the request of the State and the community

×- Not permitted

Ethiopian constitution explicitly states that the Ethiopian people and the state have exclusive right to ownership of the rural land and other natural resources presumably fish. However, except access and withdrawal rights, local community has no full ownership right on fish resource. Whereas, the State possess a full 'bundles of right' and alienation, the right to sell or lease management and exclusion rights is solely vested to the State.

The reality however is, despite the existence of the rules of the game (property right rules and proclamation/regulation) there is no restriction on access and harvesting of the resources at the user level. Therefore, every potential fishermen can access and harvest at any time, using any type of fishing gear, as he/she wish. Every fisherman is tempted to overharvest and exploited the available fish species. They have no incentives to conserve the fishery because any fish they leave is just going to be picked by the next fisherman. As a result, endemic fish species are overexploited and decreased dramatically, livelihood of fishermen are severely challenged, fish production and marketing activities of fish cooperatives were decreased drastically, competition and conflict over fish resource also increased. In recent times, the characteristics of common pool resources entitlement and inefficient rule enforcement also give incentives for many youths entering into the fishing sector as an option for their livelihood. In addition, investors are more interested on aquaculture investment in the Lake Tana fishery sector. And yet, Bureau of Agriculture who is given the responsibility to handle the overall governance of the fishery sector is not finalized preparation to start licensing of fishers and management of the resource.

## Knowledge and perception of the 'rules of the game'

For equitable, effective and efficient resource management and governance, all the various stakeholders must be part of the decision making and management processes. In a participatory approach, management decisions are more easily embraced by those who have been part of the decision-making process, and greater attention is paid to the needs and expectations of all actors (Ostrom *et al.,* 1993). Governance rules and regulations must be clear in nature and seen as appropriate by local stakeholders (Fisher *et al.,* 2010). Stakeholders can only be confident about the reliability of the resource governance process if they know what is going on their commons. As quoted by the famous Scientist Albert Einstein, '*knowing and understanding the rules of the game will help the stakeholders to play their roles better than anyone else in their respective action arena*'.

Yet, decisions which govern the use of fisheries resources should be societal and participatory where viewpoints from multiple stakeholders are balanced against potential costs and benefits of various resource uses (Scrimgeour and Wicklum, 1996). However, resource users are not well informed about and aware of the fishery proclamation and regulation. At the user level almost 95% of the respondents do not know the presence of proclamation and/or regulations. Surprisingly, from the sampled experts at district level, more than 87% were not even aware of these proclamations and regulations. This implies that, once the rules regulations enacted at higher level, either there was poor communication to the lower level to enforce the rules or the process of rulemaking (the way rules come to existence) were not participatory to involve all parties in the fishery governance system.

| **Knowledge of proclamation and/or regulation** | **District experts (N=15)** | | **Users (N=200)** | |
|---|---|---|---|---|
| | Frequency | Percent | Frequency | Percent |
| Yes | 2 | 13.3 | 11 | 5.5 |
| No | 13 | 86.7 | 189 | 94.5 |
| **Total** | **15** | **100** | **200** | **100** |

**Table 5: Knowledge about fisheries development prevention and utilization proclamation and regulation**

Source: Own survey result, 2011

Perception of resource users about the rules of the games that govern these resources could affect the management and governance of fish resource positively or negatively. It is their views of apparent benefits or costs arising from the resource system which will determine whether they will cooperate positively or act negatively towards governing the resource. Particularly conservation of natural resources is greatly affected by the perception and attitudes of the stakeholders who are involved in the process of resource management and governance. Therefore, Community members were asked to evaluate how well the existing institutional arrangements (property right of land and other resources) are clear and open to all resource users. The general perception is that institutional arrangements that are currently in place are not clear and open to all users. The Majority (80%) of the respondents who are the primary users of the CPRs were agreed that the current institutional arrangements are very poor to govern the CPRs and more than 65 percent of the respondents considered the existing rules and regulations as inefficient and not transparent for all users.

There are cases where rules and regulations are applied appropriately to administer natural resources. In some areas there were occasional arrests and detention of local farmers for fish, wetland and grazing land encroachments. First they will be judged by the community and local administrations. If they admitted not to repeat the same mistake, they will be freed with warning. However, sometimes cases will be directed to social courts and those who are found guilty will be penalized by imprisonment or money. Conversely, within the same villages or districts people who have a strong affiliation with administrators were allowed to cultivate common pool resources such as wetlands, even they were not given warning to stop their wrong doings. As a result, community members were confused of the administration procedures and for whom these rules and regulations are applied or not applied.

In summary, major challenges of Lake Tana fishery are:

Uncontrolled overfishing and loss of endemic species

Illegal fishing activities (Mesh size less than 8mm is used)

Importation of Illegal fishing gear from Sudan

Lengthy law making process to protect the declining stock of fishery resource

Stakeholders' poor knowledge and perception of legislations, proclamation and their derivatives

Mandated agencies not discharging their responsibilities (Bureau of Agriculture)

Weak institutional framework to guarantee sustainable fish resource management and development

## Conclusions and the way forward

Despite its potential contribution to the country's drive to food and livelihood security, Lake Tana fishery is threatened by a multitude of management and governance related problems. The governance interaction has to come up with a positivc outcomc in ordcr to guarantee the sustainable fish resource use, management and governance. However, if one has to ask at a glance, whether the present-day natural resource governance system of Ethiopia is appropriate for the current and anticipated future developments in Lake Tana basin or not?. The likely answer is 'no'. Because, though there are rules and regulations enacted at constitutional and collective choice arena, the implementation and enforcement of these rules and regulations at the user level is not as such satisfactory. Formal laws were enacted without considering the local context where community bylaws operated and even without the consent and consciousness of resource users. On the other hand traditional resource management and governance institutions are losing ground to effectively manage local level resource systems.

Proclamation and regulation for fisheries resource development, protection and utilization were enacted at national and regional level. However, most of the resource users and experts at a grassroots level were not aware of the existence of these rules and regulations. There were no binding legal rules that govern the behaviors of fishermen and other stakeholders at user level. Because directives to enforce rules and regulations at the operational level, were not prepared yet. Though the power to control and monitor the fishery resource management and governance issues were vested to Bureau of Agriculture, the Bureau is not discharging its mandates and responsibilities as early as possible.

Therefore, for a controlled development and proper management and governance of the expanding Lake Tana fisheries, enacting legislation and its enforcement (regulation) at the constitutional level could not be a panacea to protect the fish resource depletion; rather the process of rulemaking should be participatory, it has to be communicated with actors in the action arena (where rules interact with actors to come up with certain outcomes) and the legislation has to be enforced at the user level. Legal instruments should also lay out the procedures that must be followed by public agencies to identify and allocate rights and responsibilities within specific management instruments and agreements. Organizing local level users association with their own bylaws (crafted by them for themselves), with a minimal outsider (government, NGOs) support is imperative for sustainable fish resource production and conservation.

## References

ACP (2012) African, Caribbean and Pacific Group of states (ACP) Fish II project. Accessed online on March 13, 2012 at http://acpfish2-eu.org/index.php?page=ethiopia&hl=pt

Anteneh, W. (2005) "*The spawning migration of* Labeobarbus *(Cyprinidae: Teleostei) of Lake Tana to Dirma and Megech Rivers, Ethiopia.*" Unpublished MSc thesis, Addis Ababa University, Addis Ababa, Ethiopia.

Awulachew, S. B. Erkossa, T. Smakhtin, V. Fernando, A. (Comp.). (2009) *Improved water and land management in the Ethiopian highlands: Its impact on downstream stakeholders dependent on the Blue Nile. Intermediate Results Dissemination Workshop held at the International Livestock Research Institute (ILRI), Addis Ababa, Ethiopia, 5-6 February 2009.*

Bavinck. M, Chuenpagdee, R. Diallo, M. Van Der Heijden, P. Kooiman, J. Mahon, R. and Williams, S. (2005) Interactive fisheries governance, Delft: Eburon Publishers. 72 pp.

Bromley, D.W. (1991) Environment and Economy: Property Rights and Public Policy. Basil Blackwell, Oxford, UK

Brugha, R. and Varvasovszky, Z. (2000) Stakeholder analysis: a review. *Health policy and planning,* 15, 239.

CARE ETHIOPIA, (2011) Lesson Learned *Negede Woyto* Minority Group in Bahir Dar. Accessed Online on January 20, 2011 at http://www.care.org.et/LessonDetail.aspx?id=3

Chuenpagdee, R. (2011). An 'interactive Governance' Perspective on human and ocean systems interactions. Integrated Marine Biogeochemistry and Ecosystem Research (IMBER). Issue n°18 - September 2011 Accessed online February, 2012. At http://www.imber.info/index.php/Products/Newsletters/Issue-n-18- September-2011

Crewett, W., A. Bogale, and KORF, B. (2008) Land Tenure in Ethiopia: Continuity and Change, Shifting Rulers, and the Quest for State Control. CAPRi Working Paper 91. International Food Policy Research Institute: Washington, DC. http://dx.doi.org/10.2499/CAPRiWP91

CSA. 2003. Central Statistics Authority, Addis Ababa, Ethiopia

De Graaf, M., Van Zwieten, P. A. M., Machiels, M. A. M., Lemma, E., Wudneh, T., Dejen, E. and Sibbing, F. A. (2006) Vulnerability to a small-scale commercial fishery of Lake Tana's (Ethiopia) endemic Labeobarbus compared with African catfish and Nile tilapia: An example of recruitment overfishing? *Fisheries Research,* 82, 304-318.

De Graaf, M., Machiels, M. A. M., Wudneh, T. and Sibbing, F. A. (2004) Declining stocks of Lake Tana's endemic Barbus species flock (Pisces, Cyprinidae): natural variation or human impact? *Biological Conservation,* 116, 277-287.

FAO. © 2005-2012. Fisheries and Aquaculture topics. What is governance?. Topics Fact Sheets. Text by S.M. Garcia. In: *FAO Fisheries and Aquaculture Department* [online]. Rome. Updated 27 May 2005. [Cited 7 August 2012]. http://www.fao.org/fishery/topic/12271/en

Fisher, B., Kulindwa, K., Mwamyoka, I., Turner, R. K. & Burgess, N. D. (2010) Common pool resource management and PES: Lessons and constraints for water PES in Tanzania. *Ecological Economics,* 69, 1253-1261.

Freeman, R.E. (1984) Strategic management:a stakholder approach.pitman. Boston, MA, USA.

IIED. (2005) Stakeholders power Analysis. Power tools. International Institute for Environment and Development (IIED). Accessed on July, 2012. Online at http://www.policy- owertools.org/Tools/Understanding/docs/stakeholder_power_tool_english.pdf

Jentoft, S. (2007) Limits of governability: Institutional implications for fisheries and coastal governance. *Marine Policy,* 31, 360-370.

Kooiman, J., Bavinck, M., Chuenpagdee, R., Mahon, R. and Pullin, R. (2008) Interactive governance and governability: an introduction. *The Journal of Transdisciplinary Environmental Studies,* 7,2-11.

Kooiman, J., Maarten, B., Svein, J., and Roger, P. (Ed). (2005) *Fish for life: interactive governance for fisheries*, Amsterdam Univ Pr.

Mahon, R., Bavinck, M. and ROY, R.N. (2005) Governance in action. In: Kooiman, J., M. Bavinck, S. Jentoft and R. Pullin (eds.), 351-374.

McCartney, M.; Alemayehu, T. Shiferaw, A.; Awulachew, S. B. (2010) Evaluation of current and future water resources development in the Lake Tana Basin, Ethiopia. Colombo, Sri Lanka: International Water Management Institute. 39p. (IWMI Research Report 134). doi:10.3910/2010. 204

Nagelkerke, L.A.J. (1997) The barbs of Lake Tana, Ethiopia: morphological diversity and its implications for taxonomy, trophic resource partitioning, and fisheries. Doctoral thesis. Experimental Animal Morphology and Cell Biology, Wageningen Agricultural University, Wageningen. The Netherlands

Ramirez, R. (1999) Stakeholder analysis and conflict management. *Cultivating peace: conflict and collaboration in natural resource management*, 101-126.

Reyntjes, D., Mengistu, T., Wudneh, T. and Palin, C. (1998) Fisheries development in Ethiopia–which way now? *European Union Bulletin* 11 (1), 20–22.

Schlager, E. and Elinor, O. (1992), 'Property rights regimes and natural resources: A conceptual analysis', *Land Economics*, 68, 249–262.

Scrimgeour, G. J. and Wicklum, D. (1996) Aquatic ecosystem health and integrity: problem and potential solutions. Journal of the North American Benthological Society 15:254–261.

SMEC. (2008) "Hydrological Study of the Tana-Beles sub-basins, main report. Snowy Mountains Engineering Corporation (SMEC).." Ministry of Water Resource, Addis Ababa, Ethiopia.

Vijverbg, J., Sibbing, F. and Dejen, E. (2009) Lake Tana: Source of the Blue Nile. *The Nile*, 163-192.

Wilkes, A. (2005) The Creation of Community Rangeland Management Institutions in Dimaluo: Three Casesof Community-Based Natural Resource Management (CBNRM). Working Paper No. 11. Center for Biodiversity and Indigenous Knowledge, Yunnan

WME. (2009) No 01 Tana-Beles, Watershed Monitoring and Evaluation (WME) Project Document. July 2009.135 p.

Wondie, A. (2010) Improving management of shoreline and riparian wetland ecosystems: the case of Lake Tana catchment. *Eco hydrology and Hydrobiology,* 10, 123-132.

Wudneh, T. (1998) "*Biology and management of fish stock in Bahir Dar Gulf, Lake Tana, Ethiopia.*" PhD Thesis, Agricultural University, Wageningen, the Netherlands.

# PART IV
## REVISITING THE ROLE OF INDIGENOUS KNOWLEDGE IN HIGHER LEARNING INSTITUTION

# CHAPTER 15

## Promises and Perils of the Virtuous Knowledge Exchange Cycle in Ethiopian HEIs

*Elizabeth Ayalew*

### 1. Introduction

Society is informed and formed by its universities who constantly construct, disseminate and reconstruct knowledge. As products of the society in which they operate, these institutions in return are influenced by societal requirements. Unlike earlier times, prior to 1990s, that advocated investment in primary and secondary education, the last two decades have seen considerable progress towards the wide acceptance of the crucial role that universities play to promote economic competitiveness, regional regeneration and creation of innovation hubs for business and the community (The World Bank 2002, 2009a). This international recognition has accelerated the growth in the sector; but, in order for this growth to be qualitatively meaningful, a mechanism that allows the flow of knowledge between university and society – a virtuous cycle of knowledge exchange – needs to be in place.

This article bases its background on the international context of Higher Education (HE) whose constant change is mainly attributed to changes in the contemporary world of knowledge economies. While doing so, it identifies the challenges and opportunities the international setting offers to local universities found in developing economies. Thus, this argument brings about the essential role that the Knowledge Exchange (KE) process introduces to the expected function that Higher Education Institutions (HEIs) are believed to deliver for societal development while at the same time enhancing their internal performance. Finally, the paper addresses key concerns in the way forward to tackle challenges in the Ethiopian context of KE.

## 2. The realities of the 21st century and the context of HE in Developing Economies

While land was the main source of wealth and income in agricultural societies, capital and machinery were in the industrial society. Today, information and knowledge drive economic growth which gives HE a preeminent space in the production and dissemination of knowledge. However, for the cycle to be complete and for knowledge to bring about desired development, the two dimensions - production and dissemination - need to be supported by the third one – knowledge absorption by the society from which it is constructed (Rochford 2007). This paper argues that the HE system, particularly in developing economies, need to go beyond the traditional mission of universities - that of teaching and research - and embark on knowledge exchange to impact on economic, social and cultural aspects of life.

Knowledge economies place greater significance and accord significant priority to the production, dissemination and absorption of knowledge. While it is true that some countries produce more knowledge than others, those who produce less need to work much harder and devise ways of protecting themselves from the monopoly of it. The World Bank Report (2009a) estimates that nearly half of today's HE students live in the developing world. However, across most of the developing world, the potential of HE to promote development is being realized only marginally. Thus these developing economies will need to focus on strategizing on their knowledge systems if they aspire to bring about development through their HE systems. Although the challenge is immense, there is room for HEIs of developing economies to be able to identify their niche in this regard. While considering the scenario of HE in the contemporary world, this section opts to adopt three major areas of analysis – expansion, differentiation, and knowledge revolution– which have been taken as major stances of development of HE in World Bank's report (2000) and are highly related to concerns of quality in the system.

a. Expansion - In 1995 more than 47 million students were enrolled in HE in the developing world, up from nearly 28 million in 1980 (The World Bank 2009a). The effect of the constantly increasing number of students in HE worldwide cannot only be seen in terms of numbers. Above all, this quantitative increase has brought about noticeable changes in the demography of the adult population. There is growth in income as well as in urbanization of settlement areas thereby adding value to the growing economic importance of knowledge and skills. As a result, in most countries, HE is no longer a small enterprise for the elite; instead, it has gained universal access and recognition as a vital instrument for a nation's plans for development (World Bank 2009a). Along with expansion however, the visible imbalance in equity - between rich and poor households, urban and rural areas, men and women, and among ethnic groups - needs to be addressed.

b. Differentiation – in the traditional university structure, prominent entities are vocational studies (also known as polytechnics), teaching faculties (those institutions that grant degrees but do not conduct research), and community colleges (Lee 2007). Today, economic development is associated with a more refined division of labour, and an increasing demand for diversified knowledge and skill. Therefore, universities are undergoing the process of vertical and horizontal differentiation (World Bank 2000). While horizontal differentiation refers to the proliferating diversified field of studies driven by the increasing demand for HE; vertical differentiation refers to the increased diversity of graduates at various levels of qualification. It could be worth noting that the emergence of new institutions operated by private providers, both for-profit and no-profit purposes (philanthropic and religious), further accelerates the process. The spread of distance-learning operations is a typical and important example of differentiation.

c. Knowledge revolution- the rapid expansion of scientific capacity particularly in some economies like China, Japan, Hong Kong, Korea, etc. ; the increased number of patent applications (The World Bank 2009b) contributing to technical innovations, the increased output in published scientific papers (World Bank 2009b) coupled with advances made in Information Communication Technologies (ICTs) and internet technology have not only pushed the knowledge frontier, but have revolutionized the creation and dissemination of knowledge impacting quality of life in general.

Given the above major realities, academic institutions all over the world are now grappling with ways and means of coping up with these transformations taking place in the system. At a broader level, the fast-growing market economy, the rapid development of science and technology, and rising individual income levels and living standards have motivated the ever-increasing demand for HE opportunities. And at the national level, there are various distinctive features of countries that interpret these global changes to the context of each country. Thus, while the challenges play themselves out on a global level, academic systems function in a national environment.

More so, such interpretation gets more complicated as seen in the context of developing economies. Faced with competing demands on resources, governments of developing economies find it difficult to provide adequate funds for HE to maintain standards at the level required. In brief, Altbach and Peterson's report (1999) clearly spelt out the contemporary realities of African universities and identified the following as some of the predominant issues: the effects of demographic changes (student mobility, life-long learning, etc.) on the provision of HE; the adverse effects of a deteriorating economy on quality education; the participation of institutions of higher learning in the production and ownership of knowledge—especially in the area of information technology; and the role of universities in the political and cultural changes that are likely to occur in the 21st century.

For any significant shift in paradigm or policy direction to take place, it is essential that a comprehensive understanding of the working system of

HE in such contexts is well in place. Therefore, the following section highlights priority issues that characterize HE systems in developing economies in order to be able to identify the possible niche of meaningful change for low-income countries to consider.

**2.1. Resources in HE of Developing Economies -** Many of the problems involving HE in the context of developing economies are rooted in a lack of resources. Governments in these countries are pressed by demands for public budgets from the sector of HE. In addition, there is poor coordination of capital and operating budgets (The World Bank 2002) often reflected in situations where major new facilities are built, but then are left with no funds for operation and maintenance. HEIs in the developing world are seen as having deteriorating buildings, inadequate libraries and laboratory facilities, and scientific equipment that cannot be used for want of supplies and parts. Carrying over unspent funds for use in later years is nearly unthinkable in these institutions and wining a budget that is higher than the previous year's actual expenditure is not easy. This creates the "use-it-or-lose-it" mentality, resulting in misspending of resources.

This is further exacerbated in research universities as the pressures of expansion in teaching universities gets significant by the year. Public universities in Africa and Asia often devote up to 80 percent of their budgets to personnel and student maintenance costs (The World Bank 2009a), leaving few resources for infrastructure maintenance, libraries, equipment, or supplies—all key ingredients in maintaining a research establishment. It is not easy for HE systems in these countries to look for ways with which they can make more effective use of their physical and human capital, including access to new technologies needed to connect themselves to the global intellectual mainstream.

**2.2. Faculty and students -** Most important to consider would be quality of faculty and problems associated with students. A well-qualified and highly motivated faculty is vital to embed a more enlightened view of learning in the system in general and to augment the active intellectual engagement of students in particular. In these institutions, faculty pay is generally very low in relation to that offered by alternative professional occupations; and pay increases are governed by bureaucratic personnel systems that reward long service rather than success in teaching or research (Janjua 2011, Mwiria 2003). Moreover, politicization can also have a wider impact on the ambiance of the system. While university campuses are well remembered for being stages of riot demanding for democracy, it has also been witnessed that many such institutions have been victims of political favouritism and external interference on internal decisions.

A significant number of faculties are engaged with a divided attention (Janjua 2011) as they work part-time at several institutions. Classes are often severely over-crowded and students' living conditions are distracting with limited support from student services. In addition, many students start their studies with inadequate basic and secondary education (Sawyer 2002), combined with a lack of selection in the academic system. Their choice of field of study is dictated by exposure to opportunities and the market condition despite merit and personal inclination; and the risk of unemployment, at the end of education, is rising by the year as the number of graduates constantly increases. Generally, the providers of HE have shortcomings of various nature – the public sector being underfunded while the private (for-profit) sector focusing on addressing market-driven needs primarily.

**2.3. Governance and HE leadership** - Governments in developing economies have often strived to ensure strong relations of political accountability with leaders of HEIs (Sawyer 2002, Teferra and Altbach 2003). To this end, institutional governance arrangements have been modified to incorporate less stakeholder representation. And the resulting encroachment on academic freedom and university autonomy has been proven to have seriously affected university's identity as a research hub. Important earlier studies were undertaken to deepen understanding of the systemic peculiarities of HE, which have shed light on the sociological organization of universities

(Maitra 2007). In fact, it is imperative that governments understand the nature of such institutions, loosen their role from 'controlling' to 'supervision', and give space for faculty to recognize and respond to their changing role in knowledge-based society.

**2.4. The Status of Research in HEIs -** Research is a very expensive undertaking particularly for poor countries of the world. For example, Botswana, Burkina Faso, Namibia, Tanzania, Zambia, and Zimbabwe all spend less than 0.5 percent of GDP on research and development (Mouton 2008). Often, research in such universities is conducted by funds from external organizations as governments' budgets for research diminish by the year. Besides, the increased focus of graduate programs on teaching and the increasing engagement of HE staff in increased teaching workload, for purposes of supplementing inadequate academic salaries, have all together compromised the conduct of research and quality of output. Ultimately, as research universities become unable to find their capability to act as reference points for the rest of the education system, it rapidly becomes difficult for countries to make key decisions about the critical issues affecting them.

**2.5. University Education, Employment and Brain-drain -** Education has always played an important role in the successful integration of young men into the labour market. In the past, the process of leaving school and getting a job took place in a single step; however, the transition from education to work has become more complex nowadays. The reasons for such a bleak prospect in the reality of the job market could be many among which the rapidly expanding HEIs and their subsequent increasing supply of graduates holds a prominent position, at face value. The diverse specialization and improved qualification of new graduates, and the advancements made in automated technology with its potential risk on job security could be mentioned as additional reasons. While human right provisions of freedom of movement and choice of employment encourage the mobility of migrants to other countries increasing the risk of investment in HE, HEIs in developing nations face the risk of losing their staff to other countries (Teferra and Altbach 2003). Junior staffs are sent abroad for further training and often do not return once their studies are completed; and senior ones look for jobs in either local non-government organizations or international offices where they can be paid better.

In general, despite notable exceptions, most HEIs in developing countries suffer from deficiencies in each of the above discussed areas. While each emerging nation has its own national priority to determine what it can realistically expect its HE system to deliver, it is undeniably important that this decision must be informed by historical and comparative knowledge about the contribution of HE to social, economic, and political development. Resolving the situation is critical, but not easy. And a nation's capability to play a meaningful role in the change process, as the future unfolds to portray its unpredictable development, requires vision and projection. Every developing country needs to think strategically about how, given their inevitably limited resources, HE might best be deployed to the advantage of future generations.

One best way of strategizing on this line, which is the major theme of this article, is enhancing university function to assist HEIs become innovative rather than focusing on their traditional role of teaching and research only. What distinguishes innovative universities from the mainstream is a holistic approach to knowledge exchange with the recognition that there is a virtuous knowledge transfer and exchange cycle. The following section conceptualizes Knowledge Exchange (KE) and further identifies its specific value and the particular perspectives from which its implementation has to be seen in the context of Ethiopian HEIs.

## 3. The Rationale for Knowledge Exchange in HEIs

As knowledge becomes more important, so does HE. It has become mandatory for countries to educate more of their young people to a higher standard. A mundane example to mention would be that of first degrees becoming basic qualification for many skilled jobs these days. So also, the quality of knowledge generated within HEIs, and its availability to the wider economy, is becoming increasingly critical to national competitiveness.

For quite some time, the term 'technology transfer' has remained popular in the repertoire of relevant disciplines. Gradually, the focal vocabulary changed from 'technology transfer' to 'knowledge transfer' which, in turn, is being challenged by the concept of the more free-flowing multidimensional concept of 'knowledge exchange' between universities and society.

The term knowledge transfer appears too restrictive in terms of depicting the innovative practice of knowledge exchange, which is defined as "a set of policies and practices which enable the efficient and effective exchange and co-creation of knowledge between producers and users" (Hagen 2008:113).

This definition describes a virtuous circle of engagement between knowledge producers (typically scientists, but potentially all academics) and knowledge users (policy makers, practitioners, stakeholders, businesses, social enterprises and other publics) so that the boundaries between the producers and users ultimately become merged. Often times, researchers prefer writing their research output in a format in which they can meet their academic requirements. However, decision makers, media and consumers would rarely read them. Therefore, HEIs have to look for opportunities that create effective engagement with audience outside the academic circle thereby forming the virtuous circles between teaching, research and KE with each strand supporting and reinforcing the other.

The concept of KE cycle demonstrates a synergy and inter-connectedness between the two traditional missions of a university and involves a third component– KE. This component of the cycle provides the public (business and the community) with the knowledge, skills and competence it needs while the public can, in turn, influence the development of university's function in terms of developing the curriculum, learning methods, research, governance and public engagement. When graduates move into the world of work, they take with them the body of knowledge, skills and aptitudes acquired at university. While in the world of work, they acquire additional knowledge and skills. This is where knowledge begins to be transferred to the public; but as the graduates return to HE for further training, as many more of them do several times across their lifetime, mechanisms have to be in place for them to bring this additional knowledge to bear on the university. An adaptive university system will judiciously use this to refresh its curriculum, update its professional knowledge, feed into its research agenda, create collaborative research opportunity and develop its learning and teaching markets. Moreover, these students then return to the world of work, where the cycle is repeated.

Recently, there has been recognition of the need for open partnerships between academia and business/industry at the level of HEIs in developing economies. One of the recent World Bank's reports (World bank 2009b) is based on a study of six countries in Sub-Saharan Africa (Ghana, Kenya, Mauritius, Nigeria, South Africa, Tanzania, and Uganda) observing the relationship between economic growth and education, during 1990-2006, with focus on key drivers behind growth performance in these countries. In this study, emphasis is made on diversification of economic activities and exports and technological capabilities by strengthening ties between tertiary institutions and the business sectors. Some recommendations for policy actions are also made to help raise quality of HE to increase technology-led development in Africa. The notion of this innovation appears to be largely based on the recognition of the importance of a vibrant interchange between creativity and technology to produce commercializable inventions to boost the growth of the business sector. Generally, the focus of such initiatives seems to be focused on technology and knowledge transfer.

The KE dynamics however needs a broader framework of establishing a knowledge management strategy to underpin its responsible partnering processes within a knowledge exchange model, as has been corroborated by Hagen (2008). Accordingly, this approach gives room for the community to participate and influence the direction of the university at a wide number of interfaces:

- Strategic business alliances and collaborations;
- Consultancy, business support and expert advice;
- Joint R&D ventures with business;
- Joint research projects;
- Designing CPD/professional training (short courses for business and professional communities);
- Contracting with each other for the commercialization of Intellectual Property (IP);
- Student placements (graduate and undergraduate) in business and the community;
- Student incubators (enterprise/start-up support) jointly supervised;
- Graduate/post-graduate employment panels;
- Dissemination of ideas via publications with joint articles; and
- Co-publication and sales of software/materials.

(Hagen 2008:114)

Hagen (2008: 128) defines 'Innovative universities' as those which "seek to combine use of knowledge with industrial, business and regional growth and provide a frame for a dynamic interaction between educational, research and development, business and government sectors". As has been underlined in this definition, innovation does not take place within a single company, but involves a range of partners. As the various actors of KE interact and collaborate within the context of an innovation system, the challenge is for universities to stay competitive in a developing global market for HE. The key collaborative partners in this process are public sector bodies, local and central government, research and development institutions, other HEIs, industrial partners, Small and Medium Enterprises (SMEs) and international partners.

Kearney (2009:10) corroborates this idea by taking an integrated view of Higher Education, Research and Innovation (HERI) as the basis for social development and considers the following three dimensions important:

— Placing knowledge, including high-level scientific knowledge, at the service of development

— Converting knowledge, in all its forms, into value via applications and impact assessment

— Sharing good practice, to ensure widespread benefits.

There are quite a number of benefits that the KE process draws to the HE system. Under the general umbrella of allowing a strong relationship between university and the public, it forges a closer cooperation between science and industry to enable HE recognize the gap in the national production process. This means, the value of applied research will be strengthened. It also encourages the engagement of researchers in the field of education in studying practical approaches of learning-teaching that help transform curricular deficiencies. More importantly and at a higher level, KE introduces new framework for HE teaching and research thereby consolidating the national innovation system and increasing international visibility, competitiveness and research impact that HEIs can make.

Thus, HEIs will have better chance of improving their capability to exploit the knowledge in the economy. Through KE efforts, these institutions will be well positioned to interact with the knowledge absorbents. And the effect of this interaction is that HEIs are no longer closed systems that only serve to supply knowledge, but also receive knowledge from the interaction to be able to look inwards and align their activities with needs of industry and other public and private organizations. Ultimately, the whole interchange is expected to foster a culture of KE among these actors.

## 4. Priority issues to overcome challenges in the Context of Ethiopian HEIs

Alike many other developing economies, HEIs in Ethiopia are under great strain to accomplish their objectives. And it is high time that they redirected their focus and be more strategic in their approach in light of the escalating demands they have to respond to. All the issues raised under section two (Items 2.1 -2.5) that describe the realities of the 21st century in the context of HE, apply to the Ethiopian case also. Consequently, like many other nations, Ethiopia has taken some measures to respond to the effects of globalization as well as national demands; however, a lot remains to be done in terms of ensuring the 'fitness-for-purpose' of output from the HE system. This has been strongly reiterated by Sawyer (2002: 42) while explaining the value of maintaining advanced knowledge and research-driven universities even in poor countries:

There is no suggestion that all universities succeed in achieving this condition all the time. The point is rather that the best universities aim at doing so and, to the extent that they succeed, they fulfil their true mandate and make a unique contribution to society. Further, that every society – even the poorest – is entitled to access to such a resource if it is to derive full benefit from the world's vast reservoir of knowledge resources.

Perhaps the most critical challenge facing most institutions will be to develop the capacity for change in order to reposition themselves for a constantly changing society. This requires a concerted effort and commitment by the academic community to embark on what should be a great adventure for higher education. The World Bank (2009b) makes references to few universities in Africa who have been so far successful in revisiting their productive and relevant roles to society. Jimma University in Ethiopia is one such case that has pioneered community-based education in the country, through which it requires tertiary students in medicine, agriculture, and other professions to engage local communities, to learn from them, and to pursue problem solving with them. Community-based education is believed to enhance educational relevance to demands of the community. The community is used both as a learning environment to the students and a beneficiary agent of the educational experience. Therefore, the program is of clear benefit to both students and the community. This approach has enabled the university to offer diverse studies in inter-disciplinary areas; it has allowed a team approach to teaching; it has enhanced interaction with the community and has redirected the focus of research to be problem-based and community-demand driven (www.ju.edu.et/am/research).

It is against this background that the challenges facing Ethiopian universities and their various attempts to survive and thrive must be assessed. In this connection, the following critical points may be considered for any serious deliberation that aims to address the knowledge exchange gap seen in HE.

**4.1. The need for systematic collaboration** -The following remark (Aschcroft 2004:15) best describes the situation observed among Ethiopian HEIs as far as the issue of synergy and collaboration is concerned:

If a higher education system is defined as a set of interrelated institutions each with its own function within the system, each with its own goals, each of which makes a particular contribution to the functioning of the country, at the present time it could be said that Ethiopia has a collection of institutions rather than a system.

HEIs need to seize the initiative to work within their own selves to build and strengthen their respective institutions and facilitate full use of endogenous capacities.

One potentially effective way of doing this would be that HE leaders have to, with active support of the academic community, design and develop clear institutional policies, programs and strategies for cooperation and collaborations. Partnership and alliances amongst stakeholders is a key component of creating synergy. National and institutional policy-makers, teaching and related staff, researchers and students, stakeholders from non-governmental organizations and community groups need to build common grounds for effective change to take place. Such a synergy should be based on a consensus-building practice particularly with and between immediate members of HEIs which at the same time recognizes that HE students and staff have the right to organize themselves autonomously. This calls for ensuring conditions necessary for the exercise of academic freedom and institutional autonomy.

**4.2 Managing scarce financial resources** – As governments move toward the cost-sharing scheme in many countries all over the world, HEIs may need to develop new mechanisms that may allow them earn additional financial support from their stakeholders. In fact, KE creates additional opportunities for universities to devise ways of enhancing their financial autonomy if they work out how to leverage greater funding from non-public sources by fostering alumni organizations and establishing business partnerships. Indeed, they also need to revolutionize their financial system of to be able to use available funds more efficiently.

**4.3. Empowering HEI leaders and managers to promote a Culture of Knowledge** - implementing KE calls for owning appropriate orientation towards knowledge and its increasing role in society. Training is one such means where leadership and management capabilities can be moulded to accommodate change and innovation in universities' functions. However, such trainings should be designed as a developmental process, to address future as well as present needs so that sustainability of efforts can be maintained over time.

For HEIs to increasingly engage with the external society, they have to be more open, flexible and porous, with multiple access points and well-publicized services. Academics typically require some form of KE support in terms of changing their attitude and having adequate infrastructure for their engagement. An accommodating culture toward KE by academics is a pre-requisite for the effective exploitation of knowledge from HEIs. However, attitude change does not come for free, it has to be enhanced through targeted incentives. Most critical of all, the HE leadership should believe in the value of KE and committed to promote and implement it because its success requires a system overhaul.

**4.4. The need to improve research capacities** – The reality of policies and strategies for research in developing economies need meticulous revisiting. Such strategies must be founded on existing conditions and institutions, and preferably be drawn by in-country researchers and policy-makers. The refinement of appropriate strategies and plans, both on the generation and use of financial resources, will allow the optimal expenditure of efforts and resources and avoid spreading resources for research too thinly. Some may argue that investment on research in poor countries is a luxury, but Olsson and Mkandawire (2009:29) strongly justify the need for research and innovation in such countries as follows:

Increasing agricultural production for example is more likely to result from new knowledge than from finding better land; negotiating terms for the exploitation of natural resources; .... mineral resources attractive to foreign investors, requires scientific data collection and analysis; and local problem formulation is needed in order to choose strategies for growth, for education and for international relations, not to mention for signing international treaties and conventions and for weighing their implications.

Along with this, attention has to be given, greater than ever, to the notion of devising strategies for innovation, which are closely related to research strategies. And the role of universities, as opposed to research institutes, also needs to be clarified.

## 5. Conclusion

This article attempted to bring the issue of KE in HE to the attention of the academic public. Having outlined the prevailing dynamics of HE, both at the national and international levels, it displayed the most visible challenges for local HEIs today. Thus educational policy and reform strategies may need to go beyond improving access, or changing governance structures, given the immediate need for policy interventions that would improve the fitness-for-purpose of universities. Increasing socio-economic impact requires HEIs to become responsive and relevant agents in national and international innovation systems through close relationships with external organizations and with regular dialogue to monitor changing needs. It is also important that there is a porous and flexible HE space, providing easy access to knowledge. More importantly, leadership support is crucial to establish a strong culture of external engagement.

In a nut shell, to confine universities to a mechanical place of learning and teaching only, in the progress of society, is to diminish them; it offers only eventual disillusion as has been clearly explained in Daniel's article (2004).

## References

Altbach, P. G. and Peterson, P. M. (eds). 1999. Higher Education in the 21st Century: Global Challenge and National Response. IIE Research Report Number 21. Institute of International Education and Boston College, Center for International HE.

Ashcroft, K. (June 2004). The Massification of Higher Education: A Comparison of the UK Experience and the Emerging Ethiopian Response. The Ethiopian Journal of Higher Education, Vol 1. No.: 21-40.

Hagen, S. 2008. From tech transfer to knowledge exchange: European universities in the marketplace. Portland Press Ltd. Available at www.portlandpress.com/pp/books/online/univmark/084/. Accessed on July 11, 2011.

Janjua, P. Z. 2011. Upgrading the standard of higher education in developing countries through international cooperation. International Journal of Education Administration and Policy Studies Vol. 3(7), pp. 103-111,

Kearney, M-L., "Higher Education, Research and Innovation: Charting the Course of the Changing Dynamics of the Knowledge Society" In Meek, L., Teichler, U., & Kearney, M-L. (2009). Higher Education, Research and Innovation: Changing Dynamics – Report on the UNESCO Forum on Higher Education, Research and Knowledge 2001-2009. International Centre for Higher Education Research Kassel (INCHER-Kassel), University of Kassel, Germany. Available at http://www.uni-kassel.de/incher/. Accessed on Feb. 22, 2010.

Lee, C. "The Theory versus Practice Balance at Universities" In The Value of Knowledge: At the Interface. Morley, I. (ed.). 2007. The 1st Global Conference on The Value of Knowledge, 12th - 14th Feb., pp. 113-124. Sydney, Australia. Inter-disciplinary press, Oxford, UK.

Maitra, P. 2007. Higher Education and Global Challenges. Saurabh Publishing House, New Delhi, India.

Mouton, Johann. 2008. "Study on National Research Systems: a Meta-Review." Presented at the UNESCO Symposium on Comparative Analysis of National Research Systems, Paris, January 16.

Mwiria K. 2003. "University Governance and University-State Relations" In Teferra, D. and Altbach P. G. 2003. African Higher Education - An International Reference Handbook, pp. 32-43. Indiana University Press, Bloomington, USA.

Olsson, B. and Mkandawire, T. Compelling Rationale for a UNESCO Forum on Knowledge Systems In Meek, L., Teichler, U., and Kearney, M-L. 2009. Higher Education, Research and Innovation: Changing Dynamics - Report on the UNESCO Forum on Higher Education, Research and Knowledge 2001-2009. International Centre for Higher Education Research Kassel (INCHER-Kassel), University of Kassel, Germany. Available at http://www.uni-kassel.de/incher/. Accessed on Feb. 22, 2010.

Rochford, F. Why Universities "Oversell What We Do Least Well": Conceptions of Knowledge and the Modern University. In The Value of Knowledge: At the Interface. Morley, I. (ed.). 2007. The 1st Global Conference on The Value of Knowledge, 12th - 14th February , pp. 151-164. Sydney, Australia. Inter-disciplinary press, Oxford, UK.

Sawyerr, A. 2002. "Challenges Facing African Universities - Selected Issues"; paper presented at the 45th annual meeting of the African Studies Association, Washington DC, 5-8 Dec.

Smith, Daniel (December 2004). Universities, Paradigms and Ontological Choices. The Ethiopian Journal of Higher Education Vol. 1, No.2:61-99.

Teferra, D. and Altbach P. G. 2003. "Trends and Perspectives in African Higher Education" In Teferra, D. and Altbach P. G. 2003. African Higher Education - An International Reference Handbook, pp. 3-14. Indiana University Press, Bloomington, USA.

Teferra, D. and Altbach P. G. 2003. African Higher Education - An International Reference Handbook. Indiana University Press, Bloomington, USA.

The World Bank Report. 2009a. Accelerating Catch-up: Tertiary Education for Growth in Sub-Saharan Africa – Directions in Human Development. Washington: The World Bank.

The World Bank Report. 2009b. The Challenge of Establishing World Class Universities – Directions in Human Development. Washington: The World Bank.

The World Bank Report. 2002. Constructing Knowledge Societies: New Challenges for Tertiary Education. Washington: The World Bank

The World Bank. 2000. Higher Education in Developing Countries - Peril and Promises. Washington: The World Bank

# CHAPTER 16

## Prospects of Knowledge Sharing Among Ethiopian Institutions of Higher Learning

*Rahel Bekele and Ermias Abebe*

### Introduction

In what is widely known as the "information spectrum", knowledge is information applied with experience and judgement. In this so called knowledge age, knowledge is considered as the most critical "means of production", even more critical than the traditional ones like land, labour, and capital. Therefore, knowledge needs to be managed if it is going to yield the required results.

There are a lot of definitions of knowledge management as every writer tries to mould one that fits her/his domain of operation. According to King (2009), a very simple definition "for knowledge management could be "[the] planning, organizing, motivating, and controlling of people, processes and systems in the organization to ensure that its knowledge-related assets are improved and effectively employed".

In the context of educational institutions, Kidwell et al. (2000) state that knowledge management promises better services to stakeholders, reduced costs, shorter development cycle for products like curriculum and research, and an overall improvement in performance. In their widely cited work, "The knowledge-creating company", Nonaka and Takeuchi (1995) propose a knowledge cycle with four stages. Initially, an individual makes her own discovery at a personal level. When the personal knowledge is made explicit, it will become in the community domain. At the third stage, people will blend and reuse knowledge gained from the community knowledge space and create new knowledge. At the closing loop of the cycle, we return back to the point where the individual makes her own investigations or research to solve "personal"/unique problems.

Based on a similar "knowledge cycle", Huysman and Wit (2003), identified three types of knowledge sharing.

Knowledge retrieval: Knowledge is transferred from the organization to the individual.

Knowledge exchange: Knowledge is transferred from an individual to other individuals.

Knowledge creation: Knowledge is transferred among individuals.

Accordingly, Conway and Sligar (2002) claim that knowledge is in one of the four spaces or frames: individual, community, corporate, or public. Individuals and organizations do, therefore, have roles to facilitate the creation, transfer, and regeneration of knowledge.

One can also empirically suggest the barriers for knowledge sharing. In a case study done in Malaysia, Jain et al. (2007) identify rewards, availability of time, organizational effort, organizational culture, and lack of interaction as the five most important barriers for knowledge sharing. In some other researches individual, organizational, and technological factors are identified as the three most important categories of issues influencing the success or failure of knowledge sharing in organizations.

Alhammad et al. (2009), in their study of knowledge sharing in Jordanian Universities, concluded that "academic staff have fewer mutual relationships, team working opportunities, intentions and motivations to share their knowledge". What is more, while there is no difference in the knowledge sharing habits of females and males, younger staff are not motivated to be "creative".

Cheng et al. (2009) made a report of a case study research on the knowledge sharing practices of staff in the Multimedia University of Malaysia

(MMU). In MMU, there is a formal requirement that all academic staff have to share their research (even if it is an abstract) at least once a year on a system called ShareNet. Most of the respondents however had some misgivings about this imposition and said that they would not have used it had it not been required by the management. The researchers concluded that people-orientation should take primacy over technological orientation in creating an environment of knowledge sharing. Monetary and non-monetary incentives, and "personal expectations" like developing ones own knowledge and being recognised as 'selfless by sharing', were found to be the two most important factors in the decision of staff whether to share or not.

In a study conducted in the Indian Business School with the objective of "assessing success factors of knowledge management initiatives of academic institutions", Basu and Sengupta (2007) identify "integrated technical infrastructure, organisational culture, motivation and commitment of users and senior management support" as the four most critical success factors.

Although one observes that knowledge sharing is becoming important in higher learning institutions to support the teaching learning process and research activities, this does not seem to be successfully implemented in Ethiopian higher learning institutions. It was therefore found important to initiate a study on knowledge sharing practices. The study presented in this paper is the first step in an ongoing research work to assess existing level of staff awareness and practice of knowledge sharing among higher learning institutions in Ethiopia as well as the level of organizational and infrastructural support for knowledge sharing.

The paper is organized as follows. This introductory section briefly gave highlights of knowledge sharing as an introduction of the study. Section two presents the study objectives and background. Section three outlines the methodology employed to collect data. The fourth section presents data obtained as a result of the surveys conducted and interviews made with some selected senior faculty. This section also discusses findings based on the presented data. Conclusions and recommendations are provided in the last section.

## 2 Study Objectives and Background

Recently in Ethiopia, there seems to be an increased understanding and recognition at national level on the importance of an educated workforce to economic growth and national development, and that greater access to higher education is a pre-requisite in this regard. To this end, major and rapid initiatives are being taken to expand higher learning institutions in the country. At the time of writing this paper, there are 23 public universities in Ethiopia. The oldest University is 60 years old while the youngest ones (13 in number) are only four years old.

The challenge for these new institutions basically lies in establishing systems because of lack of organization experience. Creating curriculum, setting up research themes and groups, and even delivering courses are the main challenges in these new academic institutions not only because they lack experience but also they are short of the requisite resources. Therefore, sharing knowledge with other academic institutions is a must for many of them.

Moreover, with such expansion, the huge enrolment has outstripped the capacity of the faculty in virtually all the institutions. In the current setting, due to the overwhelming increase in the student population, most faculty members are very busy in teaching. Most of their time is consumed in preparing lecture materials, project design and examination activities. They have actually no time left for conducting research and related academic activities. Under the existing circumstance, maintaining quality is an even more challenge as compared to the expansion.

In view of the above, one duly recognizes the role of knowledge sharing as one of the areas to be explored to meet the challenges. More specifically, the study aimed to:

evaluate to what extent academic staff of Ethiopian public higher learning institutions understand and practice knowledge sharing;

assess personal, organizational, and technological barriers that limit knowledge sharing in Ethiopian institutions of higher learning; and establish ways of supporting knowledge sharing practices for the betterment of performances of the Ethiopian public higher learning institutions.

# 3. Methodology

## 3.1 Subjects of the study

The study focused on nine public universities, namely, Adama University, Addis Ababa University, Arbaminch University, Bahir Dar University, Gondar University, Haramaya University, Hawassa University, Jimma University and Mekele University, These institutions were purposely considered for the study because of their relatively long history of existence (more than 10 years) and since they have better technical infrastructure including networks, computers and software. Some of these institutions have even developed an organizational support infrastructure for knowledge sharing through the use of eLearning platforms. With the view to create awareness, develop initiatives, advocate and support IT-based knowledge sharing practices initially, the study focused on individual teaching staff from the IT Departments (as well as those working in the ICT offices) of the 9 institutions.

## 3.2 Data collection

In order to meet the objectives of the study, questionnaire surveys, interview and observation as well as informal discussions were used as data collection methods.

### Questionnaire survey

The variables considered in this preliminary study were individual, infrastructural, organizational, and technical factors. In particular, the data collection instrument contained four parts. The first part contained 5 items relating to the demographic profile of respondents. Part two comprised 15 items that elicit information on knowledge sharing practices. Part three contained 9 items pertaining to organization support regarding knowledge sharing.

Part four contained three items relating to the availability and use of technical infrastructure for knowledge sharing. The Likert Scale was considered, which asked respondents to express their level of belief/attitude on a set of statements using a three-point scale. This method seemed the most viable option to capture the responses and to gauge the perception they have of the knowledge sharing practices in their respective universities.

To expedite the data collection process and to ensure high rate of return, existing contacts were used to identify individuals to whom the questionnaires were emailed. From a total of 125 questionnaires distributed, 90 were properly filled out and returned, giving a response rate of 72%.

### (iii) Observation and informal discussions

In order to gain a better understanding of the study objectives outlined above, the questionnaire survey was supplemented by semi-structured interviews with relevant individuals. The websites of the nine institutions were also studied in order to get a deeper understanding of the knowledge sharing practices. Moreover, informal discussions with colleagues at Addis Ababa University were very important in supplementing the data collected.

Attempts were also made to consult available documents and reports pertaining to the availability and use of technological infrastructure.

## 3.3 Data analysis

The data collected from the formal and informal interview as well as observations made on the website were qualitatively categorized and thematically analysed. Since the nature of the data collected through questionnaire was quantitative, a statistical package (SPSS 16) was used to code, process and analyse the data. Data are reported by percentages of responses for all parts of the questionnaire.

# 4. Findings and Discussion

## 4.1 Respondents' profile

Following the presentation formats employed in similar studies elsewhere, the demographic variables used in this study are presented in Table 1 below.

| Respondents' profile | Classification | Percentage |
|---|---|---|
| Gender | Male | 88.9 |
| | Female | 11.1 |
| Age Group | Less than 23 years | 3.3 |
| | 23-30 years | 30.0 |
| | 31-40 years | 41.1 |
| | 41-50 years | 23.3 |
| | Above 50 | 2.2 |
| Educational Level | Bachelors | 40.0 |
| | Masters | 54.4 |
| | Doctorate | 5.6 |
| Academic Rank | Assistant Lecturer | 20.0 |
| | Lecturer | 72.2 |
| | Assistant Professor | 7.8 |
| Experience in higher learning institutions | 3-10 years | 55.6 |
| | 11-15 years | 35.6 |
| | Greater than 15 years | 8.9 |

**Table 1: Demographic profile of respondents**

As can be seen from the figures in Table 1, the majority of the respondents were male (about 89%). The highest number of respondents hold the rank of Lecturer (about 72%), with Master's degree (54%). Most of them are 31-40 years old (41.1%), with less than 15 years of experience (about 91%).

## 4.2 Knowledge sharing practices

One of the objectives of this study was to assess to what extent academic staff in the various IT Departments of Ethiopian public higher learning institutions understand the importance of knowledge sharing and practice same. The responses received on the statements related to knowledge sharing practices are presented in Table 2 below.

| Knowledge sharing practice statements | High | Medium | Low |
|---|---|---|---|
| Rate of motivation to transfer knowledge to colleagues within and outside the University | 45.6% | 36.7% | 17.8% |
| Rate of practice of documenting own work | 28.9% | 17.8% | 53.3% |
| Rate of participation in workshops, seminars, panels, etc. within own University and in other public HEIs | 36.7% | 35.6% | 27.8% |
| Frequency of knowledge sharing | 27.8% | 17.8% | 54.4% |
| Rate of use of knowledge networks such as (email, Web, Social media) to communicate with colleagues. | 27.8% | 18.9% | 53.4% |
| Rate of cooperation and helpfulness by colleagues when asked for some information or advice. | 36.7% | 17.8% | 45.6% |
| Considering knowledge sharing strength and knowledge hoarding as a weakness. | 26.7% | 18.9% | 54.5% |
| Intra-team communication and sharing of knowledge in own University. | 18.9% | 17.8% | 63.4% |
| Informal knowledge sharing practice within the University. | 45.6% | 27.8% | 26.7% |
| Frequency of sharing knowledge obtained from workshop and training to the University staff? | 18.9% | 27.8% | 53.4% |
| Frequency of sharing knowledge gained from guidelines, journals, and book to the University staff? | ---- | 53.3% | 46.7% |
| Frequency of sharing education results, research findings with own colleagues in the University? | 8.9% | 26.7% | 64.5% |

**Table 2: Knowledge sharing practices**

As observed in Table 2 above, most (about 82%) of the respondents have medium to high motivation to transfer knowledge to their colleagues, of which large number (about 46%) of respondents rated 'high'. This, however, is in marked contrast with the low rating of actual practice of sharing (as the majority rated low the frequency of sharing, intra-team communication, and hording as a weakness, about 54%, 63% and 55%, respectively). On the other hand, about 73% rated informal knowledge sharing as medium to high. Per the discussion with colleagues, a possible explanation for this may be the lack of appropriate infrastructure and administrative support (lack of events, teamwork opportunities, etc.).

With regard to the rating on knowledge sharing seen as strength and knowledge hoarding as a weakness, it is in human nature to feel insecure in sharing knowledge in professional practice and hence there is the natural tendency in every individual to hoard knowledge. This attitude may change with the provision of motivational incentives and rewarding mechanisms.

## Organizational support

The other objective of the study was to assess organizational and technological factors that limit knowledge sharing in Ethiopian institutions of higher learning. The findings from the survey are summarized in Table 3.

| Statements related to organization support | Responses | | |
|---|---|---|---|
| | Strongly Agree | Agree | Strongly Disagree |
| There is a motivational scheme by my university to encourage staff to share their knowledge within and outside the University. | 28.9% | ---- | 71.1% |
| There is a good working environment to improve your knowledge sharing attitude | 56.6% | 8.9% | 35.6% |
| Salary increment can be the incentive for better knowledge sharing practices | 46.7% | 8.9% | 44.5 |
| My university supports career development to encourage knowledge sharing | 55.6% | 8.9% | 35.6% |
| Academic and administrative promotions help improve knowledge sharing practices | 46.7% | 17.8% | 35.6% |

| | | | |
|---|---|---|---|
| Acknowledgment of one's contribution may encourage for better knowledge sharing | 46.7% | 17.8% | 35.6% |
| There is lack of technical support and immediate maintenance which obstructs work routines and communication flows. | 10% | 63.3% | 26.7% |
| In my University, there is lack of training regarding employee familiarization of new IT systems and processes. | 27.8% | 36.7% | 35.6% |
| There is a specific budget dedicated to acquire, and share knowledge in the University. | ---- | 10% | 90% |
| There is periodic plan to acquire, organize and share knowledge in the University | ---- | 27.8% | 72.2% |
| The University facilitates knowledge sharing platforms (workshops, seminars, etc.) on a regular basis. | 8.9% | 27.8% | 63.3% |
| Managers in my University encourage employees to suggest ideas for new opportunities | 20% | 35.6% | 44.5% |
| Managers in my University consult team members to make decision and solve problem | 10% | 63.3% | 26.7% |

**Table 3: Organization support**

As can be seen from Table 3, most respondents indicated strong disagreement about the availability of institutional support for knowledge sharing: encouragement (about 71%), allocation of specific budget (about 90%), periodic planning (about 72%), and facilitating platform regularly (about 63%). About 57% indicated that their universities support career development to encourage knowledge sharing. A narrow majority agreed salary increment (about 56%), promotions and acknowledgements (about 65%) can be incentives and encouragements to improve knowledge sharing practices.

## Infrastructure

With regard to availability and use of technical infrastructure for knowledge sharing, summary of responses is presented in Table 4 below.

| Statements related to infrastructure | Strongly Agree | Agree | Strongly Disagree |
|---|---|---|---|
| My University has a very up-to-date ICT infrastructure which helps knowledge sharing. | 28.9% | 44.4% | 26.7% |
| There are enough locations (Hall) within the office where staff can socialize and exchange knowledge. | 46.7% | 8.9% | 45.5% |
| Staffs do have personal home pages through which they can communicate their ideas. | 8.9% | ---- | 91.1% |

**Table 4: Infrastructure**

Responses from the survey questionnaire also revealed that the universities do not seem to make efforts to update the ICT infrastructure which helps for knowledge sharing. Findings indicated that only 10% strongly agree of the efforts being made to update the ICT infrastructure. Moreover, 44.7% of the respondents indicated that there are no enough locations within the universities where staff can socialize and exchange knowledge. In spite of the respondents being IT professionals, only 10% have personal home pages through which they can communicate their ideas.

Apart from the summary provided in Table 4, review of documents and discussions with colleagues also revealed that the various networks deployed at higher learning institutions as well as the full-fledged video conferencing facilities are all high capacity and robust networks designed to support interactive educational programs and research applications. Ever since deployment, most of these facilities are only being used for such simple and ordinary applications as email and Internet browsing. The video conferencing facilities set up in the various higher learning institutions are not put into full usage partly because of the limited band width and partly because of lack of adequate preparation on the part of the institutions.

## 4.5 Barriers to knowledge sharing

Higher Learning Institutions in Ethiopia have developed fairly advanced (for beginners) intra-institution and inter-institution network infrastructure. However, no content or information services are run on the network – limited or no digital library access, electronic journals and databases, etc. are provided. The network resources are mainly used more for web surfing and e-mail. Similarly, course teaching materials (including standard textbooks, power point presentations of lecture notes and exercises), research reports, theses by graduate students, case studies, etc. are not shared. Among the additional factors noted as barriers to knowledge sharing are: unavailability of knowledge sharing platforms (software tools, knowledge-based resources, experience sharing conversational space, structured team-based collaboration framework), and lack of knowledge sharing policies, strategies and programmes.

## 4.6 Awareness among the staff

Respondents feel very strongly about the importance of, and are motivated enough to, knowledge sharing in their work environment. Under the existing situation, knowledge sharing among the faculty takes place mostly on an informal face-to-face communication and as such mostly limited to individuals within the same institution or participating in collaborative team/group activities such as projects. Knowledge sharing through online communications, interactive workshops, virtual networks, peer coaching, best-practice review, etc. are uncommon.

In view of the constraints and extremely under-resourced environments within which the staff are operating, the increased teaching workloads, most colleagues at Addis Ababa University (that we have talked to) agreed on the significance of knowledge sharing in order to discharge their teaching and research duties and responsibilities effectively.

### 4.7 Quality Concerns

During discussions with senior faculty, among the areas cited to benefit from knowledge sharing are: reforming traditional approaches to curriculum and pedagogy, strengthening the weak research traditions by networking with more experienced faculty in other institutions overseas and the Ethiopian Diaspora. However, there is a need for a thorough study to devise strategies and mechanisms for knowledge sharing in a manner that revamps teaching and research.

One model worth considering in this connection is the consortium and collaborative networking system established at Addis Ababa University to run the IT Doctoral Program which started in February 2008. The consortium was established by bringing together academic units actively involved in the offering of graduate level ICT programs. It aimed to consolidate the scattered individual efforts at each academic unit and facilitate maximum utilization of the individual experts. This helped in strengthening the delivery of courses in the IT PhD Programs and developing quality research activities. As part of the consortium activities, national and international conferences were organized, efforts were made for joint curriculum development and mechanisms were devised for exchange of staff and research candidates. In this connection, Addis Ababa University is providing institutional support by way of covering all costs related to trips and accommodation for both instructors and students.

The International Network Joining Ethiopia in Research and Application in IT (INJERA-IT) is another initiative that could be cited as a model to tap a worldwide pool of able and willing scholars to consult and involve in the IT Doctoral Program.

## 5. Conclusion and Recommendations

This study, conducted to examine knowledge sharing among public higher learning institutions in Ethiopia, revealed the current status in knowledge sharing practices especially in relation to personal, technical, organizational factors as well as availability of infrastructure.

Based on the findings and from on own experiences and that of others, the following recommendations are forwarded for consideration.

## General

The study sampled only nine public universities. Private institutions were not considered at all. Within the nine public institutions, respondents were selected from IT related Departments and Offices. It is believed that increasing the sample size above and beyond the current focus of attention within the public institutions, as well as looking at the practices of private Universities and colleges would give a better picture.

In devising the strategies and mechanisms, due considerations need to be given to incorporate motivational incentives and reward mechanisms (related to compensations, financial rewards, performance appraisal, sponsorship for conference participation, etc.) for positive contributions in the knowledge sharing activities. Allocation in the regular budget to finance (along the lines being implemented at the IT Doctoral program at AAU) documentation, publication and circulation of own work as well as sponsoring participation in national and international workshops and conferences are strongly recommended. Equally important, in view of the findings from the survey, is to organize and conduct regular staff awareness programmes to ensure that the staff understand the benefits of knowledge sharing to advance their career and improve individual and organizational performance. Introduction of formal and informal activities to cultivate knowledge sharing can also help. It will help to organize annual forums for academic symposium, problem-solving sessions, workshops and conferences for interaction. In general, knowledge sharing should be continuously advocated. Context sensitive strategies and mechanisms need to be devised that overcome barriers.

With regard to networking, the model at the IT Doctoral program could be revamped to attract and accommodate high-level scholars of extraordinary achievements among the Ethiopian Diaspora, to provide for the human capital needed to implement a successful knowledge sharing programme.

At institutional level, management should take steps to support and motivate faculty towards the process of creating and distributing knowledge (content development). Ways and means of mobilizing funding need to be arranged, among others, for creating enabling technological facilities for capturing and disseminating knowledge, for organizing and conducting staff and student exchange programmes as well as national and international conferences and workshops which are all invaluable mechanisms in terms of providing opportunities to associate with best talent and creating relationships that stimulates learning and growth through knowledge sharing.

## Establishing communities of practice

According to Pan and Leidner (2003) and Wenger (2000), knowledge expands with the extension of social and community interactions. Knowledge contributors and seekers who share common interest areas usually come together (online or in person), forming what is generally called "communities of practice", to help each other by sharing tips, ideas and best practices. Since the critical success factor of communities of practice is very much depending on perpetual knowledge generation and sharing, cultivating communities of practice could be an effective mechanism to promote the knowledge sharing culture within institutions of higher learning in Ethiopia.

Sustained knowledge sharing in communities of practice requires continuous effort in setting a knowledge sharing agenda. This in turn requires exploring the knowledge domain, finding/recognizing gaps in the practice, and defining projects to close these gaps. In other words, this is about seeking out challenging areas for knowledge sharing, identifying opportunities to expand and deepen knowledge and relationships that promise to achieve superior results and/or mutual growth.

To this end, attempts should be made to forge networking and partnerships with collaborators and capable others in other institutions. The knowledge sharing networks may be created in some sort of formal and structured team-based collaboration within an institution or consortium of institutions to better perform the academic activities.

## Establishing enabling IT platform

IT plays an important mediating factor in knowledge sharing. The communities of practice usually produce shared repertoires of communal resource – such as documents, published research articles, case studies, technical reports, theses and related research reports, course materials, artefacts, tools, stories, websites, etc. Such repertoires which are extremely useful components for knowledge sharing in higher learning institutions need to be maintained and members of the community must have access to them. Providing computerized information system to store and retrieve such knowledge resources is critical. Electronic platforms running on the network infrastructure that provide easy to use tools for representation, storage and retrieval of such knowledge are as important.

The design process of electronic platforms should consider the application and use of such Internet technologies and services as email, listservs, chat, bulletin boards, discussion forums, web sites, and features and capabilities of related emerging access technologies of social networking tools such as: wikis, blogs, twitters/yammers, flicker, delicious and slideshare.

## Coordination of efforts

Given the limited resources at the country's disposal and to avoid "reinventing the wheel", a national coordinating body needs to be established to organize various forums for sharing of knowledge. Such endeavour naturally demands for a concerted effort and commitment from all higher learning institutions directly or indirectly involved in the development of the education sector. Among the works of the coordinating body will be promotion of staff exchange, training, dissemination of research outputs, needs assessment, review of experiences of other countries, etc. Furthermore, appropriate policies and plans might be devised by the national body for effective implementation of ICT for knowledge sharing.

## 6. References

Alhammad, L. S., Al Faori, S. and Hussan, L. S. A. (2009) 'Knowledge Sharing in the Jordanian Universities', Journal of Knowledge Management Practice, Vol. 10, No. 3. http://www.tlainc.com/articl199.htm. (accessed 14/08/11).

Basu, B and Sengupta, K. (2007) 'Assessing Success Factors of Knowledge Management Initiatives of Academic Institutions: A Case of an Indian Business School', Electronic Journal of Knowledge Management, Vol. 5, No. 3: p. 273 – 282.

Cheng, M., Ho, J. S. and Lau, P. M. (2009) 'Knowledge Sharing in Academic Institutions: A Study of Multimedia University Malaysia', Electronic Journal of Knowledge Management, Vol. 7, No. 3: p. 313 - 324.

Conway, S. and Sligar, C. (eds). (2002) Unlocking Knowledge Assets. Microsoft Press: Redmond, Washington.

Huysman, M. and Wit, D. (2003) A Critical Evaluation of Knowledge Management Practices. In Mark, V. P., Ackerman S. and Wulf, V. (eds). Sharing Expertise: Beyond Knowledge Management. The MIT Press.

Jain, K. K., Sandhu, M. S. and Sidhu, G. K. (2007). 'Knowledge Sharing Among Academic Staff: A Case Study of Business Schools in Klang Valley, Malaysia'. JASA, Vol. 2: p. 23-29.

Kidwell, J., Vander L. K. M. and Johnson, S. L. (2000) 'Applying Corporate Knowledge Management Practices in Higher Education', Educause Quarterly, Vol 4: p. 28-33.

King, W. R. (2009) Knowledge management and organizational learning. In King W. R. (ed.), Knowledge Management and Organizational Learning, volume 4 of Annals of Information Systems. Springer.

Nonaka, I. and Takeuchi, H (eds.) (1995) The Knowledge Creating Company. Oxford University Press: New York.

Pan, S. L. and Leidner, D.E. (2003) 'Bridging Communities of Practice with Information Technology in Pursuit of Global Knowledge Sharing', Journal of Strategic Information Systems, Vol. 12: p. 71-88.

Wenger, E. (2000) 'Communities of Practice and Social Learning Systems', Organization, Vol. 7, No. 2: p. 225-246.

CHAPTER 17

# Potentials of Indigenous Techniques of Knowledge Creation in Qinea Schools of Ethiopia

*Mariye Yigzaw Workineh*

## 1. Introduction

The study of knowledge is as old as human history. As a community develops its bonds through a common lived experience with shared traditions, the traditions can form societal norms and provide the members with tested solutions to problems. Over time this leads to a shared knowledge base. As such, "knowledge is socially distributed since it is constructed and entrenched in the collective action of different communities" [2, p.357].

Indigenous Knowledge (IK) is one form of knowledge that refers to the traditional knowledge that people in a given community have developed over time based on tested experience and use. IK is characterized as being developed outside the formal educational system or scientific labs, being embedded in culture, and being unique to a given society. All communities have elements of IK. Ignoring indigenous knowledge and even considering it as an obstacle to development has been the tradition of many proponents of modern knowledge. This ignorance has resulted in discrimination of people and institutions with indigenous knowledge. There is a divide between the traditional and modern worlds of knowledge management as the proponents of each view their own knowledge as complete and superior.

There is a fear that modern knowledge management practices may become another management fad that promised much and delivered little. Thus in the knowledge economy, as organizations search for ways to harness their knowledge and create new knowledge, we must recognize that knowledge has been created since the beginning of human life. While advances in technology and the formation of the modern organization have facilitated knowledge creation, looking to the past to try to understand something about the future can add new insights. Traditional knowledge creation practices may lend an additional lens from which we can enhance our modern knowledge creation practices and protect knowledge management from becoming another management fad.

Ethiopia, known for its ancient civilizations, possesses a 1,700-year tradition of elite education linked to the Orthodox Church. Among the different indigenous schools, there is a special category called the Qinea (Poetry) School where students learn to instantly create knowledge. A Qinea poem must have double meanings called wax and gold, where the hidden meaning (i.e., gold) is covered with a layer of explicit meaning (i.e., wax). It is always new and is composed instantly using different techniques. Individuals acquainted with these indigenous techniques are capable of instantly generating convincing solutions and replies to questions.

This research is motivated by the existence of understudied indigenous techniques of knowledge creation used for more than 1500 years. Studying the indigenous techniques is an opportunity for modern knowledge management to gain detailed insights on indigenous knowledge creation.

Our objective is to explore the indigenous techniques of knowledge creation and to transfer these insights to modern knowledge creation. We will present the review of relevant literature, to be followed by the research question. After describing the research approach, we will present our major findings.

## 2. Literature review

This research is about the potentials of indigenous knowledge creation techniques in the traditional poetry (Qinea) schools of Ethiopia, which involve knowledge, creativity, and poetry. We will start the review of literature related to knowledge and its management, to be followed by knowledge creation and associated processes. We will then discuss the relationships between creativity, knowledge creation, and poetry. Finally, we will describe indigenous techniques of knowledge creation in the Ethiopian poetry schools.

## 2.1. What knowledge to manage

There are varying views on the term knowledge. Knowledge can be viewed as access to information, repositories of information, sets of rules, and knowing/understanding. Knowledge can also be a state of mind, an object, a process, a condition of having access to information, or a capability. Knowledge is also viewed as the epistemology of possession which treats knowledge as something people *have or possess (*knowledge being person's 'justified true belief'; a possession of the human mind) and the epistemology of practice which treats knowledge as something people *do* [17, 18]. The widely recognized view of knowledge as having tacit and explicit dimensions is also debatable among scholars. Zack [19] says that knowledge can be declarative, procedural, and causal all of which can be made explicit. Brown and Duguid [20] claim that tacit and explicit knowledge are separate dimensions stating there is no transformation of tacit-explicit or explicit-tacit form. Hislop [21] argues that all knowledge has tacit and explicit components and that explicit knowledge has a tacit element.

Not only are there different views on the term knowledge but there are also differing strategies for its management [15]. Organizations can engage in knowledge management processes with varying steps for instance: knowledge generation, codification, transfer, and realization phases [22]; knowledge creation, validation, presentation, distribution, and application [23]; knowledge creation, storage/retrieval, transfer, and application phases [16].

However, despite these differences, knowledge is recognized as a critical resource for success and there is an increasing interest of organizations to manage their knowledge [24]. As managing knowledge is impossible without first creating it [25], knowledge creation is commonly mentioned as the initial step in knowledge management. Knowledge creation is recognized as a vital element of learning and innovation for organizational success and survival [26].

We have used in this research the definition of Nonaka [27, p.15) as a working definition of knowledge: *Knowledge is a justified personal belief that increases an individual's capacity to take effective action.* Knowledge management is also defined in this research as *the process of identifying and leveraging the collective knowledge* [28]. Knowledge as a justified personal belief is created when an individual justifies the truthfulness of his/her belief about a particular situation [29]. Therefore, knowledge creation *is the process of presenting and justifying a personal belief* where justification makes the personal belief public.

## 2.2. Knowledge creation models

The knowledge creation process is composed of *knowledge exploration* (knowledge invention and knowledge acquisition) and *knowledge exploitation* (knowledge conversion, knowledge capturing, and knowledge mapping) [30]. According to von Krogh et al. [29], the knowledge creation process has 5 steps:

*Sharing tacit knowledge*: through direct observation, imitation, experimentation and comparison, or/and joint execution (not necessarily through language).

*Creating concepts*: externalizing personal true belief (tacit knowledge) about a situation in a form of a concept (through language).

*Justifying concepts*: evaluating the created concept for consistency, uniqueness, value and benefit, and practicability.

*Building a prototype*: turning the justified belief into a tangible form (prototype) by combining the new concept with existing concepts and products/services.

*Cross leveling knowledge*: leveraging the new knowledge across organizational hierarchies by documenting, disseminating, and reinforcing the new knowledge and its prototype.

Related to this is the SECI model of Nonaka and colleagues, which is considered the standard model in the knowledge management literature. The SECI model has four modes of knowledge creation that transform the knowledge state from tacit to explicit and vice versa [27, 31]:

*Socialization (tacit to tacit)*: sharing of tacit knowledge between individuals through joint activities (not in written or verbal instructions).

*Externalization (tacit to explicit):* expression of tacit knowledge and its conversion into comprehensible forms (explicit knowledge).

*Combination (explicit to explicit):* conversion of the new explicit knowledge into more complex sets of explicit knowledge, communication and diffusion being key issues.

*Internalization (explicit to tacit):* conversion of explicit knowledge into the organization's tacit knowledge.

There are also differing factors involved in the aforementioned knowledge creation models. On the one hand, Nonaka and Konno [31] introduced the concept of four "ba's" (shared spaces for knowledge creation) that support the four modes of SECI: *Originating 'ba'* (the place for socialization), *Interacting 'ba'* (the place for externalization), *Cyber 'ba'* (the virtual world for combination), and *Exercising 'ba'* (the place for internalization) [31]. Von Krogh et al. [29], on the other hand, identified that the five-step knowledge creation process is facilitated by combinations of five enablers: installing a knowledge vision, managing conversations, mobilizing knowledge activities, creating the right context, and globalizing local knowledge.

Both the 5-steps and the 4-step spiral processes of knowledge creation have relatively similar descriptions, as summarized in the table below.

| Step | The 4-step (SECI) model | The 5-step model |
|---|---|---|
| 1 | Socialization: sharing of tacit knowledge between individuals. | Sharing of tacit knowledge. |
| 2 | Externalization: expression of tacit knowledge and its conversion into comprehensible forms. | Creating concepts: externalizing tacit knowledge. |
| 3 | - | Justifying concepts: evaluating their consistency, uniqueness, value and benefit, and practicability. |
| 4 | Combination: conversion of the new explicit knowledge into more complex sets of explicit knowledge. | Building a prototype: by combining the new concept with existing concepts. |
| 5 | Internalization: conversion of explicit knowledge into the organization's tacit knowledge. | Cross leveling knowledge: leveraging the new knowledge across organizational hierarchies. |

**Table 1. Summary of the two models of knowledge creation**

Internalization of explicit knowledge leads to the creation of tacit knowledge, making the above steps of the knowledge creation process cyclic. However, it is not clear if internalization is the only source of tacit knowledge. Related to creativity, Weisberg [32] claims that using existing knowledge could result in negative transfer to new situations (due to the possibility of being trapped into habitual modes). For a creative thinking (as involved in knowledge creation process) to produce true advances, it has to go beyond the limits of existing explicit knowledge [32].

## 2.3. Creativity, poetry, and knowledge creation

Creativity is the process of bringing into being something novel and useful [33]. For Weisberg [32], creative thinking goes beyond knowledge however knowledge provides the basic elements (building blocks) for constructing new ideas. He states that extensive domain specific knowledge, although not sufficient, is a prerequisite for creative functioning. However, creative thinking can produce true advances only if it goes beyond the limits of knowledge as using past experience could result in negative transfer to new situations [32]. Organizational creativity is the creation of a new valuable product, service, idea, procedure, or process by individuals working together in a complex social system [34]. Creativity for individuals and organizations implies doing something for the first time or creating new knowledge [34]. Thus, creativity involves the extended use of existing knowledge to generate new knowledge, breaking the limits of existing knowledge and avoiding negative transfer.

Differing views exist regarding the relationship between knowledge and poetry [35]. While one view asserts poetry as an expression of feeling with no contribution to the knowledge, another view asserts that poetry can transmit or *produce* or *embody* knowledge uniquely [35]. Poetry holds the potential to capture emotion and express the unsayable with passion, truth and intensity; it can be used to flavor organizational learning [36]. It can facilitate new organizational knowledge particularly for issues requiring a reflective space that taps into emotions and uncovers layers of thoughts and feelings [36].

Poetry can facilitate creative problem solving through its positive affect (making people feel good through humor), increasing their capacity of creative problem solving [37]. Anything that makes us feel good is likely to help us think more broadly and creatively [37]. What makes people in one culture feel good may not have the same effect in another culture.

## 2.4. Indigenous techniques of knowledge creation in Ethiopia

Indigenous Knowledge (IK) has been orally passed down for generations from person to person and is expressed in language (stories, epics, legends, tales or poetry), music (folk songs or instrumental music), spiritual activity (dance, rituals, or ceremonies), and arts and crafts [38, 39]. Unlike other forms of poems, Qinea poems combine many of these expressions such as stories, folk songs, spiritual dances, and ceremonies.

IK has enabled its holders to survive in harmony with nature [6] even in problematic situations. People all over the world have been using their own indigenous knowledge for their own situations.

Despite the clear tradition of viewing IK and institutions holding it as obstacles to development [5], IK is increasingly considered to play a key role for sustainable resource use and balanced development [39]. Development efforts fail to achieve desired objectives when they are ignorant of the local technologies, systems of knowledge, and environment [39]. IK is the local knowledge that is unique to a given culture or society and as the information base for a society that facilitates communication and decision making [40]. IK is most often tacit knowledge that resides within its holders [7].

In the Ethiopian context, the Poetry Schools (Qinea Bet) have been teaching students to actively create and share knowledge [11]. Qinea has been the oral tradition of instant knowledge creation and sharing for centuries. It was originally used for religious purposes, but now it is embodied in the culture and is often practiced by traditional singers (called Azmaris).

Qinea is the output, the knowledge in the form of poem that results from a "live" process of knowledge creation. Qinea poems are instantaneously composed to express new personal insights about special events, or ceremonies, or festivals. The person who produces the poem gathers information through observation of events (for forming the deeper meaning), retrieves from his brain relevant previous knowledge (as a cover meaning), matches this knowledge with the actual event, and produces the poem. The length of Qinea (poem) can range from 2 to 11 verses of various sizes and there are 9 different types of Qineas [11]. A composer can compose the different types of Qineas (a total of more than 33 lines) and deliver them orally within few minutes and "off the top of the head".

Through the poems, the composer becomes the knowledge creator. Combining his previous knowledge and capturing the viewers' expectations, he expresses new knowledge that best describes the truth about the event in exemplary poems. This could be related to the transformation of tacit knowledge into explicit knowledge or to what von Krogh et al. [29] describes as justifying concepts.

From the knowledge creation perspective, there is no previous research on the practices of knowledge creation in Qinea schools. The few previous research projects on Qinea are from the poetry and linguistic perspectives or from the Church treasury perspective.

## 3. Research question

Despite the existence of rich indigenous techniques related to knowledge creation, modern Ethiopia faces at least three problems with respect to knowledge creation. First, the country has lost documented knowledge of what enabled its earlier civilization [41]. Second, there has been little effort put forward by modern organizations to foster knowledge management as manifested by the absence of knowledge management strategies or initiatives. Third, the poor quality of modern education [10, 42] hasn't provided the country with the knowledge that enables it to be competitive in the modern economy.

In order to be successful, best practices must embrace local technologies, local systems of knowledge, and the local environment [39]. Thus, for Ethiopia and other similar countries to advance in the knowledge economy, they must learn to adapt best practices from other parts of the world while ensuring that they incorporate their own traditional knowledge. However, the traditional techniques of knowledge creation are not well studied, making it difficult to harmonize with modern knowledge management.

With the objective of exploring the indigenous techniques of knowledge creation and transferring these insights to modern knowledge creation, the research addresses the following questions.

*RQ1: How and why do experts in the indigenous Qinea Schools create knowledge?* Besides exploring the indigenous techniques, we will see what factors hinder or facilitate knowledge creation. This question takes our research beyond the source and state of knowledge to consider the conditions that facilitate knowledge creation, as suggested by Alavi & Leidner [16].

*RQ2: What can organizations learn from indigenous forms of knowledge creation?* In addressing this question, we will identify possible application areas of the indigenous techniques and their implications for organizational knowledge creation.

# 4. Research Approach

Given our objective to explore how indigenous techniques of knowledge creation are performed within its social context, we elected to anchor our research in an interpretive research paradigm. Accordingly, we recognized a socially constructed view of reality (i.e., ontology wise) and a subjective construction of knowledge (i.e., epistemology wise), requiring us to get 'inside' the world of those using the indigenous techniques for knowledge creation. These choices influenced our data collection and analysis methods.

## 4.1. Data source background

The data sources for this research were Qinea Schools and experts of qinea (teachers and students). The qinea schools are boarding schools; students are expected to live there for extended periods of time. In such schools, three to six students typically live in a small circular hut (with a diameter of less than 3 meters). Students within a given hut are usually at different ranks so that a student at a lower level can be assisted by a higher level student. High level students learn from even higher level students or from the master teacher.

A beginner needs to spend over four years at a school to become a teacher of qinea. The time it takes for a student to become a composer is usually less, and is based on personal effort and previous knowledge of religious thoughts of the Ethiopian Orthodox Church and its languages (Geez and Amharic). If one has such previous knowledge, he normally starts to compose qineas within few months of study at Qinea Schools.

## 4.2. Data collection methods

Data was collected via interviews, participant observation, and reviewing of secondary documents. Before entering the field, the main author tried to get a better picture about qinea from secondary sources, that is, knowledgeable persons, books containing qineas, and an Amharic documentary film on qinea schools entitled "Qinea: the philosophical foundation of the ecclesiastical education" by the Mahibere Kidusan of Ethiopian Orthodox Church [43].

He was informed by the producers that to truly understand qinea, it was necessary to visit the Chegodea Hanna Qinea School (located in West Gojjam, some 255 miles northwest of Addis Ababa, and requiring an arduous journey), and to consider other top qinea schools like Debre Tabor Savior and Zara Michael (South Gondar), and Shinbit Michael (West Gojam).

The first author was able to collect data from three top level schools (Chegodea Hanna, Shinbit Michael, and Debre Tabor Savior) and two feeder schools (Bahir Dar Giorgis and Kimer Dengay, Arbaitu Ensisa). Across these schools, interviews were conducted for five master teachers. During each qinea school visit, the researcher initially introduced himself and created rapport with the teachers. As the researcher was previously a church member, he had no difficulty establishing a trusting relationship with the teachers and students. Once trust was established (which usually took a day or two), the researcher started requesting interviews from different school teachers and students. As the interviews included both qinea teachers and students, two separate interview guides, in the local language, were created (see Appendix 1).

In addition to formal interviews, informal discussions were conducted with students. In fact, given that the researcher lived in a hut with other students during each school visit, many opportunities for informal discussions arose. Whenever the research obtained valuable information through such informal discussions, he requests for the information to be repeated and recorded. Overall, the participants were very candid, open, and comfortable through the interview process. In total, we collected 386 minutes of formal and 600 minutes of informal interviews. A summary of the data collection techniques used at each school is provided in Appendix 2.

## 4.3. Data analysis

Field notes containing key findings were written or recorded by the first author before leaving each of the Poetry schools. These field notes helped the researchers staying focused on what had been done, what had been learned, and what needed to be done next. After completing all interviews, these (and the field notes) were incorporated in a qualitative data analysis software (i.e., the Coding Analysis Toolkit, at http://cat.ucsur.pitt.edu/) for coding and analysis.

A coding scheme inspired from our review of literature was used for the initial round of coding. However, as we progressed in our coding, we ended up adding many codes for incidents that we deemed interesting and relevant, but for which the literature had not suggested a code. We thus, revised our coding scheme to a great extent over time (see the final version in Appendix 3). We stopped revising the codes as we reached saturation (classified all major incidents) and as the codes were fully supported by the data.

After we completed coding, we then created categories representing the key themes that emerged from our data sets. To do this, we first extracted all the coded segments for each code, we then summarized what these coded segments were all about, and we combined group of codes that were relevant to one another into themes.

Qinea (poem) verses and tales were also analyzed to provide insight into values, history, practices, and beliefs reflected in them. Therefore, in addition to the recorded interviews, it was necessary to select orally told or documented verses of poems and related stories to illustrate how experts used the knowledge creation techniques to justify their belief and to solve the problems they face.

The data sets were categorized into 37 codes. The codes were further grouped to seven aggregate themes and then to four major implications. The codes were further grouped to seven aggregate themes and then to four major implications, which are the major discussion points of the findings of this research. The details are presented in Table 3.

As this paper is extracted from the PhD dissertation (research-in-progress) of the first author, it will present one of the four implications (*potentials of the indigenous techniques*) and will provide only highlights of the other implications.

| Code (37) | Aggregate code (7) | Implications (4) |
|---|---|---|
| Interview ID | Biography of interviewees & Qinea schools | How experience matters in knowledge creation? |
| BI | | |
| BS | | |
| SR Beginner | Ranks of students | |
| SR Tutor | | |
| SR Corrector | | |
| SR Plunderer | | |
| Qinea Definition | Nature and methods of qinea (description) | What are the implications to knowledge creation and problem solving? |
| Qinea Sources | | |
| Prerequisite Knowledge | | |
| Methods | | |
| Qinea Techniques | | |
| Qinea Types | | |
| Wax and Gold | | |
| Validity measures | Validity measures | |
| Qinea vs poem | | |
| Reward system | | |
| E Knowledge | Enabling factors to study/teach qinea | What motivates individuals for knowledge creation? |
| E Freedom | | |
| E Satisfaction | | |
| E Challenge | | |
| E Religion | | |
| B Economic | Barriers to study/teach qinea | |
| B Academic | | |

| | | |
|---|---|---|
| QP Unsayable | Qinea Potentials: Practical areas where qinea has been applied by the indigenous community | What are the potentials of the indigenous techniques; how and where can we apply them? |
| QP Instant creativity | | |
| QP Creative problem solving | | |
| QP Predicting the future | | |
| QP Creative education | | |
| QP Social problem solving | | |
| QP Personal problem solving | | |
| QP Utilizing Geez documents | | |
| QP Religious services | | |
| QP Entertainment | | |
| QP Tolerance | | |
| QP Ethics | | |
| QP Inspiration | | |

**Table 3. Summary of the codes and emerging themes**

## 5. Findings on the potentials of the indigenous techniques

Knowledge of qinea has been serving the indigenous community for centuries in every aspect of their life. The interviewees' reply as to for what purposes they have been using qinea techniques resulted in a list of application areas: expressing the "unsayable", instantly creating replies or solutions to problems, predicting the future, utilizing Geez documents, creative education, entertainment, tolerance, inspiration (for moving the mass), fulfilling religious services, and promoting ethics. These areas are practiced among qinea people, but they remain largely unused in modern Ethiopia.

### 5.1. Expressing the "unsayable"

Qinea composers have been using their qinea knowledge for systematically saying the unsayable or touching the untouchable (especially challenging the authorities for their wrong doings which otherwise is unsayable in the Ethiopian tradition) in a way that is pleasant to the authorities too. For example, Aleqa Gebre Hanna (1814-1898), an Ethiopian legend for humorous creative replies and criticisms, was one of the honored persons of Emperor Menelik II (1844-1913). While he was faced with an illness, Empress Taitu sent him oxen meat of one leg (which was considered as disrespectfulness). Aleqa expressed his anger at the Empress by saying: "*If the generous Empress gives one leg for the nobles, what will be left for the King?*" In saying 'giving one leg', the plain meaning is related to her generosity in giving one oxen leg (meat), but the hidden meaning is that she gives her own leg up for the novels (publicizing the unsayable rumor that the Empress commits adultery with novels).

Composers do not hesitate from criticizing the bad decisions or acts of authorities through their qineas. As stated by a teacher, "*One can make a serious criticism against injustice covering with the wax (layer meaning) to let the qinea look like in favor of the target person, or may present a hidden praise or appreciation but covered with something that officially seems a blaspheme.*" The following Sillasea qinea of Aleqa Zewolde can be mentioned here as an example:[66]

*Lion, the son of lion;*
*My lord! Your arm broke all your enemies;*
*You the true breaker lion; the king of lions;*
*With your arm broken all the sticks;*

66 Translated from Oromifa (with some Amharic):

አርካ ኬቴ ሁንዱማ ጨብሴ፣

ሴንሜ ጎፍታኮ ኢጆሌ ሴንሜ፣

አሁንም የእውነት ግሩም አንበሳ የአንበሳ ኮርማ የቢያ፣

በክንድህ ሰበርኸው የሁሉን ሶማያ::

ዱጉማ ጎፍታኮ ዱጉማ እሲ መና ኬቲ ገቢያ (ገደ ቢያ)፣

ሀሬ ጉደቴ ፈርዳን ቆሊያ፣

ማሎ ማሎ ጎፍታ ኪያ!

መና ፈርዳ ቆደቴ አህያ::

*True my lord, true in your house;*
*Donkey gets the honor which was denied to horses.*
*Why, my lord, why?*
*The donkey inherited the house of the horses?!*

This qinea was composed during Emperor Menelik II (1889-1913) for praising and criticizing the Emperor. Like a king of lions, Menelik came to the throne conquering all his enemies. As a muscle person breaks sticks easily, Menelik defeated the wishes of his enemies. This comparison of Menelik with king of lions is honor. However, as a horse is better in many aspects than a donkey, it is not a wise decision to honor donkeys and disregard horses. In a similar way, the composer criticizes Menelik for disregarding wise persons and honoring inappropriate persons in his throne.
As expressed by a respondent,

*Through qinea, you can systematically complain about a problem and force authorities to change their decision in your behalf. It is possible to produce a qinea that shows the right and wrong doings of a person.*

The story of a known qinea teacher in Delanta (North Wollo) called Sirak confirms this. The church administrator, who has not a positive attitude towards the qinea teacher, assigned a monthly salary of Birr 30 for the qinea teacher and Birr 60 even for the ordinary priests and deacons. The teacher expressed his sorrow by composing the following qinea[67]:

*While donkeys and mules cost sixty,*
*Sirak, the god of Delanta was sold for thirty.*

What the composer calls donkeys and mules are deacons and priests. He associated himself with Christ (God) and the administrator with Judah who accepted a price of 30 silver coins for betraying Christ to his accusers (Math

67 Translated from Geez:
አእዱግ ወበቅል እንዘ ያወጽኡ ሥድሳ፤

ሲራክ አምላከ ዋድላ ተሰይጠ በሠላሳ::

26:14-16). It was (and still is) not common to criticize spiritual leaders and publically complain on their decisions.

Such challenges are not limited to kings and religious leaders. It is free in qinea to question even God by asking the 'unquestionable' as reflected in the following short qinea of Aleqa Gebre Hanna[68]:

*You [God] who says 'love your enemies',*
*Why do you punish sinners in hell?*

The composer challenges God Who ordered us to love our enemies, why should not He also love His enemies (sinners) and avoid punishing them in Hell. Sinners, contrary to the righteous, do not perform in accordance with God's will and are considered as His enemies. It is written that God will pay them the price of their bad doings during the Day of Judgment (Rev. 22:12). The price for sinners after death is expected to be thrown into hell. In another teaching, Christ ordered his followers to love even their enemies (Math. 5:44). Although societies have accepted these two teachings, the composer of this qinea challenges God that He who orders us to love our enemies, should also love His enemies (sinners).

## 5.2. Instantly creating replies or solutions to problems

Qinea people compose large poems and provide convincing solutions instantly right at the time the situation demands. For example, while Aleqa Gebre Hanna (19th C.) was crossing the Nile valley on foot (going from Addis Ababa to Gondar), the travelers with him told him that dangerous thieves were around. As there was no time to think or discuss farther, Aleqa thought in terms of what he had: church umbrellas and clothes (taking to rural churches) and a wooden plate (for preparing flax). He then quickly covered the plate with the church clothes, put it on his head like an Arc, and

68 Translated from Geez:
ኮንኖኃጥአንኩሎሙኢይደልወከምንት፤

አፍቅሩጸላእትክሙእንዘትብልአንት።

ordered the travelers to hold umbrellas over the 'Arc'. When the burglars reached his group, they asked whose Arc it was, Aleqa instantly replied: "St. Gorge of the Flax Processing." All churches in Ethiopia have Arcs named after Saints, Angels, and God Himself and it is common to have an adjective of the Arc (like 'The Listening Mary'). The instant reply didn't give time for the thieves to suspect he was not telling the truth; they instead showed the safe way to the travelers respecting the 'Arc'. What Aleqa did here is to find a solution from what he already has. This is against the experience of many who search for a solution to a problem far from what they have/know.

Another teacher, as he was teaching about the Church doctrine of baptizing children in their 40$^{th}$ days of birth for baby boys and 80$^{th}$ day of birth for baby girls, was asked the day of baptizing a child with two sexes (asexual). The answer was not simple as there was no rule stating on baptizing such a child (if there is). The teacher thought that if there is such a baby, s/he has to use one of the organs for urination and replied that the baby's sex will be determined with what s/he urinates. This was incorporated in the Church's law of baptism. Such 'automaticity' is the life of qinea masters.

In relation to this, Weisberg [32] states that creative accomplishment is gained through automaticity of skills, which comes through years of immersion for practicing the skills. A person with such automaticity of skills does not need to think about how to express his ideas, but produces novelty as the ideas become available [32]. Deep immersion is important to perfect a skill and to develop heuristics [32].

The above examples show how the qinea people solve problems or react to problematic situations through their automatic reply. This helps them to minimize the chance of occurrence of problems. However, once a problem occurs, the qinea persons make fun and joke on it by composing poems. The qinea education process and their life style there help students to learn how to find solutions to their problems or how to make fun out of their problems; making their problems powerless by reducing psychological consequences. The statement of a teacher confirms this:

"*We have to have a strong mentality that tolerates every problem we may face. The study of qinea requires great endeavor, concentration and peace of mind. When you first have the determination, the problems you face will be powerless and your study will be much better.*

Thus, composing qineas on problems is practiced as a means to minimize the psychological effect of problems and share personal experience. A master teacher said:

*We get relief when we compose qinea on our problems. Disclosing our problem reduces its strength and has a psychological importance. Knowledge of qinea helps us to minimize the incidence of problems specially in interacting with other persons. Qinea knowledge helps us to gain attraction, to normalize bad-tempered situations, and to get a judgment in your favor.*

Another said: "*Qineas can touch a problem systematically and give relief. It is like gently touching the surrounding of a wound on your body.*" As an example, a teacher who already lost his sight at early age also lost his hands and feet due to gangrene. He shared his tested experience in such problematic situation through, among others, the following Mewodis qinea[69]:

*Oh God;*
*Diseases of flesh came to infect me on your behalf;*
*While star/my flesh drops off,*
*Sun and moon/my eyes hided not to see you crucified;*

69 Translated from Geez:

ኦ አምላኪ!

ህማማተ ሥጋ በእንተ ህማምከ መጽኡ በከመ ይህመመው ዚአየ፣

እስመ አመ ወድቀ ምድረ ኮከበ ሰማይ ሥጋየ፣

ፀሐይ ወወርህ ዓእይንትየ ኢይርእዩ ሥነ ስቅለትከ ተሰወሩ ዝየ፣

ብርያልሂ ዘፈንዎሙ ደዌያተ ሥጋ ጸርየ፣

እህዞኒ እሲሮሙ እደውየ ወእገርየ፣

እንዘይመጽኡ ሰርከ እንበለ ይርእይ ዓይንየ፣

በሰላም ወበጠኢና መጠነ ህለውኩ ጸጊቢየ::

> We get relief when we compose Qinea on our problems. Disclosing our problem reduces its strength and has a psychological importance.

*My enemies; Satan-sent diseases of flesh,*
*Held me tying my feet and hand;*
*Always coming while my eyes are closed;*
*Held me tying my feet and hand,*
*Enjoying with peace and health, as much as I lived.*

While Christ was crucified, the sun and moon became dark, stars dropped … (Luke 23:45). This composer relates Christ's pain in being crucified with his own pain in losing his hands and feet; stars which drop from the sky with the cutoff of his body; the darkness of the sun and moon with his blind eyes. The poem expresses as sun and moon became dark not to see Christ's body, his eyes were blind not to see his slashed body, and he faced with serious health problems repeatedly that lead him to lose his eyes, then his hands, and then his feet.

This way, a qinea person jokes on and extracts fun from his problems. This is related to the capacity of poetry in facilitating creative problem solving through its positive affect (making people feel good through humor), increasing their capacity of creative problem solving as described by Isen et al. [37].

## 5.3. Predicting the future

Qinea persons exercise predicting the future by predicting what poem another person is going to compose. As stated by a respondent,

*It is a common practice in Qine communities; we call it 'neteka' or 'zerefa' (plundering). When a person begins to compose, we can predict what he is going to say from the first two or three lines.*

The term 'Neteka' means 'taking' or 'plundering' a qinea under composition from the mouth of its composer. As a composer starts his qinea, a plundering composer looks at the style and the story from the first lines as stated by a respondent:

*The prediction is made based on carefully looking at how one starts his qinea (based on the story of the day and the associations the composer is presenting in the first verses). The wise plunderer (who is also a composer) instantly finishes the qinea under formation in the same way as the composer was going to say.*

This capacity is used to have a skill to easily predict a person's reaction to dialogues, to facilitate communication, and to convenience others. This is supported by the following statements of respondents:

*If a person has a good knowledge of qinea, he can facilitate smooth communication, make a better prediction of what a person is going to say, and turn things in his own way.*

*A qinea person has better chance of controlling situations. The stories, books, and lessons from qineas teach us to clearly imagine things and their consequences ahead of their occurrence so that we can better lead ourselves on the right track.*

*Qinea enables you to read the ideas of other persons and to choose your actions accordingly....*

As practical evidence, let's see how Aleqa Gebrehanna used the creativity of qinea in problematic situations. Aleqa cursed a priest who was not good for him saying "Ahiya," meaning 'donkey'. The priest brought the case to court and Aleqa was asked why he said so. He then instantly replied: "My lord, he heard me saying 'Ahiya Sherahiya' in my prayer and it seemed him that I cursed him because he is illiterate. This made Aleqa free from the charge. Aleqa fore sought at the beginning what will happen and what his answers should be. This quality is acquired through qinea as stated by a student:

*I now see things brightly through qinea as it is a key to understand everything. That is why I prefer to study qinea. Some fathers say there is nothing like qinea that enables us to have a better knowledge about the future.*

## 5.4. Utilizing Geez documents

Geez is among the few languages used for writing in the ancient world. There are Biblical evidences showing that Ethiopians have books of Prophecy in their language since the times of Old Testaments, (Acts 8:26-30). The country's documented knowledge before the $16^{th}$ BC is all in Geez. As respondent states, "Geez is the language with which many of Ethiopian history and knowledge is documented."

Today, vast documents in Geez are found today not only in Ethiopia and Eritrea, but all over the world. For example, there were 2980 Geez manuscripts until 1981 in the central libraries of Europe; France (854), Britain (701), and Germany (411) taking the lion's share [44]. Despite its restriction to Church services in its origin (Ethiopia), Geez is offered in renowned universities in the world including Harvard and Oxford [45]. Many Western scholars who studied Geez produced their studies about the language and translated many Geez documents to Latin, French, English, Italian, and other languages [45]. A qinea teacher expressed the influence of Geez even in the Bible itself stating:

*There are many Geez words in the Bible that are still in use in other languages. For example, the first man Adam has no meaning in other languages, but in Geez. Adam means attractive or pleasant and derived from the root verb Ademe (meaning became attractive and pleasant).*

As the Geez language is the medium of instruction in qinea schools, students join these schools with a purpose of knowing Geez and then understanding the vast collections. A student said:

*We come to qinea School to understand the knowledge in books and hymens in Geez. I can say qinea schools are special in giving knowledge.*

The sayings of other respondents confirm with this:

*The Geez books, prayers, and songs need to be translated to the local languages for the people to understand what the church says. This requires knowledge of qinea. Thus, qinea is useful for translating the vast Geez collections.*

*Qinea opens our eye to clearly understand the scriptures and everything else. As a direction of a road leads to a desired destination, qinea is also a clear direction of the scriptures.*

*Qinea is a key tool to understand books. For a person who has the knowledge of qinea, it will be easy to study the bible and other books.*

## 5.5. Creative education

Creative education enables students to develop the wisdom of predicting the consequences of their actions. An interviewee said:

*The stories, books, and lessons from qineas teach us to clearly imagine things and their consequences ahead of their occurrence so that we can better lead ourselves on the right track.*

In the Qinea Schools, students are trained to compose as many as 40 verses of poems at a speed of a normal talk. A creative education sharpens students as stated by a master teacher:

*Qinea sharpens our brain. Matured qinea students do not need to repeatedly read or listen to catch it. For example, when I compose full qineas (up to 40 verses) continuously, the students catch it at ones. I do not repeat even such long qineas, but they are sharp to recall it and repeat after me for lower level students. Even when they read a book, they need to read it only once and they have it in their brain. This is a useful technique that can be used for the country's modern education.*

Following are some aspects unique to the academic environment in Qinea Schools:

Challenge: Interviewees confirmed that compared with other teachings which are based on transferring the knowledge in books through the teacher, qinea requires to always compose new poems. This creativity differentiates qinea from other fields and makes it difficult for students.

Attraction power of the study: The creativity required in qinea is not only a challenge, but also what attracts students to study it. As a respondent stated, *"Nothing gives as much challenge and pleasure as qinea." A master teacher also said: "Qinea creates an emotion of happiness and refreshment. It is attractive and I wish I hear nothing else than qinea."*

Iterative incremental study: The students' progress in creativity in terms of the complexity of the qinea they compose and the speed of creativity. In terms of complexity, students start to compose simple qineas and progress to complex ones (11 verses and the complete types) iteratively. Regarding the speed of creativity, students begin to create by counting/thinking what to say and finally start to automatically generate all the different types of qineas at ones without previous thinking. One should do the composition at a speed of a normal talk.

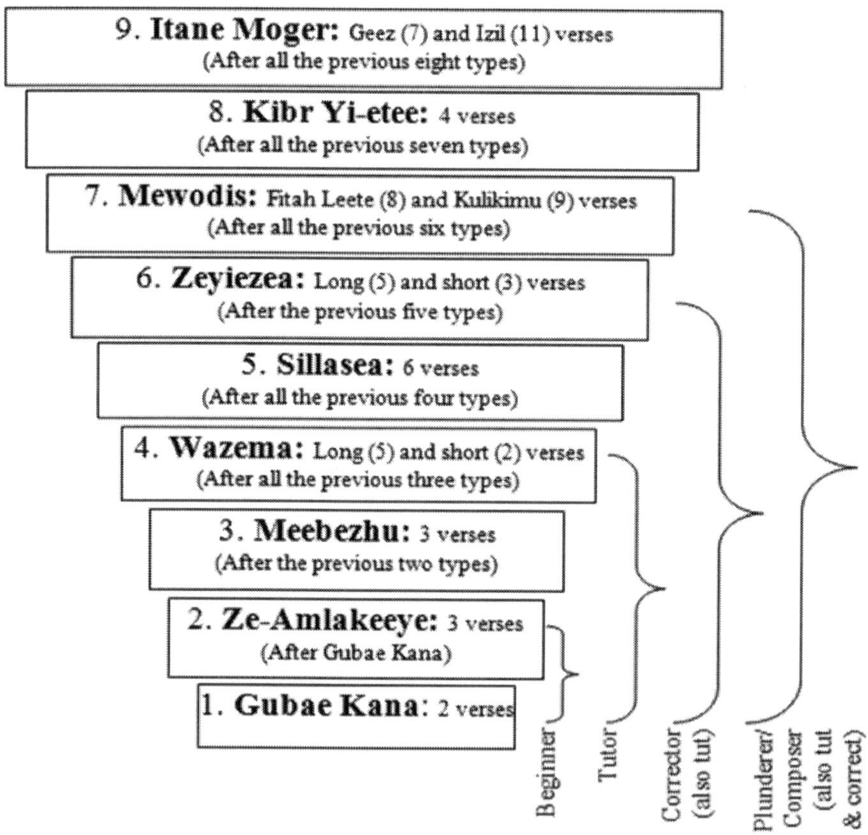


**Figure 1. Iterative incremental education and student ranks**

Dual role of students: a student is a teacher at the same time; he studies and gets perfection by teaching what he knew. He needs to be equally engaged in assisting other students and continuing his study. A teacher stated: *"A student has to make an uninterrupted effort to know, and to help others know what he knew."* It is more than a peer study. A student lives with a best friend of him, but also his friend is his co-teacher. The senior student who coaches his friend is gaining a teaching experience and perfection in his lower level studies. Such a friend tutorship creates a more relaxed academic environment as stated by a senior student: *"Students, especially beginners feel relaxed to present their new poems to senior students when compared with presenting to the main teacher."* Supporting this idea, a beginner student said:
"This approach is very convenient for us; we can ask whatever question we want because they are our friends. It is easier for us to interact with our friend teachers than with the master teacher."

Time distribution: the students invest almost all of their study time for creating their own qineas and almost no time for writing and memorizing composed qineas. A teacher confirmed this by saying:

*"Because we can always create qineas, there is no tradition of writing them. Besides, students have to focus on creating new ones rather than emphasizing on recording qineas."*

Strong relationships: There is strong friendliness and brotherhood relationship among students. As qinea schools are boarding schools, they live together day and night and they care to one another. The relationship between students and their master teacher is also parental. The master teacher is their father of knowledge and their guardian. In the same way as a teacher loves his succeeding students more than his biological children, students also love their teacher more than their biological father.

Academic freedom: students join schools any time, study at their personal pace, and leave the school by their own decision (even no dismissal for academic failure). This flexibility gives students more freedom. Another aspect of academic freedom is the freedom to compose qineas on any issues.

Despite being one of the religious institutions, students are free to compose qineas of non-religious issue including sex. For example, many students recall the following anonymous Gubae Kana poem:

*The upper and lower fire of vagina remains hot,*
*For penis/wood is inserted in it.*

The poem states that in a similar way to fire with sufficient wood in it, the female sex organ (vagina) remains hot to cool down hot penis. Traditionally, we know that fire is cooled down with water. Here, we are told that hot penis cools down with hot vagina. The warmth of the vagina is an indication that it will sexually function and continue to be the source of giving birth to new life. If the female sex organ loses its hotness, it ceases to sexually function and facilitate birth.
Strong discipline and self-motivation: As a teacher said,

*"Students are usually disciplined and learn how to behave from senior students. The school environment is peaceful. As they are self-motivated for the study, they are fully committed to know more and have no other business here."*

Despite severe economic problems, the students are highly motivated to acquire the knowledge. A student justifies why:

*The love for knowledge of qinea ties you and when you know more, you will love it more. We are now economically poor, poorer than everybody else. But we have the knowledge that money cannot buy. Students in modern education attend schools expecting to gain economic benefits later on. In Qinea Schools, we learn for the sake of knowledge, not for money or cloth. We know all this will pass as time goes, but knowledge and peace of mind live with forever.*

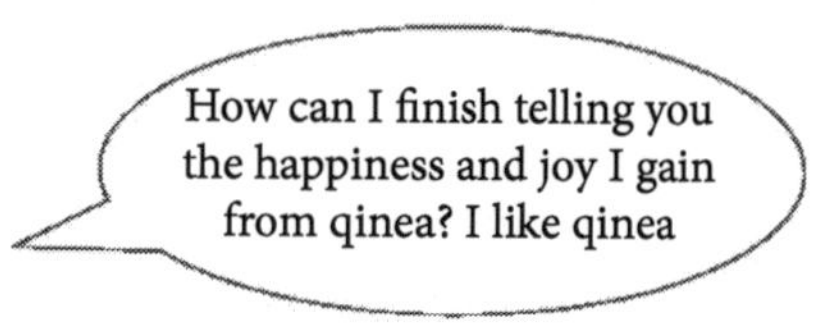

## 5.6. Entertainment

The way composers present qinea poems are pleasant to the composers themselves, to the listeners, and even to those who do not understand the poems. A teacher shared her experience by saying:

*"Qinea creates an emotion of happiness and refreshment. When I compose qineas in front of farmers who do not even know the language, I see how they feel and they frequently reward me even without understanding the meaning. It is attractive and I wish I hear nothing else than qinea."*

Another teacher stated:

*"How can I finish telling you the happiness and joy I gain from qinea? I like qinea more than a delicious food. The secret of qinea is a good refresher for our brain. It never fads and I never get staffed with qinea."*

Qinea helps to extract fun even from problematic situations. As a teacher said,

*"Qinea is a means to create pleasure within sadness. It touches the problem systematically and gives relief. It is like gently touching the surrounding of a wound on your body. It serves in the times of joy and sorrow, at wedding and funeral, for the kings or for the farmers."*

A composer student confirms this:

*"I get great pleasure and relief when composing qinea. It gives me mental satisfaction. I feel happy to study qinea.... Qinea expresses the joy and sorrow, the hunger and satisfaction, the warmth and coldness, and every other thing."*

Baym [46, p.33] states: "Because humor relies on group norms, knowledge, practices, and problems, it provides a way to deal with problematic issues within the group, to generate unique identities and to create group identity and solidarity." Thus, the availability of such a humorous culture as qinea indicates the potential to problem solving and to strengthen identity and solidarity. Besides, as fun is a function of linguistic, geographical, diachronic, socio cultural and personal boundaries [47], the same things are not equally funny among different cultures. We can say qinea helps to make fun in the linguistic, geographical, diachronic, socio cultural and personal boundaries of Ethiopia. Especially the composition of qinea in Wax and Gold form (a cover meaning and a deeper secret meaning) is unique in its ambiguities. It is very entertaining to be able to understand the deeper meaning (secret) from layers of meanings, covered with the Wax (cover meaning). This is related to what Chiaro [47] calls as the three interacting systems that make up the competence required in producing a joke: linguistic, socio-cultural, and poetic. In order to deal with the ambiguities and hidden traps of a language, both the producer and the recipient of a joke are required to recognize instances of broken linguistic rules and to have a high standard of linguistic knowledge [47].

## 5.7. Tolerance

Qinea teachers and students exercise the power of tolerance by passing through two challenges: the poor economic life that leads to starvation (economic challenge) and the complex hierarchy of qineas (academic challenge). Once they passed through these challenges as students, they can tolerate problems and accept ordinary life after becoming a master in the knowledge of qinea.

For example, Kifle Yohans, a renowned qinea teacher in Gondar, managed to gain the recognition and respect of the early 18th century kings (Eyasu and Dawit). But, as he was not in a good relationship with the priests who serve the kings rather than God, he hated the town life, went to a rural place where he was unknown, and was hired as a shepherd for a farmer. Once upon a time, he was ordered by his employer to take food to a nearby church for a ceremony. There were knowledgeable persons at the church who identified him and they were amazed of looking the great teacher leading a shepherd's life, dressing worn and ordinary cloths. They asked him why he preferred to live as such and as a reply, he forwarded the following Mewodis qinea[70]:

*Being a shepherd is better than being a priest;*
*Cattles house is also preferred to a big temple;*
*Because the temple has become a den of robbers;*
*The evil on priests is not on the shepherds.*
*In order to see the proof of this;*
*The Sun (Christ), who never becomes dark, came out of the manger of cattle;*
*Shepherds played with High Lord of Angles.*
*Again before this, from the place of shepherds, the tale of good news is heard;*
*Son of Anberem [Moses] for chronicle and son of Jesse [David] for kingship were called.*

The composer showed here in what ways his position as a shepherd was better than as a priest. He compares the priests with shepherds and the temple with manger and preferred the shepherds and the manger over the priests and the temple, justifying his belief with Biblical facts. The fact that Angels song with shepherds on the birth of Christ [Luke 2:8-12], that Moses was called for leading Israel from Egypt while he was serving as a shepherd at his father-in-law [Exodus 3:1-10], and that David the shepherd was also called for succeeding King Saul [1Sam.16:11-13] are presented to show how being

70 Translated from Geez:

እምከዊነ ካህን ይኄይስ ከዊነ ኖላዊ፤

ወኀለ እንስሳ ትትበደር እምቤተ መቅደስ ዐቢይ፡፡

እስመ ቤተመቅደስ ኮነት ቤተ ፈያታይ፤

ወበላዕለ ኖሎት ኢሀለወ በላዕለ ካህናት ዘሀሎ እከይ፤

ትእምርተ ዝኒ ከመ ንርአይ፤

በኀለ እንስሳ ሠረቀ ዘኢየዐርብ ፀሐይ፤

ወኖሎት ተዛውዑ ምስለ ሐራ ልዑል ሰማይ፡፡

ዓዲ እምቅድመ ዝንቱ እመካነ ኖሎት ዜና ሰናይ፤

ወልደ እንበረም ለምስፍና ወለቅብዐ መንግሥት ወልደ ዕሤይ፡፡

a shepherd is better than being a priest. Besides, because the priests changed the temple of God into a den of robbers (similar analogue with Christ on Mat.21:13), he prefers the manger to the temple. He concretizes his personal belief with Christ's preference to be born in a manger rather than in a temple (Luke 2:12).

With all the knowledge he had and being respected even by the kings, he had the capacity to change himself from a great teacher to a shepherd. That is why a respondent said: *"Knowledge of qinea helps you to pass the times of sadness and to have a power of tolerance."*

Regarding the academic challenge, a teacher said:

*"Learning qinea demands much tolerance, effort, and time to pass through the different stages of studentship: from starting as a beginner to intermediate level and then to expert level. Even after reaching a composer, a good student needs to study more on the different methods of thought and on increasing his perfection".*

Another teacher confirmed this stating:

*"Throughout the process, whenever a student goes to a next higher level, he composes it together with all the previous forms iteratively. ...Even after becoming a composer, a student stays further serving as assistant teacher and studying the different schools of taught to increase his methodological knowledge. Studying qinea is a difficult process. The knowledge is not acquired from the teacher alone, but from the creativity of the student himself. He improves his knowledge and skill by daily creating qineas, presenting and justifying it to his teacher, accepting criticisms, and making improvements. This is done for years and is a challenging task that tests ones patience and endurance".*

## 5.8. Inspiration (for moving the mass)

With his vocal style, word selection, and body movement, a good composer attracts attention more than any other show. When we see such a composer acting like a hero in war field, it will be a must for us to be inspired, unconsciously move our body, and applause. Going back in history to 1891, five years before the Ethio-Italian war at Adwa, the following Asere Nigus qinea[71] of Aleqa Lemma is an example of inspiring Emperor Menelik to march for Italians at Adwa:

*Menelik, do as much as your arm can do on Rome,*
*For its harvest time is approaching to reap.*
*Again, think of what is in the Gospel;*
*As the Word (Christ) said,*
*There is no fruit on fig tree/Rome.*

For the wax, Jesus said in Gospel (Luke 13:6-9; Math 7:20) that if a fig tree cannot bear fruit, its owner will cut it down and throw it into fire. Relating Rome (Italy) with this fruitless fig tree (indicating the uselessness of Italy for Ethiopia), the composer inspires the Emperor of Ethiopia to fight and conquer (reap) Rome. He also confirms that the time has come for Rome's downfall.

Historically, this qinea was composed at Addis Ababa Trinity Church on July 19, 1891 (before the Ethio-Italian war at Adwa). Menelik and Italy were not in quarrel during this time. However, many people considered this inspiring and prophetic qinea as a prophecy and expected that war between Ethiopia and Italy is inevitable. As expected, on the 5th year of this qinea, the Italians fought with Ethiopia at Adwa and were defeated by Emperor Menelik II.

A teacher said:

71 Translated from the Geez:

ምኒልክ ግበር ላዕለ ሮምያ መጠነ እዴከ ትክል፤

እምጣነ ለአጺድ በጽሐ ማእረራ ወመዋዕሊሃ ለኃጕል፤

ዓዲ ተዘከር ውስተ ወንጌል፤

ላዕለ ሮምያ በለስ ኢሀሎ አስካል፤

ከመ ይቤ ወልደ ያሬድ ቃል።

*"When I compose qineas in front of farmers who do not even know the language, you see how they feel and they frequently reward me even without understanding the meaning".*

Whenever there are major events in the Church, qineas are presented to establish and cement shared sentiments among the mass, which is the source of their unity and inspiration. As the Ethio-Eritrean war (1998-2000) brakes out, a composer called Merigeta Fisiha Kidane (Dessie) produced the following Sahilke qinea[72], indicating the current situation and what Ethiopian should do:

*This untiring country said:*
*I don't want to be recalled in the name of Essaiyas;*
*For there is a short man called after my name.*

If a man asks a married wise woman for extra marital relations, she rejects the offer by saying that she has her own man. In a similar fashion, when President Esayas of Ertiria invaded Ethiopia in late 1990's, Ethiopia made a right decision in collaborating with the existing government forces (despite internal differences). The short man of Ethiopia refers to the then Prime Minister Melles Zenawi.

This qinea was composed while Ethiopia and Eritrea were in border fighting. As the war began, there was a belief that the Ethiopian governing body (EPRDF) failed to win the heart of its people and the Ethiopians might not collaborate with EPRDF to fight with Eritrea. The poem iterates this situation and shows the way how Ethiopians should act to the war. There are several qineas related to this and other major events to inspire the people.

72 Translated from the Geez:
ትቤ ሀገር ዘኢትጻሙ፣

ኢይፈቅድ እንበር ለኢሳይያስ በስሙ፣

በስምየ እስመ ሀሎ አሐዱ ወሬዛ ዘሐጺር ቆሙ::

## 6. Findings on the process, barriers and enablers of knowledge creation

One of the major questions in this research was *"How and why do experts in the indigenous Qinea Schools create knowledge."* The '*how*' aspect is related to the process of the indigenous knowledge creation. The '*why*' aspect is related to identifying the barriers and enablers (motivating factors).

### 6.1. Indigenous knowledge creation process

The biographic data of informants show that the composer students and master teachers have studied qinea for at least four years. Beginner students study Geez morphology and the teacher's poems under the guidance of a tutor before producing a single new poem a day. Tutoring students can produce relatively less complex types of poems within shorter periods of time. Corrector students produce almost all types of qineas within a short time. Composer students also called Plunderers, produce all types of poems instantly "off the top of the head" and can even predict the poem of others under formation. They are like the master teacher except methodological knowledge.

The process of composing qinea poems is always initiated by a triggering event. When a person is called to compose at such events, he quickly scans the situation (key participants of the event) for the deeper meaning and relates it to previous knowledge in his brain for the cover meaning. Combining his previous knowledge with the information of the ceremonial events and capturing the viewers' expectations, he expresses new knowledge that best describes the truth about the event in exemplary poems. The process is depicted in the following figure.

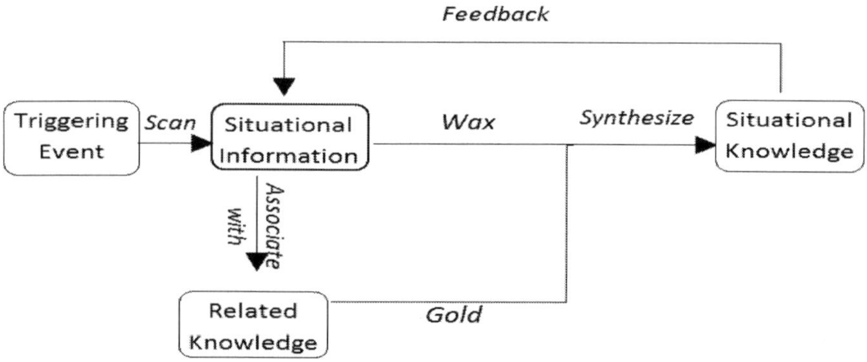


**Figure 2. Processes of knowledge creation in Qinea Schools**

The general process of composing qinea poems resembles the socialization and externalization processes in the SECI model. Socialization in SECI involves the sharing of tacit knowledge between individuals through physical proximity (being together or living in the same environment) [27, 31]. Similarly, qinea students and teachers live together (have physical proximity) and share, in poetic forms, their personal belief (tacit knowledge) about a situation. The difference is that the knowledge sharing in qinea is in verbal form, contrary to the socialization of SECI model which doesn't necessarily involve written or verbal instructions. This articulation of tacit knowledge through dialogue where the composer explains why he said everything in his poem is related to externalization of SECI model. When a person composes a poem, he at the same time creates the concepts in language and the knowledgeable audience evaluates his poem from different perspectives (originality, appropriateness of language, size, and meaning layering in wax and gold). This is related to the first three steps in the 5-step model of von Krogh et al. [29]: *sharing tacit knowledge, creating concepts, and justifying concepts.*

## 6.2. Barriers

Economic and academic problems were reported by almost all respondents as the major barriers to studying qinea. The economic barriers are widespread. As almost all students went far across regions to study qinea, families do not provide economic assistance. Thus, students usually get food by begging the villagers around the Qinea Schools. As most Qinea Schools are located in rural areas, the villagers are farmers with their own economic problems and it is difficult for them to give food to the students. Academic barriers included the problem of understanding the subject matter (manifested by inability of creating qineas as desired) and the shortage of text books that support the study. Compared to the economic barriers, the academic barriers are more harmful and unmanageable.

## 6.3. Enablers

Qinea students and teachers enjoy devoting their entire life to the study of qinea, despite their severe economic problems. We have to ask here why they study qinea under such conditions. Interviewees described the primary motivating factors for studying the Qinea as: the Challenge (of creating new qineas sharpens the brain), Knowledge (the love for knowledge in Qinea Schools), Freedom (of creativity), Satisfaction (derived from composing and listening to good qineas), and Religion (providing better religious services). Knowledge was mentioned as the dominant motivator to study qinea. Students claim that qinea is the source of knowledge that money cannot buy and that such knowledge enables them to predict the future, curve the direction of things to their convenience, lead a preferred lifestyle, understand things easily, make wise judgments, and lead followers to desired destinations. Knowledge is mentioned as a source of other motivating factors for studying qinea (religious fulfillment, satisfaction, and freedom of creativity).

We can relate the enablers with the four ba's of Nonaka and colleagues [31]. The Qinea Schools have good Originating 'ba' allowing students to informally share problems and opinions with one another. The Qinea presentation sessions (allowing a student to present new qineas and get feedback daily) can be considered as interacting 'ba'. The difference is that in the interacting 'ba', the people selected to engage in dialogue have a relatively balanced mix of knowledge. In Qinea Schools, the dialogue is between students and correctors (having unbalanced knowledge and capabilities) and the objective is primarily to develop the knowledge construction capacity of the student while contributing to the new knowledge.

## 7. Findings on implications of indigenous knowledge techniques

Although there are different techniques of qinea, the predominant one is expressing new ideas covering familiar ones in a form of wax and gold. The wax and gold tradition takes qinea far beyond poetic expressions. It is like defining something new in terms of familiar concepts, as shown in the following short poem of a student:

*To liberate Adam/student from illiteracy/Hell;*
*Savior true God/Qinea of Yaread was born in Ephrata.*

This poem associates students with Adam, illiteracy with sin/hell, qinea of the master teacher (Yaread) with Christ, and the Qinea School with Ephrata where Christ was born. This way, a qinea composer approaches a new situation through familiar situations. Thus, when he wants to share his experience and personal belief, he presents it as gold covering familiar concepts (wax). This makes it easy for listeners to understand the composer's new concepts.

This tradition implies that instead of thinking directly about creating new knowledge (or solution to a problem), it is better to start with what we know and try to associate it with existing knowledge or solutions. This is related to the analogy (metaphor) creativity technique. Couger et al. [48] state that analogies help to find new insight and approaches to the nature and solutions to IS problems. Metaphors are powerful tools of creativity which facilitate the description of new situations through reference to familiar ones, thereby allowing conceptual leaps [49].

## Conclusion

The objective of this research was to explore the indigenous techniques of knowledge creation and to transfer these insights to modern knowledge creation. *We have raised two questions: How and why do experts in the indig-*

*enous Qinea Schools create knowledge,* and *what can organizations learn from indigenous forms of knowledge creation.*

In addressing the how and why, we have seen that qinea techniques have been used by the community for such purposes as: expressing the "unsayable", instantly creating replies or solutions to problems, predicting the future, utilizing Geez documents, creative education, entertainment, tolerance, and inspiration (for moving the mass). Despite that the indigenous communities are enjoying these, such potentials are absent from the modern education of Ethiopia. This calls for further research to leveraged the potentials of the indigenous techniques to the best use of modern Ethiopia. As the capacity of creating new knowledge and converting it into products and services is key for the progress of a country, qinea can be considered as a reach base of knowledge creation in the context of Ethiopia. However, its application by the community is limited to the previously mentioned areas, and future researches focusing on converting the knowledge into tangible products and services are most welcomed.

We have also seen that the general process of composing qinea poems resembles the socialization and externalization processes in the SECI model with minor differences. However, once the tacit knowledge is externalized in poetic forms, we do not see the combination step of SECI and the knowledge is not often converted to tangible forms (products or services). The indigenous techniques are more focused on tacit knowledge creation. What we learnt is that it has unique methods for creating and evaluating tacit knowledge. Even the existence of multilayered formal procedures of validation does not hamper the justification process.

Couger et al. [48] explain creativity as a dynamic phenomenon encompassed of four highly interactive components: motivated persons, facilitating processes, product (resulting from creative works) and press (positively influencing environment). This description matches with the features of the Qinea Schools. The students' dedication for the study of qinea against all odds (serious economic problems) evidences that they have intrinsic (internal) motivation. The Qinea School environment and the educational processes facilitate students to daily produce new products (poems).

Further work is recommended to model the indigenous techniques, test it with simulated problems, and compare the results with SECI and other models of knowledge creation to see their effectiveness in creating knowledge for solving a given situation/problem. Further works are also required to exploit the various potentials of the indigenous techniques of knowledge creation and to shape their applicability in specific focused domain areas.

## 9. Reference

Agrawal. A. (2004), "Indigenous and scientific knowledge: some critical comments," IK Monitor 3(3), pp.1-9.

Agrawal, A. (1994), "Dismantling the Divide between Indigenous and Scientific Knowledge," Workshop in political theory and policy analysis, 11/18/1994; Indiana University, USA.

Alavi, M. & Leidner, D.E. (2001) "Review: Knowledge management and knowledge management systems: Conceptual Foundations and Research Framework," *MIS Quarterly 25(1),* pp.*107-136.*

Baym, N. K., (1995) "The Performance of Humor in Computer-Mediated Communication," Journal of Computer-Mediated Communication, 1995, pp. Publisher's official version: http://dx.doi.org/10.1111/j.1083-6101.1995.tb00327.x Open Access version: http://hdl.handle.net/1808/871

Bhatt, G.D. (2001) "Information technology; Knowledge management; Knowledge management systems; Knowledge workers; Interaction," Journal of Knowledge Management 5(1), *pp.*68-75.

Boven, K. & Morohashi, J. (2002), "Best Practices using Indigenous Knowledge," A joint publication by Nuffic, The Hague, The Netherlands, and UNESCO/MOST, Paris. http://www.unesco.org/most/bpikpub.htm

Brown, J.S.& Daguid, P. (1998) "Organizing Knowledge," California Management Review, 40(3), pp.90-111.

Chiaro, D. (1995) "The Language of Jokes: Analyzing Verbal Play," London, Routledge.

Couger, J.D. Higgins, L.F. and S McIntyre, C. (1993) "(Un)structured creativity in information systems organizations," *MIS Quarterly;* Dec 1993, 17(4), pp.375-397.

Cook, S. & Brown, J. (1999) "Bridging epistemologies: the generative dance between organizational knowledge and organizational knowing," *Organization Science* 10(4), pp.381- 400.

Fahey, L. & Prusak, L. (1998), "The Eleven Deadliest Sins of Knowledge Management," *California Management Review, 40(3), pp.265-276.*

Gashe, A., (2011) "The Wisdom Compass to Eternal Life," Xlibris Corporation, USA.

Grenier, L. (1998) "Working with indigenous knowledge: A guide for researchers," International Development Research Centre (Canada).

Grisoni, L. (2007) "Cooking up a storm: Flavouring Organizational Learning with Poetry," Proceedings of OLKC 2007 – "Learning Fusion", pp. 341-356.

Gorjestani, N. (2000), "Indigenous Knowledge for Development: Opportunities and Challenges," Indigenous Knowledge for Development Program (The World Bank), UNCTAD Conference on Traditional Knowledge, *Geneva*

Grover, V. and Davenport, T.H. (2001) "General perspectives on knowledge management: Fostering a research agenda," Journal of Management Information18(1), pp.5-21.

Hislop, D. (2002) "Mission impossible? Communicating and sharing knowledge via information technology," *Journal of Information Technology 17*(3), pp.165-177.

Hirst, G. (1997), "Context as a Spurious Concept," AAAI Fall Symposium on Context in Knowledge Representation and Natural Language, Cambridge, Massachusetts, 8 November 1997.

Isen, A.M. Daubman, K.A. and Nowicki, G.P. (1987) "Positive Affect Facilitates Creative Problem Solving," Journal of Personality and Social Psychology, Vol. 52. No. 6, pp.1122-1131.

Kebede, M. (2008) Radicalism and Cultural Dislocation in Ethiopia: 1960-1974, University of Rochester Press.

Lewis, T. (2006) "Creativity: A framework for the design/problem solving discourse in technology education", Journal of Technology Education, 17(1), pp.36-53.

Levine, D.N. (1972), Wax and Gold: Tradition and Innovation in Ethiopian Culture; University of Chicago, USA.

McQueen, R.J. (1998), "Four Views of Knowledge and Knowledge Management," Proceedings of AIS'98, University of Waikato, pp. 609-611.

Malhotra, Y. (2000) "Knowledge Management for E-Business Performance: Advancing Information Strategy to 'Internet Time'," Information Strategy: The Executive's Journal 16(4), pp.5-16.

Moteleb, A.A. Woodman, M. Galal-Edeen, G.H. and Bakry, W.M. "Ontology of Knowledge Management Systems Elements" accessed on July 3, 2009 from http://www.iseing.org/emcis/EMCIS2005/pdfs/34.pdf

Mahibere Kidusan, (2009) "Qinea: the philosophical foundation of the ecclesiastical education", Documentary film.

Nkrumah, G. (2003) "Older than Egypt is Ethiopia," Al-Ahram Weekly 21-27, August 2003, Issue No. 652 http://weekly.ahram.org.eg/2003/652/heritage.htm

Newell, S. & Swan, J. (2002) "Managing Knowledge Work" in *Knowledge Work and Knowledge Workers*. eds S. Newell, H. Scarbrough, J. Swan, J, and M. Robertson London: Palgrave.

Nonaka, I. (1994) "A dynamic theory of organizational knowledge creation," *Organization Science,* 5(1), pp.14-37.

Nonaka, I. and Konno, N. (1998) "The Concept of "Ba": Building a foundation for knowledge creation," California Management Review 40(3), pp.40-54.

Nonaka, I. (2005), "Knowledge Management: Critical Perspectives on Business and Management," Taylor & Francis Publishers.

Ocholla, D. (2007), "Marginalized Knowledge: An Agenda for Indigenous Knowledge Development and Integration with Other Forms of Knowledge," IRIE (International Review of Information Ethics), Vol.7 (09/2007).

Patke, R.S. (2006) "Poetic Knowledge: Theory, Culture & Society," SAGE Publications 23(2–3), , pp.199–222.

Puga, D. and Trefler, D. (2003) "Knowledge creation and control in organizations," *University of Toronto; NBER* Working Paper 9121, November 2003.

Prusak, L. (2001), "Where did knowledge management come from?," IBM Systems Journal 40 (4), pp.1002-1007.

Puri, S.K. (2007), "Integrating scientific with indigenous knowledge: constructing knowledge alliances for land management in India," MIS Quarterly 31(2), pp. 355-379.

Rahman, A. (2000), "Development of an integrated traditional and scientific knowledge base: a mechanism for accessing, benefit sharing, and documenting traditional knowledge for sustainable socio-economic development and poverty alleviation," UNCTAD expert meeting on systems and national experiences for protecting traditional knowledge, innovation and practices, Geneva.

Saint, W. (2004) "Higher Education in Ethiopia: The Vision and Its Challenges," JHEA/RESA 2(3), pp.83–113.

Schmidt, L. (2009) "Traditional knowledge: valuing folklore," Managing Intellectual Property, the Global IP Resource, November 1, 2009.

Soo, C.W. Devinney T.M., and Midgley, D.F. (1999) "The Process of Knowledge Creation in Organizations; Centre for Corporate Change," Australian Graduate School of Management; The University of New South Wales, Australia,

Sternberg, R.J. and O'hara, L.A. (1999) "Creativity and Intelligence," in Handbook of Creativity, R.J. Sternberg, Ed., Cambridge University Press, pp.251

UNESCO, (1983) "Note on Ethiopian Cultural Objects Abroad," United Nations Educational Scientific and Cultural Organization (UNESCO), Intergovernmental Committee for promoting the return of cultural property of its countries of origin or its restitution in case of illicit appropriation, Third Session, Istanbul, May 1983, pp.9-13.

von Krogh, G. (1998) "Care in Knowledge Creation," California Management Review 40(3), pp.133-153.

von Krogh, Ichijo, G. K. and Nonaka, I. (2000) "Enabling Knowledge Creation: how to unlock the mystery of tacit knowledge and release the power of innovation," Oxford University Press, US.

Warren, D.M. and Rajasekaran, B. (1993) "Putting local knowledge to good use," *International Agricultural Development* 13(4), pp.8-10.

Weisberg, R. W. (1999) "Creativity and knowledge: A challenge to theories," in Handbook of Creativity, R.J. Sternberg, Ed., Cambridge University Press. pp.226-250.

Woodman, R.W. Sawyer, J.E. and Griffin, R.W. (1993) "Toward a Theory of Organizational Creativity," The Academy of Management Review, Vol. 18, No. 2, Apr. pp.293-321.

Workneh, H.M. (1971) ጥንታዊየኢትዮጵያሥርዓተትምህርት (Ancient Educational system of Ethiopia); Birhanina Selam Printing Press, Addis Ababa, 40th anniversary of HS.

Zack, M.H. (1999) "Managing Codified Knowledge," Sloan Management Review, Summer *1999.*

# CHAPTER 18

# Intellectual Property Rights: A Focus on Medicinal Plant Products from Africa

*Fekadu Fullas and Mammo Muchie*

## Background

Whether it is an inventor of a machine, an author of a book, or a composer of music, each owns his (her) work. Such work cannot be copied or purchased without taking into account the rights of the owner. IPRs are awarded to individuals or organizations for inventions and creative works. The premise of such rights and reward is to encourage further innovations and creativity to benefit society. The legal means of protection of these rights are through copyrights, industrial designs, geographical indications and trademarks (Lodha, 2007).

Knowledge and evaluation based on scientific facts do not automatically replace traditional beliefs. Some traditional medical practices disappear over time, while others get transformed and revitalized. Thus, medical pluralism and hybridization follow. New hybrid practices get tangled with (in collusion or harmony) cosmopolitan health practices introduced and promoted by multinational corporations. In this interaction, conflict of interest and rights is played out between multinational corporations and state institutions on one side, and local rural plant collectors, traders and traditional healers on the other side. Property rights become the center of controversy. There is a constant incompatibility of interests between indigenous property rights and international patent laws in this day and age of globalization of commerce. It has been argued that enclosure of knowledge hampers the progress of science and society as a whole. Nevertheless, the issue is so delicate that Non-Governmental Organizations (NGOs) may one-sidedly promote the interest of multinational corporations by sidelining the rights of the providers of indigenous knowledge (IK) (Reihling, 2008). In such cases, indigenous councils representing local communities are better positioned to work out an alternative method of sharing benefits potentially acceptable to both sides (*vide infra*, *Hoodia* story).

In general, products of nature cannot be patented in the form they are found in nature. Although, patent rules may impose limitations of invention on natural products, claim protection may be secured for extracts and formulations of traditional medicines that are processed by a novel and non-obvious method. Some Chinese herbal extracts have been granted in this manner (Correa, 2002). On the other hand, natural products in their purified form are patentable, provided that they are not previously known. The fact that no one has taken a patent on a commercial substance that has utility does not mean that the latter is patentable. For example, table salt has been "in the public domain" since ancient times and therefore is not patentable. "*Utility*" is an important aspect of a patent claim. If a substance is novel but doesn't have utility, then it is not patentable. The hardest aspect of a patent determination is the "*non-obviousness*" side of a claim. If an idea (a discovery) would be obvious to a person "*ordinarily skilled in the art*" at the time of the invention, then it is not patentable. The "*ordinarily skilled in the art*" person is one who has a general educational background and experience in the particular field of the invention (Mitscher et al, 1994)

There are some people who think that patenting traditional knowledge (TK) is inappropriate for the "common good." On the other hand, many researchers are prone to disseminating potentially proprietary TK in publications (grants, papers, dissertations, brochures, herbarium annotations, the Internet, etc.) and verbal presentations at meetings. All these information outlets make TK "*prior art*" and therefore not patentable. In academia, where the "*publish or perish*" motto drives academic promotions, scientists have a tendency to present TK as a part of their presentations or publications. Suggestions have been made that researchers should withhold information on the origin of the plant materials and ethnomedical uses for the sake of protecting TK. If the information eventually leads to a commercial product, such information can be revealed after making proper agreement with the parties imparting the knowledge. Professional organizations and journals should draw up strict guidelines for presentations and manuscripts which involve TK (Weiss & Eisner, 1998). The US patent law stipulates that a patent cannot be granted to a claim if evidence exists that such information is available in the USA or a foreign country in the form of orally transmitted TK, or as documented or published information. In the same manner, the European Patent Law under a broad "*State of Art*" category rejects a patent application, if it determines that the public has prior knowledge of the claim in an oral or written form, thus effectively placing TK in the public domain (Elvin-Lewis, 2007).

## What are IPRs?

IPRs are the legal rights that result from intellectual activity in the literary, industrial and scientific fields. They vary widely across countries. The USA, Japan and some European countries produce a large number of products and technologies and thus have many intellectual property laws to protect their inventions. Technology protectionism grew in scope with globalization as developing countries opened up their doors to international commerce (Teljeur, 2003).

IPRs regimes can be placed into two major categories: copyright and industrial property rights. Copyrights cover literary, musical, photographic, artistic and scientific works; maps, technical drawings and computer programs, whereas industrial intellectual property rights include inventions, designs, trade and service marks, commercial patents, industrial designs and geographical "*appellations*." Technological innovations are protected by patents. An innovation has to meet certain patentability criteria; it should be novel and "*non-obvious*" and should have utility. A patent is a document issued by a government that entitles the inventor exclusive rights for a limited period of time. At the end of the patent period, the invention will be in a public domain. Inventions differ from scientific discoveries in that the latter involve natural laws and phenomena. A third category is known as *sui generis* "of its own kind" protection, which is neither in the copyright nor in the patent realms. It is a specific mode of protection that applies in individual countries, for example, to medicinal plants and their products. In order to appreciate how patent rights are distributed, it is instructive to glean the spread of patent applications. In 2001 alone, about 90,000 new patent applications were filed with the World Intellectual Property Organization (WIPO); 42% were American, 13% German, 10% Japanese, 6% British and 4% French. All developing countries combined accounted for only 3.5% of the applications. Technology exporting countries claim that intellectual property protections will have spillover benefits, whereas developing countries argue to the contrary. The latter claim that the protection engenders increases in the cost of patented products, in addition to causing decreased access to technology (Teljeur, 2003)

## International IPRs Regimes

Some concerns arose out of knowledge or suspicion that some scientists, research institutions or commercial enterprises take biological samples from developing countries to develop pharmaceuticals without due permission, or without working out mutually acceptable terms. An additional concern was the misappropriation of indigenous intellectual properties, which would form the basis for developing drugs. On the other hand, it has been argued that exaggerated accusations of biopiracy of intellectual property rights may create tight restrictions on research. Conflations and divergence of interests in Intellectual Property Rights occur between international organizations and institutions and communities in developing countries. The on-going conflicts between the Convention on Biological Diversity (CBD) and Trade Related

Aspects of Intellectual Property Rights (TRIPS) espoused by the World Trade Organization (WTO) reflect this situation. For example TRIPS agreements do not require that patent applicants consult with indigenous communities or governments about patenting a product obtained from that country. There is also no stipulation which would require sharing benefits with indigenous people (Hamilton, 2003).

## World Trade Organization/TRIPS

WTO was established in 1994, and the Agreement of Trade Related Aspects of Intellectual Property Rights (TRIPS Agreement) evolved to define minimal intellectual property standards and their enforcement (Elvin-Lewis, 2007). Since 1995, WTO has required its member countries to comply with TRIPS. At the Seattle WTO meeting of 1999, the African group vigorously opposed the patenting of life forms, and argued in favor of protecting community rights over agricultural and biological heritage (Limson, 2002). The WTO and TRIPS do not recognize the collective rights of indigenous communities (Reihling, 2008). In 2001, the Fourth Ministerial Conference in Doha (Qatar) formulated the Doha Declaration in order to clarify the TRIPS agreement within the context of CBD. The TRIPS council was instructed to give due protection to traditional knowledge, and also to contribute to the promotion of technological innovation and transfer of technology in a manner conducive to social and economic welfare of developing countries (Elvin-Lewis, 2007). However, there are areas of divergence between CBD and TRIPS Agreement. In the case of patentable inventions for biological material, the TRIPS Agreement does not clearly require disclosure of the source of the material for the invention, nor delineates the need to acquire prior informed consent of the country of origin of the material. Although the agreement appears to provide a *sui generis* protection of TK, it does not lay out the details. The aim, however, appears to ensure plant breeder rights (PBRs) which are not covered by standard IPRs, rather than recognizing TK and sharing benefits with indigenous people. The International Union for the Protection of New Varieties of Plants (UPOV) Conventions have items that may protect PRBs. UPOV is formed by 30 countries (most of them developed) primarily to provide IPRs on new varieties of plants in member countries for market. The provisions in UPOV enable plant breeders to collect royalty fees from farmers in poor countries who want to use and harvest the new varieties the breeders may develop. The TRIPS Agreement falls short of protecting and compensating TK in a clear and transparent manner (Lodha, 2007).

## Convention on Biological Diversity (CBD)

The United Nations CBD Framework is an international agreement signed by 150 countries at the 1992 Rio de Janeiro Earth Summit. The CBD is anchored in the recognition that developed countries can contribute systematically to the preservation of world biodiversity. However, it also stipulates the rights of local communities, and states need to protect and conserve biological resources. CBD mentions that programs and policies should reflect conservation, sustainable use, and sharing of benefits that emanate from the use of these resources (Lodha, 2007). In its objectives, CBD states that biological diversity will be conserved, use will be sustainable, and benefits will be equitably shared, including technology transfer and appropriate funding. The short term non-monetary benefits include employment in Research and Development (R & D) and training, capacity building, acquisition of equipment and supplies, scientific exchange, supply of research results, technology transfer, infrastructure and research collaboration. Monetary benefits during research and development (so called "process" benefits) include access fees in the form of lump sums or milestone payments. These payments usually form the basis of trust funds for the immediate benefits of the stakeholders. Long term benefits include income generating schemes and village development projects in local communities (Rukanigra, 2001).

## The African Perspective

The African Model Law for regulating biological resources in the continent was put in place in 2000 (Ekpere, 2000). The objective of the law was to preserve the rights of indigenous African communities to uphold the conditions set forth in the Convention on Biological Diversity (CBD) to define terms when Northern corporations access these resources. The Model Law arose out of the need to protect indigenous and local community knowledge systems, particularly in light of the Northern patent system of claim allowable in the TRIPS agreement. It was particularly aimed at safeguarding "*the interests of the rather uninformed and thus rather gullible South, and to a lesser extent also the public of the North, in this unfolding scene of totally new kinds of claim to the resources (especially biodiversity) and intellectual achievements, through IPRs, of indigenous and local communities of the South but, to a lesser degree, even of the North*" (Oh, 2009). The Model was originally developed by Ethiopia and endorsed by African Union members. It formulated a "*sui generis*" system of protection of the rights of local communities, farmers, and breeders, as well as for the regulation of local biological resources. The Law provides, *inter alias*, for acquiring prior informed consent (PIC) to access biological resources, and a fair and equitable sharing of benefits arising from the exploitation of these resources (Ekpere, 2000; Oh, 2009).

In a technical report (Andrews et al, 1996), the International Union of Pure and Applied Chemistry (IUPAC) has drafted a sample contractual agreement that contracting parties may sign onto in order to extract mutual optimal benefits from research and development of plant medicinal products. The draft outlines steps that the partners can take in medicinal plant sample acquisitions and developing products. The suggested contracts stipulate a 2% of Gross Net Sales royalty to be paid to a developing country, should the product reach marketable status. Crude extracts are more valuable than unprocessed samples. About 50% of the 2% royalty is paid by the sponsoring company to the collector directly and the remainder is paid to a designated foundation representing the country of origin to fund conservation and sustainable cultivation of targeted biota. IUPAC recommends that a company make a prior agreement with a supplier to pay a specified amount of money per kilogram of sample of a plant, an extract or purified natural product. What necessitates agreement is that companies in the developed countries have the funds and the expertise, but not the biota. Developing countries have rich biota, but not adequate funds and expertise, hence the need for collaboration. The largest unexplored source of natural products lies in the subtropical and tropical regions of the world. In this region, ethnobotanical medicine has flourished, and a vast knowledge has built up over the course of many generations (Andrews et al, 1996).

Knowledge about traditional medicine may be preserved within a family, known to specialized healers, to a certain gender, or shared among members of a community, clans or tribes. For example, in sub-Saharan Africa, such information is closely guarded, so that the practitioners' livelihood is not compromised. In Madagascar, traditional knowledge is claimed by the state, considered as possession of indigenous groups or tribes in South Africa, or ascribed to specific healers in Tanzania. However, as a whole management of TK related to genetic resources is still an evolving process. Petty patents are present in African countries such as Ethiopia, Kenya, Cameroon, Central Africa, Chad, Congo, Dahomey, Cote d'Ivoire, Malagasy, Mauritania, Niger, Senegal and Upper Volta. Petty patents in these countries are relatively easy to acquire and less expensive. They are ideal for circumscribed botanicals which require lower standard of inventiveness (Elvin-Lewis, 2007).

## South-South Collaboration

In the ideal world, IPR would belong to all humanity. However, development of the pharmaceutical industry in the West has been fuelled by IPR protection of market economy in the form of patent monopolies (Patwardhan, 2005). The resources, technology and expertise to exploit biodiversity of medicinal plant products exist in the North (USA, Europe and Canada); on the other hand, rich biodiversity abounds in the South (developing countries) where such conditions lack. It has been claimed that too many stringent regulations in the South might be restrictive to research. This has been true even for local scientists. For example, such regulations have been considered to have hampered Northern scientists to probe into the rich Southern rich forests of Brazil and Malaysia to explore medicinal plants. Because of lack of definitive laws, it has claimed that it is nearly impossible to get a permit to collect plant specimens in Brazil (Hamilton, 2003). However, these complaints may not be justifiable, considering past experiences of countries in the South. Lingering resentments exist in developing countries over past uncompensated exports of genetic materials (Weiss & Eisner, 1998).

One view claimed that exchange of plant specimens even within African countries was difficult for fear of bio-piracy. In one instance, it was reported that it was easier for taxonomists in Uganda to exchange specimens with the United Kingdom than with the neighboring Kenya, even when such practice had nothing to do with bio-prospecting or bio-piracy (Hamilton, 2003). Despite some difficulties, collaboration within developing countries (South-South) and with developed countries (South-North) may spur development of natural medicinal products.

Tropical and subtropical Africa is home to 40, 000 to 45,000 higher plant species, and of these only 1% has been studied for their medicinal potential. It is estimated that about 50 modern drugs have their origin in tropical plants. The 1% deforestation rate is the highest in the world. Although Africa harbors 5,000 medicinal plants, only 83 out of the 1,100 leading commercial medicinal plants have entered international market. One reason cited for this low number is lack of standards. To remedy this situation, the African Association of Medicinal Plants Standards has been established for the sole purpose of preparing Herbal Pharmacopoeia. Capacity building networks such as NAPRECA, WANNPRES, ALNAP, NABSA and AFASSA have been established in Africa (Gurib-Fakim, 2009).

In Africa, research on the chemistry and biological activities of plants is constrained by lack of financial resources and inadequate equipment, manpower and technical resources. Therefore, natural product researchers turn to the assistance of laboratories in developed countries. In countries where natural product research exits, it is heavily academic. Most of the ongoing researches are carried out with funding from international donor agencies. A concerted approach to develop phytomedicines at the industrial scale is lacking. This problem has been compounded by lack of strong national policies and appropriate industrial technology (Rukanigra, 2001). However, sub-regional (South-South) cooperation and capacity building in medicinal plant research has been going on in Africa.

Natural Products Research Network for Eastern and Central Africa (NAPRECA) was established in 1984 to promote and coordinate natural products research in the sub-region. The founding scientists were from Ethiopia, Kenya, Sudan and Zimbabwe. Over the years, the membership grew to include additionally Tanzania, Rwanda, Uganda, Madagascar, Botswana, D.R. Congo, Cameroon in that order. It has held 13 symposia meetings, with the venues rotating between member states. In the first two years, they were held annually, but since 1989 they have been conducted biennially. In addition, NAPRECA has held several pre-symposia workshops and short summer school programs to aid in research capacity building. The funding for the network has come from the International Program in Chemical Sciences (IPICS), Sweden, the German Academic Exchange Service (DAAD), the International Foundation of Science (IFS) and the Organization for the Prohibition of Chemical Weapons (OPCW) (Midiwo, 2009). NAPRECA encourages member countries to protect their intellectual property rights, discoveries and inventions through registration of patents. However, adequate expertise and resources are lacking for handling such patents. Obviously, discoveries patented in developing countries may not easily attract the interest the western world companies (NAPRECA, 2003).

Another network in the sub-region, African Laboratory for Natural Products (ALNAP) is headquartered in Addis Ababa, Ethiopia. It links laboratories in five neighboring countries in the sub-region: Uganda, Rwanda, Ethiopia, D.R. Congo and Burundi. The network member laboratories make their facilities available for use and training to researchers and graduate students in other laboratories within the ALNAP network. In addition to availing analytical instruments, ALNAP has also developed a natural database (acronym, NADA) on African plants highlighting their chemistry, uses and other aspects. The data base has also been released on a CDROM. The Swedish organization IPICS has been the major funding agency for the effort of ALNAP (Dagne, 2009).

In the western part of Africa, the Western Africa Network of Natural Products Research Scientists (WANNPRES) was officially established in 2002 in Ghana. The seed for the network was planted early in 1997 by scientists from Cameroon, Ghana and Nigeria. Since the first successful scientific meeting held in Accra, Ghana in 2004 which attracted a sizable 80 participants from the sub-region, the network has held three similar conferences (Addy, 2005; History, 2008). The network has national coordinators in Benin, Burkina Faso, Cameroon, Cote d'Ivoire, Ghana, Guinea, Mali, Niger, Nigeria, Senegal and Togo (Structure, 2008). As stated in the Constitution, the network strives to "*share ideas, information and resources, both human and material*" in order to add *"value to natural products for industrial development.*" In the first scientific meeting, there was a presentation that was devoted to IPR issues and the OAU Model Law, which encourages *sui generis* to be applied for the protection of TK (Addy, 2005). This effort in the western sub-region of Africa parallels the already successful NAPRECA in the east and central sub-region. At the founding of NAPRECA, it was stated that the Network was established on geographic and historic considerations (Dagne, 1996). It would be prudent to establish institutionalized collaborations between the sub-regions, for example arranging a joint symposium, like the American Society of Pharmacognosy (ASP) does, by holding joint international meetings with similar societies in Europe.

In 2002, the Third World Academy of Sciences (TWAS) recognized the University of Botswana as a south Center of Excellence for natural products research. The center is called Network for Analytical and Bioassay Services in Africa (NABSA). Ten African countries (Burkina Faso, Cameroon, Nigeria, Ethiopia, Kenya, Congo, South Africa, Sudan, Tanzania and Zimbabwe) have benefited from this arrangement. In 2002, the Medical Research Council (MRC) of South Africa collaborated with NABSA to investigate the South African plant *Dicoma anomala* for its chemical constituents and biological activity. After enlisting the cooperation of a Sudanese student and a senior scientist from Addis Ababa University (Ethiopia), the structures of the active anti-plasmodial compounds were elucidated. A full patent was secured through MRC lawyers in South Africa (Abegaz, 2009). The successful collaborations achieved in this research may serve as a model for other African researchers to follow.

An important natural product research network that is potentially useful to Africa as a whole is AFASSA (acronym for *Africa, Asia, South America*). The founding of this network was prompted by the need to make advanced laboratory facilities within the South available to researchers in other countries in the region, which lack such facilities. So far, this arrangement has had limited success. It has helped a few researchers to travel and carry out research in other laboratories in the South, despite the preference of some students to do their research in laboratories found in the North. The preference stems from regular instrument malfunctions and related problems in laboratories in the South. Travel arrangement, local living accommodation, and lack of funding are some of the constraints that AFASSA has been facing. Furthermore, regular and structured symposia have not been held. It has been suggested that a sustained South-South-North cooperation (with a North partner/funding agency) rather than a South-South cooperation alone, might alleviate the constraints. Over the years of its existence, AFASSA has benefited from the financial support it received from IPICS, IFS and OCPCW (Kuma, 2009)

## North-South Collaboration

The National Cancer Institute (NCI), USA, one of the bioprospectors in Africa, has been criticized for providing unreasonably low upfront benefits, and for not guaranteeing share of royalties. To circumvent this problem, the International Cooperative Biodiversity Group (ICBG) program was put together by three US agencies---the National Institutes of Health (NIH), the National Science Foundation (NSF) and the Agency for International Development (AID). The purpose was to help build local capacity, while bioprospecting. The other purpose was to help local scientists identify rich areas of biodiversity. Other bioprospections have established trust funds to pay off royalties (Mukerjee, 1996). Through funding from International ICBG program at NIH, the Virginia Polytechnic Institute and State University, Blacksburg (Virginia) has investigated several plants from the rainforests of Madagascar. In recent years, in place of big pharmaceutical manufacturers (Big Pharma), smaller companies such as Galileo Pharmaceuticals, SelectX Pharmaceuticals in the USA, Ecopia Biosciences in Canada, and Meril Lion Pharma in Singapore are showing continued interest in developing drugs from plants. The plants that are selected for investigation need to be collected in approved and legal ways with the preservation of the rights of all parties including indigenous people. The acquisition of collection permits have to be transparent and open. In the case of Madagascar, the Missouri Botanical Garden carried out the botanical collections and conducted biodiversity surveys. Conservation International-Madagascar was involved in economic development, biodiversity conservation and benefit sharing activities. The *Center National d' Application des RecherchesPhamaceutiques* prepared extracts, and was involved in screening for anti-malarial activity and isolation of compounds with anti-malarial activities. The *Center National de L'Environment* made collections of marine organisms, and was involved in some isolation work. The industrial partners were Eisai Research Institute for bioassay, drug discovery and development, and Dow Agro-sciences for agrochemical discovery. The offshoot of this cooperation was the launching of small-scale village projects in the area of plant collections, such as construction of an agricultural warehouse, renovation of a primary school, construction of a bridge over a river, promotion of eco-tourism and helping farmers to be self-sufficient. The projects were selected by the respective villages, and upfront compensation came from industrial partners. The Madagascar-ICGB Project resulted in the isolation of several bioactive compounds (Kingston, 2000).

## Some Examples of North-South Conflicts

North America and Europe have provided patents to companies for plant products that originated in South Africa. The German company Finkelberg, a chief producer of commercial plant extracts, was granted a patent on an extract from the South African plant *Harpagophytum procumbens* (Devil's Claw). The purified extracts are allegedly used for the treatment of asthma, gastrointestinal ailment, inflammatory conditions and rheumatism. The Khosian in the Kalahari knew the healing properties of the plant long before the patent was granted. However, this was never acknowledged in the patent, and no benefit was provided to the indigenous people, despite the fact that the plant product was sold widely in Europe. The irony of it all is a 50 gram of the plant extract formulation would sell for 25 Euros, while the San in Namibia were paid 30 to 40 cents/kilogram of the plant material collected, and even worse, others were paid for food and drink to harvest the plant (Reihling, 2008). Another controversy concerns the case of *Pelargoniumsidodes* and *P. reniforme*. The first plant is found in South African Eastern cape, while the latter grows throughout the cape regions, KwaZulu-Natal and Lesotho. The plants have antibacterial, antifungal and antitubercular properties. Both plants had been used by the Xhosa, Zulu and Basotho people to treat infectious diseases. However, the company Iso Arzneimethel did not arrange any Access and Benefit-Sharing (ABS) agreement with the indigenous people, when it was granted patents in 2003 and 2005 (Reihling, 2008).

The cactus plant *Hoodia* had been used by the Khosian and San people for hundreds of years to stave off hunger and thirst (Limson, 2002; Reihling, 2008). In 1996, the local parastatal Council for Scientific and Industrial Research (CSIR) in South Africa was studying the toxic effects of plants used by the Bushmen (San). One of these plants, *Hoodia gordonii* was found to be not only non-toxic, but also caused weight loss in animals. On follow-up studies, CSIR isolated a compound coded P57 (presumably the 57$^{th}$ product investigated by the Council) from the plant, and patented the finding. In 2001, CSIR provided a license to a London-based company, Phytopharm to further develop the product. The CSIR informed Phytopharm that the tribe that discovered the use of the cactus plant had died out, and that there were no obligations to pay royalties. This revelation outraged the Bushmen knowing that Phytopharm could potentially make billions of dollars by exploiting their knowledge about the use of the plant (Hoodia Advice, 2005; Reihling, 2008). The San people, through the Working Group of Indigenous Minorities in Southern Africa (WIMSA), the South African Council, and the San Institute of South Africa (SASI) claimed to be the true owners of the knowledge. In 2003, a deal was struck with CSIR, whereby benefit-sharing agreement was reached---the San people would get 6% of the financial benefits received by CSIR from its industrial partners. If P57 would be a success, the San people were to receive 0.03% to 1.2% of the net sales of the products to be paid into a trust fund shared by the San councils in Namibia, Botswana and South Africa (Reihling, 2008). Phytopharm made an agreement to transfer its rights to Pfizer to develop a P57-based diet pill. After a few years of unsuccessful attempt to make a synthetic version of P57, Pfizer pulled out of the agreement. In 2004, Phytopharm entered another deal with Unilever to produce *Hoodia* in the latter's diet food production line. Phytopharm still holds a patent on P57 and its weight-losing properties, but not on the plant itself. Anyone can sell the plant without reference to P57 and without claim that it reduces weight (Hoodia Advice, 2005). Currently, there is a thriving world-wide market of *Hoodia* herbal products as dietary supplements (Der-Marderosian & Beutler, 2007)

Schistosomiasis affects 200-300 people in tropical and sub-tropical parts of Africa, the Middle East, the Far East, the Caribbean islands and South America (Goldsmith, 1991). Therefore, a search for a cheaper non-synthetic alternative natural mulluscicide that was applicable in developing countries seemed to be an attractive endeavor. The soap berry plant *endod*, or *Phytolacca dodecandra*, is used as a natural soap in Ethiopia. In 1964, Dr. Akiliu Lemma noticed that where women washed their clothes with *endod* berries, the bilharzia transmitting snails died. *Endod* turned out to be a good natural candidate, requiring no sophisticated technology to process. The berries were easy to harvest while still unripe, which were then dried in the shade and ground to powder. Initially, funding for a rigorous research was difficult (Rukanigra, 2001). The World Health Organization (WHO) denied funds to the Ethiopian scientists to develop *endod* on the pretext of lack of safety data for the plant (Goldsmith, 1991). Dr. Aklilu collaborated with the Tropical Plant Products Institute in London to whom he provided a sample of *endod.* In the meantime, Dr. Aklilu and his colleagues were intensifying their research and putting out several publications on the use, agro-botanical and other aspects of *endod.* To his surprise, while on a visit to London, he found out his British colleagues secured a patent without including him in the process (Addis Tribune, 24 April 1997; Mukerjee, 1996). However, he continued collaborating with Stanford Research Institute (SRI) to elucidate the chemistry of the plant, which led to a joint patent. SRI donated its share of the patent to a non-profit foundation (*Endod* Foundation). Upon the insistence and diplomacy of Dr. Aklilu, the researchers in London followed suit and donated their patent to the Foundation (Mukerjee, 1996). The University of Toledo, Ohio (USA) picked up some interest in the plant when it found out that *endod* berries can potentially remove zebra mussels that plugged up hydroelectric power pipes in Mid-West USA. After conducting some preliminary studies based on the work and observations of Dr. Aklilu, the University was quick to acquire a patent for this application (Rukanigra, 2001).

The Zimbabwe National Healers' Association submitted a sample of the plant *Swartzier madagascariensis* to the University of Zimbabwe to be tested for its effectiveness in patients with *Candida* fungal infection, foot rot and oral thrush. The University in turn sent a sample to the University of Lausanne in Switzerland. After some research, the University, in collaboration with a drug company in the USA, patented the plant product. This move caused a big controversy between the Healers' Association in Zimbabwe and the Swiss University (Correa, 2002).

Another plant of some controversy in Africa is the Neem tree (*Azadirachta indica*). It is a plant principally grown widely in India and the surrounding areas (DerMarderosian & Beutler, 2009). It has been reported that about 70 patents have been taken up by western corporations on various applications of neem products. It has been used by indigenous peoples for thousands of years; however, neem has now been a subject of Intellectual Property Rights controversies. Western companies have even allegedly claimed patents in some parts of Kenya for some of the uses of the neem plant. Neem has been dubbed "*the tree of the 21$^{st}$ century*," "*the village pharmacy*," and "*the curer of all ailments*," due to its importance and multiple medicinal uses. Over 135 compounds have been reported from the plant (Jere-Malanda, 2003). Neem oil has been shown in vitro to be active against chloroquine and pyrimethamine-resistant malaria causing *Plasmodium falciparum* (DerMarderosian & Beutler, 2009). The European Patent Office granted W.R. Grace & Co a patent for the discovery of the fungicidal properties of neem oil. The Government of India challenged the patent on the grounds that the very claim was part of Indian traditional knowledge system, and thus lacks novelty (originality) and inventiveness. The patent was eventually rejected (Lodha, 2007). Although it was challenged, another patent taken by the same company was stayed—a method developed by the company to stabilize the bioactive constituents in neem oil. In the latter case, both the owners of the patent and Indian farmers and growers of the neem tree benefited (Elvin-Lewis, 2007). Although not directly relevant to Africa, of note is a patent claim by University of Mississippi researchers in the USA on the wound healing properties of turmeric (*Curcuma longa*). The Indian Council of Scientific and Industrial Research (ICSIR) challenged the claim on the basis that turmeric has been used in India for the same purpose for thousands of years, and that the patent lacked novelty. The Council produced "*prior art*" citations written in Sankrist, Urdu and Hindu to substantiate the challenge, upon which the patent claimed by the University of Mississippi was revoked (Hamilton, 2003; Elvin-Lewis, 2003).

In some cases, western corporations exploit publicly available traditional medicine knowledge without acquiring IPRs, and thereby without sharing the benefits with the traditional knowledge owners in developing countries. Such was the case with the plant Devil's Claw from Namibia and South Africa mentioned earlier, with *Prunus africana* from equatorial Africa and many others that are popular in the western market (Correa, 2002). In 1999, Cameroon earned US $ 700,000 from the export of *Prunus africana*, whose extract is used widely to treat benign prostate hyperplasia. On the other hand, western pharmaceutical companies earned US $200 million from the same export, which exemplifies the disproportionate benefits (Musa, 2005).

## The Future of IPRs in Africa

Weiss and Eisner (1998) eloquently summarize how developing countries need to strategize in order to extract optimal benefits from IPRs:

"*A developing country that wishes a significant share in the intellectual property deriving from a successful development must be prepared to make a significant financial investment, to share the risk of failure and to delay receipt of cash income. A developing country may seek quicker returns, with fewer risks and delays, by undertaking to develop new phytomedicines (herbal extracts sold directly for medicinal purposes), personal care products and nutriceuticals (food supplements).*"

This path may have to be followed by African countries to exploit their intellectual properties maximally. However, the effort requires a consolidated and unified approach. Plant collectors (botanists) should make their collections available to local chemists, instead of shipping them to foreign companies for money, or before entering into ill-defined contractual agreements with foreign companies. Chemists in local research universities, in turn, should perhaps defer publication of their findings, so that new and bioactive constituents of plants can be commercially exploited. Otherwise, patent rights and royalty income can be forfeited (Weiss & Eisner, 1998). Internationalization of patents and the push for enforcing IPRs by big Pharma in the North have made such an approach all the more reasonable. Cognizant of such a situation, countries like Brazil have intensified their research in traditional medicine (AFASSA Coordinating Group, 2003). African countries need to do the same.

## Conclusions

IPRs and patent rights that emanate from them have been held disproportionately by individuals, universities and multinational corporations in the North. This situation is true for products that have been developed based on TK (yet, unacknowledged) and natural resources in the South. To tip this imbalance in favor of developing countries, capacity building, technology expansion and launching local industries are indispensable. As shown in this article, one way to achieve this goal is to cooperate among laboratories, institutions and research groups within the South. This has been demonstrated, *albeit* on a limited scale, by the creation and activities of NAPRECA, WANNPRES, ALNAP, NABSA and AFASSA. However, more inter-network regional collaboration is needed to strengthen information exchange, research capability, resource utilization, natural product development and eventually tangible home-grown IPRs.

## References

Abegaz, B. (2009), 'Networking to build regional capacity in Africa: The efforts of NABSA,' in Kiselman, C. (ed.), *Proceedings of International Conference (Interregional Cooperation to Strengthen Basic Sciences in Developing Countries)*, Addis Ababa, 1-4 Sept, 2009, 313-320.

Addis Tribune (1997), 'Dr Aklilu Lemma-Ethiopia's Late Great Scientist,' 24 April.

Addy, M.E. (2005), 'Western African Network of Natural Products Research Scientists (WANNPRES). First Scientific Meeting, August 15-20, 2004. Accra, Ghana: Report,' in *African Journal of Traditional, Complementary and Alternative Medicine,* 2(2): 177-205.

AFASSA Coordinating Group (2003), 'Developing countries and natural products research: A Draft Blue Print,' Addis Ababa, September 1-3.

Andrews, P.R., Borris, R., Dagne, E. et al (1996), 'Preservation and Utilization of natural biodiversity in context of search for economically valuable medicinal biota (IUPAC Recommendations)' in *Pure and Applied Chemistry*, 68(2): 2325-2332.

Andrews, P.R., Borris, R., Dagne, E. et al (1996), 'General features of contracts for natural product collaborations (Technical Report),' in *Pure and Applied Chemistry*, 68(2): 2333-2337.

Constitution of the West African Network of Natural Products Research Scientists (WANNPRES) (2008), Available at: http://wannpres.net/html/structure.htm (Accessed on: 27 July 2011).

Correa, C.M. (2002), '*Protection and Promotion of Traditional Medicine. Implications for Public Health in Developing Countries*,' August 2002, Available at : http://www.apps.who.int/medicinedocs/pdf/s4917e/s4197.pdf (Accessed on: 27 July 2011).

Dagne, E. (1996), 'NAPRECA and Its Role in the Dissemination of Information on Natural Products Research in Africa,' in *Bridge Builders. National Academy of Sciences*, Washington, D.C., USA.

Dagne, E. (2009), 'Linking African Laboratories in Five Neighboring African Countries; ALNAP,' in Kiselman, C. (ed.), *Proceedings of International Conference (Interregional Cooperation to Strengthen Basic Sciences in Developing Countries)*, Addis Ababa, 1-4 Sept, 2009, 291-298.

DerMarderosian, A., Beutler, J. A. (eds.) (2009), 'Neem. March 2009,' in *The Review of Natural Products*, St. Louis, Missouri: Wolters Kluwer Health

DerMarderosian, A., Beutler, J. A. (eds.) (2009), 'Hoodia, February 2007,' in *The Review of Natural Products*, St Louis, Missouri: Wolters Kluwer Health.

Ekpere, J.A. (2000), '*The OAU Model Law. The Protection of the Rights of Local Communities, Farmers and Breeders and for the Regulation of Access to Biological Resources*,' Lagos: Organization of African Unity Scientific, Technical and Research Commission, Available at: http://www.blauen-institut.ch/tx_blu/tp/tp/t_OAU.model_law.pdf (Accessed on: 27 July 2011).

Elvin-Lewis, M. (2007), 'Evolving concepts related to achieving benefit sharing for custodians of traditional medicine,' in *African Journal of Traditional, Complementary and Alternative Medicine*, 4(4): 443-468.

Goldsmith, M.F. (1991), 'Out of Africa comes the fruit of long research: Possible self-reliant control of Schistosomiasis,' in *Journal of American Medical Association*, 255 (20): 2650-2651.

Gurib-Fakim, A. (2009), 'Natural Product Networks in Africa and their contribution to cooperation in economic development,' in Kiselman,

C. (ed.), *Proceedings of International conference (Interregional Cooperation to Strengthen Basic Sciences in Developing Countries)*, Addis Ababa, 1-4 Sept, 2009, 103-111.

Hamilton, A. (2003), '*Medicinal Plants and Conservation: Issues and Approaches*,' International Plants Conservation Unit/WWF-UK, Godalming, Surrey.

History (2008), Available at: http://wannpres.net/html/history.htm (Accessed on: 27 July 2011).

Hoodia Advice (2005), '*The History of Hoodia*,' Available at: http://www.hoodia-advice.org/hoodia-history.htm. (Accessed on: 27 July 2011).

Jere-Malanda, R. (2003), 'Biopiracy: Neem, the wonder tree: a classic example of biopiracy from which Africa has a lot to learn-the blatant piracy of the neem tree, dubbed by the UN as the three of the 21st Century,' in *New African*, December 1.

Kingston, D.G.I. (2006), 'Successful Drug Discovery from Natural Products: Methods and Results,' in Midiwo, J.O, Abiy Yenesew, Solomon Derese (eds.), *11th NAPRECA Symposium Book of Proceedings*, Nairobi: NAPRECA Publication, 224-232

Kuma, V. (2009), 'Challenges and Constraints in South-South Collaboration. The Experience of AFASSA,' in Kiselman, C. (ed.), *Proceedings of International Conference (Interregional Cooperation to Strengthen Basic Sciences in Developing Countries)*, Addis Ababa, 1-4 Sept, 2009, 307-311.

Limson, J. (2002), '*Focus in biopiracy in Africa. Science in Africa*,' Available at: http://www.scineceinafrica.co.za/2002/Spetember/biopiracy.htm. (Accessed on: 27 July 2011).

Lodha, R. (2007), '*Intellectual Property Rights, Biodiversity and Traditional Knowledge*,' Jaipur, India: CUTS International.

Midiwo, J.O. (2009), 'Natural Products Research Network for Eastern and Central Africa (NAPRECA),' in Kiselman, C. (ed.), *Proceedings of International Conference (Interregional Cooperation to Strengthen Basic Sciences in Developing Countries)*, Addis Ababa, 1-4 Sept, 2009, 288-290.

Mitscher, L.A., Crowley, S.R., Plattner, J.J. (1994), 'Chemical Patents. Profiting from your inventions,' in *Analytical Chemistry*, 66: 575A-581A.

Mukerjee, M. (1996), '*Profiting from biodiversity. Scientific American 1996 (reproduced)*,' Available at http://www.dhushara.com/book/diversit/extra/prof.htm (Accessed on: 27 July 2011).

Musa, J. (2005), 'Cameroon. The magic of medicinal plants,' in *New African*, 58-59, February, 2005.

NAPRECA (2003), '*Network Consolidation Program. Strategic and Action Plan, 2004-2009*,' The Coordinating Office, Department of Chemistry, University of Dar es Salaam, Tanzania, July.

Oh, C. (1999), '*The OAU draft law and convention: A model for protecting community rights and access*,' Third World Resurgence No. 106, June, Available at: http://www.twnside.org.sg/title/co 1-cn.htm (Accessed on: 27 July 2011).

Patwardhan, B. (2005), '*Traditional Medicine: A novel approach for available, accessible and affordable health care*,' World Health Organization, Regional Office for South-East Asia, Pyongyang, DPR Korea.

Reihling, H.C.W. (2008), 'Bioprospecting the African renaissance: The New Value of *muthi* in South Africa,' in *Journal of Ethnobiology and Ethnomedicine*, 4:9.

Rukangira, E. (2001), 'The African Herbal Industry: Constraints and Challenges,' in *Erboristeria Domani*, August 2001.

Structure (2008), Available at: http://wannpress.net/html/structure.htm (Accessed on: 27 July 2011).

Teljeur, E. (2003), '*Intellectual Property Rights in South Africa: An Economic Review of Policy and Impact*,' Available at: http://www.tips.org.za/files/Teljeur_IPRs_2003.PDF (Accessed on: 27 July 2011).

Weiss, C., Eisner, T. (1998), 'Partnerships for value-added through bioprospecting,' in *Technology in Society*, 20: 481-498.

# CONCLUSION

*Mammo Muchie, Mentesnot Mengesha and Amare Desta*

Knowledge, learning, innovation, competence building systems (KLiCS) are critical to anchor development and structural transformation by promoting the wellbeing of people, society and nature at the same time. The unity of knowledge, learning, innovation and competence building can promote the ecological wellbeing of nature, peoples' social relations and the spiritual world together. When new knowledge is created, it is not necessary to do it through epstimicide of indigenous knowledge. On the contrary the indigenous knowledge can be a knowledge resource to revamp the sources of the different cultures and societies to open the imagination to the virtue of nurturing different ways of knowing. There is not only one way of knowing. The scientific way is one way, but there are other ways of knowing. There is even a revealational way of knowing. Unifying the different ways of knowing requires that one way of knowing should not dominate. Epistemic convergence can lead to the emergence of new ways of comprehending the complex interaction of the social, natural and spiritual worlds.

One of the difficulties in Africa is that the knowledge that was created, invented, innovated and produced has faced epstimcide over the period of destruction that Africa experienced for over 500 years. The positive data that Africans created as the originators of humanity, science, mathematics, knowledge, religion, languages and astronomy have been rejected. Still today there is racism which has made Africans to feel they are not knowledge creators but its consumers. Knowledge created in Africa by Africans is not easily recognised. Knowledge created from outside is likely to be accepted and used. This has left Africa to copy rather than be the true creators of knowledge.

The knowledge system in Africa to this day lacks indigenous authenticity and identity. This is in spite of historical records which show that, before Africa's subjection to slavery and the colonial technology of rule, there were recorded histories of indigenous higher learning in Timbuktu and other parts of Africa. This higher learning from the pre-colonial and pre-slavery period was historically ruptured and disrupted by the colonial-imperial project in Africa that also took the form of enslaving whole communities and forcibly evicting them for the purposes of slave labour in the New World of the Americas.

"The evidence of Africa's greatness is everywhere for our children to see. They only see [images of] the architecture of the Nubian civilisation and the great pyramids. They also know very little about the artistic creativity of the ancient Nok in Nigeria – a civilisation that predated Jesus by centuries. Timbuktu is another, and there are the bronze works of Benin. People often forget or simply do not know that the first university in the world was African – the Al Karaouine University – founded by an African woman in Fez, Morocco in 859 AD. I am not the one saying so. The Guinness Book of World Records says so. And the second university in the world was also African – Al Azhar in Egypt, built in 969 AD, which still exists today. Europe saw the lecture halls of a university a full 229 years after Africa; the University of Bologna in 1088 AD".( http://newafricanmagazine.com/need-africa-centred-innovations-systems-prof-mammo-muchie/)

The contributions of this pre and post-15th century varied knowledge systems have not been acknowledged or recognised. In fact, they have been bypassed, ignored, and suppressed rather than being used to serve as building blocs for the contemporary KLICS development in Africa. Thus, instead of continuity, what has taken place is discontinuity in KLICS. This context of historical rupture and the erasure of pre-existing knowledge has to be rectified and justice done means that knowledge remains an issue that cannot be ignored.

Africa's contribution to knowledge before the 15th century has not been explored and its contemporary relevance for revitalising the knowledge world has not been appreciated. The first university was formed in Africa in the 9th century A.D. The significance and knowledge and even the information is not well known to this day.(New Africa March, 2015)

Even during the 20th century Africa, modern higher education institutions were established when colonialism began to concede political defeat. European powers sent their armies and missionaries when colonising Africa. Later, when the colonial powers recognised that they may have to relinquish direct control, they realised that they had to 'export' a form of their universities to help train civil servants who would be necessary to discharge the administrative burden of running the state in the post-colonial period. They began to establish universities in their African colonies, though these largely took the form of setting up one university to serve a number of their colonies. Thus, three years after the fifth historic Pan-African Congress meeting in Manchester in 1945 at which the Congress passed a resolution and declared to the world the need to liberate Africa by any means necessary, Makerere University was set up as the East African University - not as the Ugandan University as it later (in 1963) became known. In Nigeria, in 1949, Ibadan and Legon were set up as sites for universities. In the French-ruled part of Africa, closer links between the University of Dakar and the University of Bordeaux were established, and Belgium established a link between Université Lovanium in Congo (now University of Kinshasa in the Democratic Republic of Congo) and the Université Catholique de Louvain4 . The problem is that the modern higher education system in Africa is an export from the former colonizers of Africa and their allies, and has been imported by Africa. From its inception, higher education has not been an African home-grown institution designed to serve the needs, aspirations, well-being and development of the African people despite the fact universities like humanity originated in Africa.. Just as the post-colonial state became largely a legacy of colonialism, the universities also became part of the institutions transmitted through the colonial channel.

The real question relates to the problem of how to make African higher education serve the needs and aspirations of the people of Africa. That colonial powers built universities and left them behind in African countries may not be problem in and of itself. The wholesale acceptance of the universities made in the image of the colonial world uncritically, and without interrogating the need to match them to the transformation challenges of African economies, remains an issue. There has been little or no systematic criticism of the impact of the 'export' of universities to Africa. The negative and inferiorising impact it has had continues to damage the processes of knowledge production. Africans continue to rely on external sources of knowledge lacking the self- creation of sustainable and self-worth enhancing knowledge that is also of relevance to African development and values.. That Africa has to modernise has meant that it has to catch up with the former colonial powers and not anchor its knowledge on what was created within Africa. In education, it has meant that universities are expected to 'catch up' to those in the former colonies.

The series of conferences we initiated by the NEsglobal and the Ee-JRIF open access journal editors precisely wishes to address this challenge from knowledge transfer to knowledge exchange. Knowledge exchange acknowledges that Africa has knowledge worthy of exchanging and it should not be dependent on transfer of knowledge. This series of books we will publish have the purpose and vision to create a paradigm shift to re-anchor in the African knowledge system by revivifying KLICS to re-map the education contour/landscape in Africa.

The shift from knowledge transfer to knowledge exchange is critical to create a unity of knowledge that has become one of the grand ideas of intellectual history and the history of scientific thought. The trend is to build the unity of knowledge by linking disciplines and the human, social and natural sciences. This trend will be reinforced by redressing the historic injustice of committing epistemological vice against Africa. The Knowledge transfer is loaded with a deficit of epistemic virtue. The only way epistemological virtue can be nurtured is when the knowledge exchange epistemological disobedience is fully recognised, appreciated and accepted.

This book starts a new journey for the unity of knowledge building on how the history of nature and humanity evolved in order to examine how a new knowledge that can identify and remove the epistemic vice of knowledge transfer into the epistemic virtue of knowledge exchange can be created. The knowledge exchange can be the foundation for creating a unity of knowledge based on a new paradigm , modes and interdisciplinary that builds on the human, social and natural sciences.

It is with great interest we open this new paradigm of knowledge exchange to liberate knowledge originating in Africa from the historical damage it has gone through. We welcome all to engage and contribute to the new liberated unity of knowledge.

# ABOUT THE AUTHORS

**Ermias Abebe** (MSc) is lecturer at the School of Information Science in Addis Ababa University. He also works as consultant in various information system development activities. His main area of interest includes software design and management and Natural Language Processing.

**Elizabeth Ayalew** (PhD) is an Assistant Professor at the School of Information Sciences, College of Natural Sciences, at Addis Ababa University. She has her PhD in Information Systems (Addis Ababa University); M.Ed. in Information Communication Technology in Education (University of Pretoria and Addis Ababa University-joint); MA in English Language Education (University of Warwick, the U.K), and BA in Foreign Languages and Literature (Addis Ababa University).

**Bedru B. Balana** is a Research Economist at the International Water Management Institute (IWMI). Prior to joining IWMI he was a Lecturer/Assistant Professor of Economics at Mekelle University, Ethiopia (1995-2007) and Environmental Economist at the James Hutton Institute, UK (2008-2014). Bedru obtained his Bachelor's Degree in Economics (1994) from Addis Ababa University (Ethiopia); Master's Degree in Economics (2000) from Punjab University-Chandigarh (India); and PhD (2007) in Bioscience Engineering (Agri. & Environ. Economics) from the Catholic University of Leuven (Belgium). Bedru's main research areas include integrated economic analysis of environmental & natural resources management options; agri-environmental policy analysis; environmental valuation; and economics of ecosystem services and livelihoods. Bedru published more than 35 peer-reviewed scientific journal articles and technical papers.

**Rahel Bekele** (PhD) is associate professor at the School of Information Science in Addis Ababa University. She holds a PhD in Computer Science from Department of Informatik, University of Hamburg, Germany and her research interests are mainly in the area of knowledge management, eLearning systems, software usability and mobile application in health. She has worked with several organizations including the United Nations, the World Bank and the Nile Basin Initiative as consultant in knowledge management, Gender related issues, and software related projects as well as Mobile use to improve Maternal and Child Health in Ethiopia.

**Tewabech Bishaw** is a Public Health Specialist with over 40 years of national and international health development work. She is a Founder and Managing Director of Alliance for Brain-Gain and Innovative Development (ABIDE). In 2013 Dr Tewabech was nominated as an inspiring Ethiopian and recognized for her pioneering role in public health. She was also among the top Ten Ethiopian Women of Excellence in 2012 by the Association of Women in Business. Internationally, and included in the top 100 Women Leaders in Global Public Health through a campaign organized by the Global Health Program of the Graduate Institute, Geneva. Tewabech holds PhD and Master of Public Health from Loma Linda University School of Health, Ca. USA

**Demese Chanyalew** is an Agricultural Economist and currently the General Manager and founder of a consultancy firm - DeMar Ethio-Afric PLC based in Addis Ababa, Ethiopia. He has previously worked for different institutions in the USA, Kenya and Ethiopia. He has a PhD from Kansas, and MSc from Oklahoma State Universities, USA; and a BSc from Addis Ababa University, Alemaya College of Agriculture.

**Nicholas Chisholm** (PhD) is a Senior Lecturer in Food Business & Development at University College Cork, Ireland. His research focusses on food insecurity and hunger, and also sustainable natural resource management. He has a particular theoretical and conceptual interest in complex development issues which occur at the intersection of different theoretical field to comprehend the interaction of environmental economics, institutional analysis and bio-physical processes in the context of sustainable development.

**Hailemichael Teshome Demissie** is Senior Research Fellow and Head of the Inclusive Bio-economy Programme at the African Centre for Technology Studies (ACTS) - an inter-governmental think tank based in Nairobi, Kenya. He holds PhD from King's College London and MSc from the University of Warwick in the UK. His research specialty is around the law and technology interface and the law, ethics, policy and regulation of emerging technologies and their convergence. He has published several articles on nanotechnology, biotechnology and other technologies and their impact on economic development in Africa and the developing world.

**Amare Desta** is a Senior Lecturer and a Course Director for Business Information Systems at the London South Bank University. He also taught at the LSE, Goldsmiths College of University of London, University of East London, the Open University, Thames Valley University, & at Maritime Greenwich College. Dr Amare is a co-founder and chair of GKEN (Global Knowledge Exchange Network) since it is foundered in 2011. He earned his PhD from The London School of Economics (LSE) in Information Systems and Master of Education (MEd) from University of Cambridge, MSc from LSE and BSc (Hons) degree in Computer Science from ARU, Cambridge, UK.

**Pat Enright** is a Lecturer in the Department of Food Business and Development at National University of Ireland Cork. Prior to this role, he has held various management positions in a major Irish food company. He has an MA in rural development from the University of Alberta in Canada, a BA and Diploma in Dairy Science from University College Cork and a PhD in economic geography from the University of Leicester.

**Zeremariam Fre** is a Lecturer at Development Planning Unit, University College London (UCL) since 2002 lecturing on Development Administration and Planning (DAP) and Environment and Sustainable Development. He has MSc and PhD in Agricultural extension and rural Development University of Reading in the UK and Founder director of PENHA with 30 years of development work experience in 13 countries mostly in East Africa, Far East and the Europe. Conducted research and consultancy work for a number of international organisations.

**Fekadu Fullas** is a Research Preceptor for Post-graduate Year I (PGY1) Pharmacy Residents, at Unity Point Health-St Luke's - Inpatient Pharmacy, Sioux City, Iowa, USA. Fekadu has a PhD in Pharmacognosy from College of Pharmacy, University of Illinois at Chicago, Illinois. He has published scientific papers in several international journals.

**Semahagn G. Gashu** is a visiting assistant professor at Institute of Human Rights, University of Connecticut, USA. Semahagn holds PhD from Goettingen in Germany LLM (Masters of Law) from Amsterdam, Holland and another Masters of Law from Goettingen, Germany.

**Yared Haile-Meskel** is a Managing Director of a consulting Company. He earned a BSc degree from Addis Ababa University, MSc from University of North London, and MPhil from University of Southampton and Executive MBA from University of Bath. Yared had worked as R&D Engineer in automotive and defense industry in UK for 16 years in UK before setting up a consulting company. Yared had carried out research in scientific and social related fields to published 14 papers in peer reviewed journals.

**Tadesse Biru Kersmo** is a Part-time Lecturer at Greenwich University, Department of Business Development where he teaches Introduction to Management and Entrepreneurship Development. He holds a PhD in Social and Cultural Sciences from the Polish Academy of Sciences, Warsaw, Poland and MSc Degree in Economic Theories from Kiev Taras Shvshenko University, Kiev, Ukraine.

**Dessalegn Molla Ketema** is an Assistant Professor in the Department of Rural Development & Agricultural Extension at College of Agriculture & Environmental Sciences, Bahir Dar University. Previously he has been working as a socio-economics, research & extension-farmers linkage researcher at Amhara Regional Agricultural Research Institute (ARARI). He has a PhD in Rural Development from University College Cork (Ireland), MSc in Rural Development and BSc in Agricultural Extension from Haramaya University, Ethiopia.

**Medicine Magocha** has extensive work experience in both public and private sectors and worked as head of department, district chairperson, lecturer, editor, researcher and executive head of finance at Rupare High School, Better Schools Programme of Zimbabwe. Medicine holds a Doctorate degree in Business Leadership from University of South Africa Graduate School of Business Leadership. Currently he is completing a Doctorate of Technology in Agriculture with Tshwane University of Technology.

**Mammo Muchie** holds a DPhil in Science, Technology, Innovation for Development (STI4D) from the University of Sussex. He is currently a rated DST/NRF Research Professor of Innovation Studies at Tshwane University of Technology. He is also TMDC Rsenior Research Fellow in Oxford University, UK, and Adjunct Professor in ASTU, Ethiopiac. He is a member of the African Academy of Science and the South African Academy of Sciences. He is also the chair of the Network of Ethiopian Scholars that initated also the GKEN (Global Knowledge Exchange Network). He is founding chief editor of the open access journal Ee-JRIF and the African Journal of Science, Technology, Innovation and Development (AJSTID). Since 1985, Prof. Muchie has produced over 365 publications, including books, chapters in books, research papers in internationally accredited journals and entries in institutional publications.

**Bereket T. Tsegay** is a researcher at the London based non-for-profit research institute the Pastoral and Environmental Network in the Horn of Africa (PENHA). He is also a PhD candidate in Development Studies at the School of Oriental and African Studies (SOAS), University of London working on the issue of Green Economy in Sub-Saharan Africa. His research has focused on policy analysis, natural resource governance, food security, pastoralism, and dynamics of institutions in development, climate change resilience and green economy, risk management as well as social protection in Africa, Far East and the Europe. Conducted research and consultancy work for a number of international organisations.

**Mariye Yigzaw Workineh** was a lecturer at Addis Ababa University and at various private colleges in ICT and related subjects due to his many years of professional training and management experience in the industry. He was a PhD Candidate at AAU IT Doctoral programme and held MSc. degree in Information Science and B.A. degree in Management from Addis Ababa University. His research interest in (indigenous) knowledge management with special focus on indigenous knowledge creation and sharing was really fascinating and his work was admired by many professionals in his field.

***Sadly Mariye passed away before his time due to short illness in Gonder Ethiopia in 2014.***

# INDEX

## J



## K



## L



## M



## N

## Q



## P



## R



## S



## T



## V

## W